Advances in Molecular Biophysics

Advances in Molecular Biophysics

Editor: Keira O'Donnell

www.callistoreference.com

Callisto Reference,
118-35 Queens Blvd., Suite 400,
Forest Hills, NY 11375, USA

Visit us on the World Wide Web at:
www.callistoreference.com

© Callisto Reference, 2020

This book contains information obtained from authentic and highly regarded sources. Copyright for all individual chapters remain with the respective authors as indicated. All chapters are published with permission under the Creative Commons Attribution License or equivalent. A wide variety of references are listed. Permission and sources are indicated; for detailed attributions, please refer to the permissions page and list of contributors. Reasonable efforts have been made to publish reliable data and information, but the authors, editors and publisher cannot assume any responsibility for the validity of all materials or the consequences of their use.

ISBN: 978-1-64116-278-4 (Hardback)

Trademark Notice: Registered trademark of products or corporate names are used only for explanation and identification without intent to infringe.

Cataloging-in-Publication Data

Advances in molecular biophysics / edited by Keira O'Donnell.
   p. cm.
Includes bibliographical references and index.
ISBN 978-1-64116-278-4
1. Molecular biology. 2. Biophysics. 3. Biomolecules. I. O'Donnell, Keira.
QH506 .A38 2020
572.8--dc23

Table of Contents

Preface VII

Chapter 1 **Visualizing the Ensemble Structures of Protein Complexes using Chemical Cross-Linking Coupled with Mass Spectrometry** 1
Zhou Gong, Yue-He Ding, Xu Dong, Na Liu, E. Erquan Zhang, Meng-Qiu Dong and Chun Tang

Chapter 2 **Roles of H3K36-specific histone methyltransferases in transcription: antagonizing silencing and safeguarding transcription fidelity** 13
Chang Huang and Bing Zhu

Chapter 3 **Raman spectra of the GFP-like fluorescent proteins** 21
Ye Yuan, Dianbing Wang, Jibin Zhang, Ji Liu, Jian Chen and Xian-En Zhang

Chapter 4 ***In situ* protein micro-crystal fabrication by cryo-FIB for electron diffraction** 29
Xinmei Li, Shuangbo Zhang, Jianguo Zhang and Fei Sun

Chapter 5 **Mapping disulfide bonds from sub-micrograms of purified proteins or micrograms of complex protein mixtures** 38
Shan Lu, Yong Cao, Sheng-Bo Fan, Zhen-Lin Chen, Run-Qian Fang, Si-Min He and Meng-Qiu Dong

Chapter 6 **Determining the target protein localization in 3D using the combination of FIB-SEM and APEX2** 52
Yang Shi, Li Wang, Jianguo Zhang, Yujia Zhai and Fei Sun

Chapter 7 **Radioligand saturation binding for quantitative analysis of ligand-receptor interactions** 60
Chengyan Dong, Zhaofei Liu and Fan Wang

Chapter 8 **Using 3dRPC for RNA–protein complex structure prediction** 68
Yangyu Huang, Haotian Li and Yi Xiao

Chapter 9 **Skeletal intramyocellular lipid metabolism and insulin resistance** 73
Yiran Li, Shimeng Xu, Xuelin Zhang, Zongchun Yi and Simon Cichello

Chapter 10 **Protocol for analyzing protein ensemble structures from chemical cross-links using DynaXL** 82
Zhou Gong, Zhu Liu, Xu Dong, Yue-He Ding, Meng-Qiu Dong and Chun Tang

Chapter 11 **Symmetry-mismatch reconstruction of genomes and associated proteins within icosahedral viruses using cryo-EM** 91
Xiaowu Li, Hongrong Liu and Lingpeng Cheng

Chapter 12	**Choosing proper fluorescent dyes, proteins, and imaging techniques to study mitochondrial dynamics in mammalian cells**	99
	Xingguo Liu, Liang Yang, Qi Long, David Weaver and György Hajnóczky	
Chapter 13	**Crystal structures of MdfA complexed with acetylcholine and inhibitor reserpine**	108
	Ming Liu, Jie Heng, Yuan Gao and Xianping Wang	
Chapter 14	**Structural roles of lipid molecules in the assembly of plant PSII–LHCII supercomplex**	115
	Xin Sheng, Xiuying Liu, Peng Cao, Mei Li and Zhenfeng Liu	
Chapter 15	**Dissection of structural dynamics of chromatin fibers by single-molecule magnetic tweezers**	130
	Xue Xiao, Liping Dong, Yi-Zhou Wang, Peng-Ye Wang, Ming Li, Guohong Li, Ping Chen and Wei Li	
Chapter 16	**Identification of small ORF-encoded peptides in mouse serum**	141
	Yaqin Deng, Adekunle Toyin Bamigbade, Mirza Ahmed Hammad, Shimeng Xu and Pingsheng Liu	
Chapter 17	**Class C G protein-coupled receptors: reviving old couples with new partners**	152
	Thor C. Møller, David Moreno-Delgado, Jean-Philippe Pin and Julie Kniazeff	
Chapter 18	**Significant expansion and red-shifting of fluorescent protein chromophore determined through computational design and genetic code expansion**	159
	Li Wang, Xian Chen, Xuzhen Guo, Jiasong Li, Qi Liu, Fuying Kang, Xudong Wang, Cheng Hu, Haiping Liu, Weimin Gong, Wei Zhuang, Xiaohong Liu and Jiangyun Wang	
Chapter 19	**Docking-based inverse virtual screening: methods, applications, and challenges**	172
	Xianjin Xu, Marshal Huang and Xiaoqin Zou	
Chapter 20	**Evaluation of RNA secondary structure prediction for both base-pairing and topology**	188
	Yunjie Zhao, Jun Wang, Chen Zeng and Yi Xiao	
Chapter 21	**Regulation of metabolism by the Mediator complex**	198
	Dou Yeon Youn, Alus M. Xiaoli, Jeffrey E. Pessin and Fajun Yang	
Chapter 22	**MetaDP: a comprehensive web server for disease prediction of 16S rRNA metagenomic datasets**	207
	Xilin Xu, Aiping Wu, Xinlei Zhang, Mingming Su, Taijiao Jiang and Zhe-Ming Yuan	
Chapter 23	**A new dimethyl labeling-based SID-MRM-MS method and its application to three proteases involved in insulin maturation**	217
	Dongwan Cheng, Li Zheng, Junjie Hou, Jifeng Wang, Peng Xue, Fuquan Yang and Tao Xu	

Permissions

List of Contributors

Index

Preface

Molecular biophysics is an interdisciplinary domain of research which combines the principles of physics, chemistry, mathematics, engineering and biology. It aims to explain bimolecular systems and biological functions in terms of dynamic behaviour, molecular structure and structural organization. This field seeks to understand biological processes and functions at different levels of complexity, such as from viruses to small living systems, and from single molecules to supramolecular structures. The study of allosteric interactions, molecular forces, molecular associations, cable theory and Brownian motion are also addressed by molecular biophysics. Procedures that enable imaging and manipulation of living structures aid research in molecular biophysics. Spectroscopic techniques comprising of laser Raman, FT-NMR, FT-infrared, spin label electron spin resonance, etc. are widely used to understand structural dynamics of biomolecules and their intermolecular interactions. This book is a compilation of chapters that discuss the most vital concepts and emerging trends in the field of molecular biophysics. The various advancements in this field are glanced at and their applications as well as ramifications are looked at in detail. Students, researchers, experts and all associated with molecular biophysics will benefit alike from this book.

The researches compiled throughout the book are authentic and of high quality, combining several disciplines and from very diverse regions from around the world. Drawing on the contributions of many researchers from diverse countries, the book's objective is to provide the readers with the latest achievements in the area of research. This book will surely be a source of knowledge to all interested and researching the field.

In the end, I would like to express my deep sense of gratitude to all the authors for meeting the set deadlines in completing and submitting their research chapters. I would also like to thank the publisher for the support offered to us throughout the course of the book. Finally, I extend my sincere thanks to my family for being a constant source of inspiration and encouragement.

Editor

Visualizing the Ensemble Structures of Protein Complexes using Chemical Cross-Linking Coupled with Mass Spectrometry

Zhou Gong[1], Yue-He Ding[2], Xu Dong[1], Na Liu[2], E. Erquan Zhang[2], Meng-Qiu Dong[2✉], Chun Tang[1✉]

[1] CAS Key Laboratory of Magnetic Resonance in Biological Systems, State Key Laboratory of Magnetic Resonance and Atomic Molecular Physics, Wuhan Institute of Physics and Mathematics, Chinese Academy of Sciences, Wuhan 430071, China
[2] National Institute of Biological Sciences, Beijing 102206, China

Graphical Abstract

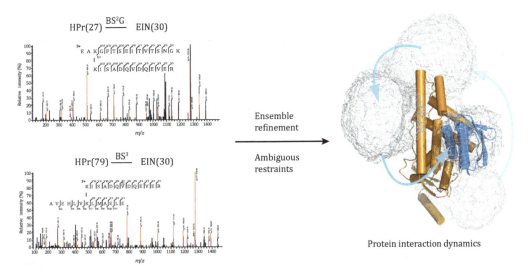

Abstract Chemical cross-linking coupled with mass spectrometry (CXMS) identifies protein residues that are close in space, and has been increasingly used for modeling the structures of protein complexes. Here we show that a single structure is usually sufficient to account for the intermolecular cross-links identified for a stable complex with sub-μmol/L binding affinity. In contrast, we show that the distance between two cross-linked residues in the different subunits of a transient or fleeting complex may exceed the maximum length of the cross-linker used, and the cross-links cannot be fully accounted for with a unique complex structure. We further show that the seemingly incompatible cross-links identified with high confidence arise from alternative modes of protein-protein interactions. By converting the intermolecular cross-links to ambiguous distance restraints, we established a rigid-body simulated annealing refinement protocol to seek the minimum set of conformers collectively satisfying the CXMS data. Hence we demonstrate that CXMS allows the depiction of the ensemble structures of protein complexes and elucidates the interaction dynamics for transient and fleeting complexes.

Zhou Gong and Yue-He Ding have contributed equally to this work.

✉ Correspondence: dongmengqiu@nibs.ac.cn (M.-Q. Dong), tanglab@wipm.ac.cn (C. Tang)

Keywords Protein–protein interaction, Encounter complex, Fleeting complex, Ensemble refinement, Ambiguous distance restraint

INTRODUCTION

A protein interacts with other proteins to perform its function. The binding affinity or K_D value between two proteins ranges over ten orders of magnitude, and the resulting complex can be stable, transient or fleeting (Jones and Thornton 1996; Nooren and Thornton 2003). Examples of stable complexes include enzyme/enzyme inhibitor and antigen/antibody (Kastritis et al. 2011), while transient and fleeting complexes are often involved in cell signaling. Transient complexes are those with K_D values greater than 1 μmol/L, whereas fleeting complexes are three–four orders of magnitude weaker with K_D values in mmol/L (Vinogradova and Qin 2012; Xing et al. 2014; Liu et al. 2016).

Two transiently interacting proteins not only form a stereospecific complex, they can also form a series of nonspecific encounter complexes (Tang et al. 2006; Fawzi et al. 2010; Schilder and Ubbink 2013). Encounter complexes are important structural intermediates, and facilitate the formation of the stereospecific complex. Yet, encounter complexes constitute only a minor population of the total complex, and are difficult to study (Berg et al. 1981; Schreiber and Fersht 1996; Gabdoulline and Wade 2002). With the K_D value in mmol/L, the distinction between specific and non-specific complexes starts to blur, and the subunits in a fleeting complex often adopt a variety of conformations (Tang et al. 2008; Liu et al. 2012). As such, to characterize the structure of a protein complex, especially a transient or fleeting complex, it often requires an ensemble description to recapitulate the different conformational states.

Chemical cross-linking of proteins coupled with mass spectrometry analysis (CXMS) is an emerging technique to investigate protein-protein interactions (Rappsilber 2011; Herzog et al. 2012; Kalisman et al. 2012; Lasker et al. 2012; Walzthoeni et al. 2013; Politis et al. 2014). Amine-specific homo-bifunctional cross-linkers, including bis-sulfosuccinimidyl suberate (BS^3) and bis-sulfosuccinimidyl glutarate (BS^2G), are commonly used. Recently, carboxylate-specific cross-linkers reactive towards glutamate or aspartate residues, such as pimelic acid dihydrazide (PDH; Leitner et al. 2014), were added to the CXMS toolbox. In theory, two primary amine groups (either lysine side chain or protein N-terminus) or two carboxylate groups (either glutamate or aspartate side chains) that are close in space can be covalently linked. The cross-linked residues can be identified with the use of a database search engine (Rinner et al. 2008; Yang et al. 2012), and each intermolecular cross-link can be converted to a distance restraint for modeling the complex structure (Rappsilber 2011; Kalisman et al. 2012; Walzthoeni et al. 2013; Schmidt and Robinson 2014).

As CXMS has been increasingly used for the structural characterization of protein complexes, two technical issues have become apparent (Rappsilber 2011; Merkley et al. 2014). First, only a fraction of the cross-links expected from the known structure of a protein complex are experimentally observed. This can be due to low accessibility and reactivity of the involved residues (Leitner et al. 2014). Second and more intriguingly, for a subset of cross-links, the theoretical distance between two cross-linked residues, as calculated from the specific complex structure, sometimes exceeds the maximum length of the cross-linker (Kahraman et al. 2013). Incorrect identification of cross-linked peptides has been blamed for such discrepancies (Zheng et al. 2011; Kalisman et al. 2012). Yet, with the most stringent criteria that essentially eliminate false identifications, sometimes there remain cross-links violating the distance limits (Lossl et al. 2014). So what are the origins of these "incompatible" cross-links?

CXMS data have been recently implemented in ROSETTA software package for modeling protein complex structures (Kahraman et al. 2013; Lossl et al. 2014). The approach aims to obtain a single structure that satisfies CXMS restraints and has the lowest ROSETTA energy score, and is suited for characterizing stable complex structures. Nevertheless, as transient and fleeting complexes can adopt a multitude of conformational states, a single-conformation representation may not suffice. Here we show that the highly reliable but seemingly incompatible cross-links arise from alternative modes of protein–protein interactions. We present a rigid-body refinement protocol against all the experimental cross-links, and show that an ensemble representation comprising multiple conformers of the complex is often required when characterizing transient and fleeting complexes.

RESULTS

Refinement of the stable complex structure

To refine against intermolecular CXMS restraints, we treated each subunit as a rigid body. Any two cross-linked lysine residues were restrained to have their C_α-C_α distance to be less than the maximum length of the corresponding cross-linker using a square-well pseudo-energy potential. BS^3 and BS^2G covalently link lysine residues <24 Å and <20 Å apart, respectively, as measured from C_α to C_α atoms (Lee 2009; Kahraman et al. 2011). Cross-links may also involve protein N-terminus; when fully extended, the maximum C_α-C_α distance between an N-terminal residue and a lysine is 15 Å for BS^2G and 19 Å for BS^3.

We then assessed the refinement protocol on the complex between trypsin and bovine pancreatic trypsin inhibitor (BPTI), a stable complex with a K_D value of ~60 fmol/L (Marquart et al. 1983; Kastritis et al. 2011). Based on the known structure of the complex (PDB code 2PTC), there can be a maximum of 17 theoretical inter-subunit lysine-lysine cross-links with BS^3 cross-linking reagent (Table S1). Starting from the structures for the free proteins (PDB codes 4GUX and 1JV8, for trypsin and BPTI, respectively), we fixed the coordinates of trypsin and allowed BPTI to freely rotate and translate as a rigid body. With simulated annealing, we refined the complex structure against the CXMS restraints, with additional van der Waals repulsive term employed. Calculating one structure takes less than 2 min on a single core of Intel Xenon 5620 CPU. Repeating the calculation from different starting positions for the two subunits afforded a set of highly converged structures with overall root-mean-square deviation (RMSD) for backbone heavy atoms almost 0 Å. Importantly, the RMS difference between the CXMS model and the crystal structure was only 0.54 Å (Fig. 1).

Further assessment of the rigid-body refinement protocol

In practice, however, it is rare to have as many as 17 intermolecular cross-links for a complex with the size of trypsin/BPTI (281 residues total and 18 lysine residues). Often, only a few cross-links can be experientially identified. To assess how robust the refinement protocol is with fewer CXMS restraints, we obtained CXMS data from the published studies (Herzog et al. 2012; Kahraman et al. 2013) for the complex between protein phosphatase 2A catalytic subunit (PP2Ac) and immunoglobulin binding protein 1 (IGBP1). PP2Ac and IGBP1 interact with each other with a K_D value of ~300 nmol/L (Jiang et al. 2013), and six intermolecular cross-links were identified between Lys^{28}-Lys^{158}, Lys^{33}-Lys^{166}, Lys^{35}-Lys^{163}, Lys^{40}-Lys^{158}, Lys^{40}-Lys^{163}, and Lys^{40}-Lys^{166} (from PP2Ac to IGBP1) (Herzog et al. 2012). Starting from the structures for free PP2Ac (PDB code 2NYL) and IGBP1 (PDB code 3QC1) proteins, we obtained their complex structures by refining against the CXMS distance restraints. The probabilistic distribution was computed for PP2Ac with respect to IGBP1 in all the structural models and was shown as atomic probability map (Schwieters and Clore 2002), which encompassed the known complex structure (Fig. 2A). Importantly, the overall backbone RMS difference between the CXMS models and the crystal structure for PP2Ac/IGBP1 complex was as small as 2.8 Å (Fig. 2B) (Jiang et al. 2013).

Then what is the minimum number of intermolecular cross-links needed to model the complex structure? With the use of three experimental cross-links involving PP2Ac Lys^{40} (Lys^{40}-Lys^{158}, Lys^{40}-Lys^{163}, and Lys^{40}-Lys^{166}), the resulting structures took up similar positions (Fig. S1A) as the structures calculated using the full set of CXMS restraints, though a bit more scattered. With only one CXMS restraint, for example from PP2Ac Lys^{35} to IGBP1 Lys^{163}, the modeling still afforded a set of CXMS models that are similar to those calculated with the full set of experimental CXMS restraints (Fig. S1B). Thus, the more CXMS restraints were incorporated, the more converged the resulting models were. We also performed the structural refinement using five out of the six cross-links, and then back-calculated the C_α-C_α distance for the unused cross-link. Except for the cross-link between PP2Ac Lys^{28} and IGBP1 Lys^{158}, the calculated distances are mostly within the maximum length stipulated by the corresponding cross-linker (Table S2). Thus, the cross-link between PP2Ac Lys^{28} and IGBP1 Lys^{158} afforded a key restraint about the complex structure, and owing to the sparsity of the intermolecular cross-links, this cross-link is not redundantly provided by other cross-links.

Using CXMS, we characterized the complex between CDK9 and Cyclin-T1. This complex is responsible for transcription elongation, and its two subunits interact with each other at a K_D value of ~300 nmol/L (Baumli et al. 2008). We focused our attention on the intermolecular cross-links that were identified twice or more, for which the probability of being observed by random chance was below 10^{-8} for at least one instance and below 10^{-3} for additional instances (a false discovery rate cutoff of 0.05, an E-value cutoff

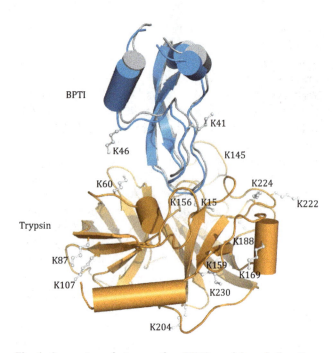

Fig. 1 Comparison between the CXMS model and the X-ray structure for the complex between trypsin and BPTI. The two structures are superimposed by trypsin (*orange cartoon*), and BPTI in the CXMS model and in the crystal structure (PDB code 2PTC) are colored *gray* and *blue*, respectively. The CXMS model was obtained by refining against 17 theoretical inter-molecular cross-links. The RMS difference of backbone heavy atoms between the two complex structures is 0.54 Å. Lysine residues involved are *labeled*

rate of 10^{-3}, spectral count ≥ 2, and the best E-value cutoff of 10^{-8}). With these stringent criteria, it would be unlikely that the cross-links were identified by random chance, and the remaining cross-links should be correctly assigned. Three intermolecular cross-links were identified for CDK9/Cyclin-T1 (Table 1) and the corresponding MS2 spectra are shown in Fig. S2. For each, the two linked lysine residues were found within the maximum length of the cross-linker, as calculated from the known structure of the complex (Baumli et al. 2008).

We treated each subunit in CDK9/Cyclin-T1 as a rigid body, and refined against the intermolecular CXMS distance restraints: two cross-linked lysine residues were restrained to have their C_α-C_α distance to be less than the maximum length of the corresponding cross-linker using a square-well energy potential. Since each intermolecular cross-link was observed with both BS^2G and BS^3 cross-linkers (Table 1), we restrained the C_α-C_α distance to be shorter than the length of BS^2G (20 Å for lysine-lysine cross-links and 15 Å for lysine-protein N terminus cross-links). In the refinement, the coordinates for one subunit, CDK9, were fixed, while the other subunit, Cyclin-T1, was grouped as a rigid body, given full translational and rotational freedoms. A single intermolecular CXMS restraint was readily satisfied, but the

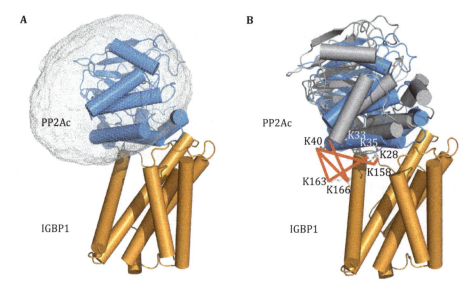

Fig. 2 CXMS model obtained for the complex between PP2Ac and IGBP1. **A** The distribution of PP2Ac with respect to IGBP1 (*orange cartoon*) is shown as atomic probability map, plotted at 30% threshold and shown as *gray meshes*. **B** The RMS difference between the CXMS model (*gray cartoon* for PP2Ac) and the crystal structure of the complex (PDB code 4IYP) can be as small as 2.8 Å. With PP2Ac superimposed, IGBP1 in the crystal structure is shown as *blue cartoon*. Cross-linked lysine residues are *labeled* and the intermolecular cross-links are shown as *red lines*

resulting complex model was poorly converged, with Cyclin-T1 dangling along one side of CDK9 (Fig. S3). As Lys^{74} and Lys^{144} are adjacent to each other in CDK9, cross-links of Cyclin-T1 Lys^{6} to these two residues provided redundant information about the complex structure. Cyclin-T1 Lys^{100} and CDK9 Lys^{56} are located at the other side of the complex; as a result, the refinement against the corresponding cross-link restraint afforded a different but overlapping distribution of the complex. With all three restraints used, a narrower distribution was obtained (Fig. 3A). Significantly, the structural models based on CXMS restraints encompassed the known crystal structure of CDK9/Cyclin-T1, and the pairwise RMS difference between the CXMS model and the PDB structure was as small as 2.86 Å (Fig. 3B). Thus, we show that the CDK9/Cyclin-T1 complex can be modeled as a single conformer, based on sparse CXMS distance restraints.

CXMS analyses of transient and fleeting complexes

We then performed CXMS analysis for EIN/HPr and ubiquitin homodimeric complexes using BS^2G and BS^3. EIN and HPr are involved in signal transduction for bacterial sugar uptake and interact with each other with a K_D value of ~ 7 μmol/L (Suh et al. 2007). Ubiquitin is an important signaling protein in cell and can noncovalently dimerize with a K_D value of ~ 5 mmol/L (Liu et al. 2012). Using the same stringent criteria described above, intermolecular cross-links for the two complexes are also presented in Table 1, and the corresponding MS2 spectra are shown in Figs. S4 and S5. A total of 13 intermolecular cross-links were identified for EIN/HPr, but only one of them (EIN Lys^{58} to HPr Lys^{24}) was found consistent with the stereospecific complex structure (Garrett et al. 1999). For validation, we also performed

Table 1 Intermolecular cross-links observed for transient and fleeting protein complexes

Cross-linked pairs	BS^2G	BS^3	Total spectra	Best E-value[a]	C_α-C_α (Å)[b]	Remarks[c]
Cyclin-T1(6)–CDK9(144)	13	15	28	1.7×10^{-10}	18.5	–
Cyclin-T1(6)–CDK9(74)	9	35	44	3.7×10^{-11}	11.2	–
Cyclin-T1(100)–CDK9(56)	30	25	55	2.5×10^{-13}	9.3	–
EIN(1)–HPr(24)[d]	26	0	26	2.4×10^{-18}	49.1	EC-III
EIN(1)–HPr(49)	23	14	37	1.3×10^{-32}	45	EC-III
EIN(29)–HPr(1)[e]	0	18	18	9.4×10^{-16}	49.1	EC-II
EIN(30)–HPr(1)[e]	0	7	7	1.1×10^{-18}	46.9	EC-II
EIN(30)–HPr(24)	25	74	99	2.1×10^{-24}	36.8	EC-II
EIN(30)–HPr(27)	31	12	43	1.2×10^{-23}	35.3	EC-II
EIN(30)–HPr(49)	3	1	4	7.8×10^{-42}	29.9	EC-II
EIN(30)–HPr(79)[e]	0	10	10	2.2×10^{-10}	49.5	EC-II
EIN(49)–HPr(24)[d]	15	0	15	8.4×10^{-21}	22.3	EC-I
EIN(49)–HPr(49)[e]	0	36	36	2.0×10^{-25}	26.5	EC-I
EIN(49)–HPr(72)[d]	8	0	8	1.3×10^{-18}	35.2	EC-I
EIN(58)–HPr(24)	1	13	14	2.5×10^{-26}	15.4	SC
EIN(238)–HPr(24)[d]	9	0	9	2.5×10^{-18}	56.1	EC-III
Ub(6)–Ub(48)	24	24	48	5.1×10^{-17}	–	–
Ub(6)–Ub(63)	23	2	25	4.5×10^{-17}	–	–
Ub(11)–Ub(48)	3	83	86	1.6×10^{-24}	–	–
Ub(29)–Ub(48)	15	22	37	1.8×10^{-12}	–	–
Ub(33)–Ub(48)	174	78	95	7.5×10^{-24}	–	–
Ub(48)–Ub(48)	67	103	170	3.5×10^{-18}	–	–
Ub(63)–Ub(63)	8	0	8	3.3×10^{-19}	–	–

[a] The best E-value among all the MS2 spectra for each cross-link. E-value is the probability of observing the cross-link by chance

[b] Distance was calculated from the known stereospecific complex structure, PDB accession code 3EZA. No uniquely defined structure is available for the ubiquitin dimer

[c] Designations of the conformational clusters for EIN–HPr complexes

[d] Observed with BS^2G, but not with BS^3

[e] Observed with BS^3, but not with BS^2G

CXMS analysis for EIN/HPr using PDH (Leitner et al. 2014) as the cross-linking reagent.

In order to identify intermolecular cross-links between two ubiquitin subunits in a ubiquitin homodimer, we performed CXMS analysis on a mixture of ^{14}N-labeled (natural isotope abundance) and ^{15}N-labeled ubiquitin proteins (Liu et al. 2012). The cross-links between ^{14}N- and ^{15}N-labeled peptides with characteristic MS1 spectra (Fig. S6) should only arise from intermolecular interactions (Taverner et al. 2002). In this way, we identified a total of seven intermolecular cross-links for the ubiquitin homodimer.

Ensemble structure refinement of protein encounter complexes

To account for the experimental cross-links and to model the structure of EIN/HPr complex, we fixed the position of EIN and treated HPr as a rigid body given rotational and translational freedoms. The intermolecular cross-links could not be satisfied with a single-conformer representation of the complex, as the restraints were consistently violated with an average violation >8 Å (Fig. 4A). This means that in addition to the stereospecific complex, HPr sampled a multitude of conformations with respect to EIN, which were captured by cross-linking. Thus, we invoked ensemble representation for the complex—with EIN fixed, HPr was represented as multiple conformers. We treated each intermolecular cross-link as an ambiguous restraint (Nilges 1995), and defined the CXMS energy averaged over all the conformers in the ensemble with a steep dependence on the C_α-C_α distance. In this way, a CXMS restraint could be satisfied providing that it was accounted for by at least one conformer in the ensemble. The ensemble refinement showed that a minimum of four conformers was required to fully satisfy the intermolecular CXMS restraints with an average distance violation close to 0 Å (Fig. 4A). Too large an ensemble size, however, would lead to over-fitting. When using five conformers to represent the complex, HPr in the additional conformers were found scattering around, making no contribution to the CXMS energy (Fig. S7).

Using a spherical coordinate system, we projected the positions of HPr with respect to EIN in the CXMS models to lower dimensions. In the 2D plot, HPr was found in four distinct clusters (Fig. 4B), thus explaining the requirement of four conformers in the ensemble. One cluster (SC) contained conformers overlapping with the known complex structure, and therefore accounted for the stereospecific EIN/HPr interactions. HPr was positioned away from the specific interface with EIN in the other three clusters (EC-I, EC-II and EC-III), which represented non-specific interactions between EIN and HPr. Each cluster of conformers accounted for multiple intermolecular cross-links (Table 1).

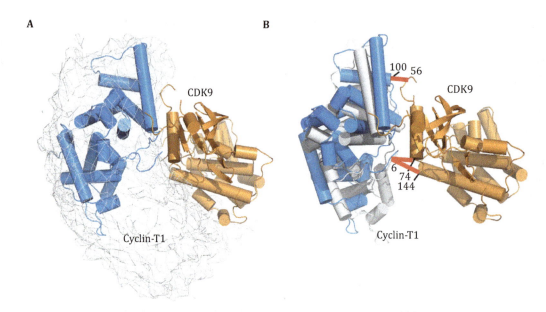

Fig. 3 Structural model for the CDK9/Cyclin-T1 complex refined against intermolecular CXMS restraints. **A** The distribution of Cyclin-T1 with respect to CDK9 (*orange cartoon*) is represented as an atomic probability map plotted at a 10% threshold (*gray mesh*). **B** A selected CXMS model, shown as *orange* and *gray cartoon* for CDK9 and Cyclin-T1, respectively. For comparison, the CDK9 of the crystal structure (PDB code 3BLH) is superimposed, and the Cyclin-T1 crystal structure is shown as a *blue cartoon*. The root-mean-square deviation between the two complex structures is 2.86 Å. Each set of two cross-linked residues is denoted with a *red bar*

We could cross-validate the ensemble structure modeled from lysine-lysine cross-links with the CXMS restraints from a different cross-linking reagent, PDH (Leitner et al. 2014). For a pair of PDH cross-linked glutamate residues, the C_α-C_α distance should be less than 22 Å. With high confidence, the PDH cross-links were identified between EIN Glu[41] and HPr Glu[85] and between EIN Glu[67] and HPr Glu[85] (Fig. S8). Calculated from the stereospecific complex structure (Garrett et al. 1999), the C_α-C_α distances for these two pairs of residues were 41.2 and 12.9 Å, respectively. Clearly, the cross-link between EIN Glu[41] and HPr Glu[85] could not be accounted for with the stereospecific complex structure alone. In the four-conformer ensemble structure modeled from BS^2G/BS^3 CXMS data, however, the averaged C_α-C_α distance between EIN Glu[41] and HPr Glu[85] was 23.1 ± 4.9 Å.

Previously, EIN/HPr complex has been characterized with paramagnetic nuclear magnetic resonance (NMR), and it was shown that EIN and HPr form a multitude of encounter complexes, which facilitate the formation of the stereospecific complex (Tang et al. 2006; Fawzi et al. 2010). Protein encounter complexes are of low occupancies and short lifetimes. Previous NMR studies

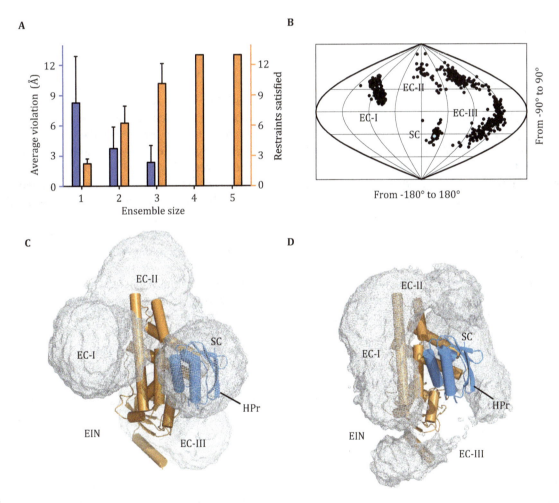

Fig. 4 Ensemble refinement for the complex structure between EIN and HPr. **A** Average violation of CXMS distance restraint (*blue axis* on the *left*) and the number of the satisfied restraints (*orange axis* on the *right*) versus the number of conformers representing the complex. With four or more conformers, all CXMS restraints can be satisfied. **B** *Spherical coordinates* for the four-conformer ensemble structures showing the distribution of HPr with respect to EIN. In each ensemble structure, the HPr is found in four clusters, namely EC-I, EC-II, EC-III, and SC. For comparison, the structure for EIN/HPr stereospecific complex (PDB code 3EZA) is indicated as a *cyan dot*. **C** Atomic probability map of the distribution of HPr with respect to EIN in the ensemble structure refined against intermolecular CXMS restraints. The difference clusters of CXMS conformers are *labeled*. **D** Atomic probability map of the distribution of HPr with respect to EIN in the ensemble structure refined against intermolecular PRE data. The NMR ensemble was calculated based on the previously published data (Tang et al. 2006). EIN is fixed and shown as *orange cartoon*, the distribution of HPr is shown as *gray meshes* and plotted at 20% threshold. For comparison, the stereospecific complex structure is superimposed, with HPr shown as *blue cartoon*, and the four clusters are also marked

estimated that encounter complexes made up less than 10% of the total EIN/HPr complex, thus putting the apparent K_D value for the encounter interactions >10 mmol/L (Fawzi et al. 2010). Importantly, the distribution of HPr relative to EIN modeled on the basis of CXMS data (Fig. 4C) resembles the EIN/HPr encounter complexes previously depicted using NMR spectroscopy (Fig. 4D).

Ensemble structure refinement of a fleeting complex

Performing CXMS experiments on an equimolar mixture of ^{15}N- and ^{14}N-labeled ubiquitin proteins, we identified five inter-molecular cross-links. We fixed the coordinates for one ubiquitin, and allowed the other one to move. A single conformation for the ubiquitin dimer failed to satisfy all the restraints, with average violations ~2 Å. Hence we represented the ubiquitin dimer with two, three, and four conformers, with C_2 non-crystallographic symmetry enforced for each pair of ubiquitin dimer. The CXMS restraints could be satisfied with an $N = 2$ ensemble. Increasing the size of the ensemble did not improve the agreement between experimental and calculated C_α-C_α distances, and the additional conformers in the $N = 3$ and 4 ensemble scattered around with respect to its dimer partner (Fig. S9). Thus, the $N = 2$ ensemble was sufficient to describe the dynamic interactions between two ubiquitin proteins.

In the CXMS models, the two ubiquitins adopt a variety of orientations (Fig. 5A), characteristic of fleeting protein-protein interactions (Liu et al. 2016). This also explains why Lys^{48} in one ubiquitin was able to cross-link to five different lysine residues, except for Lys^{27} and Lys^{63}, in the other ubiquitin. Importantly, the two subunits interacted at the β-sheet region in the CXMS models, and the distribution of the CXMS models was in good agreement with a previous NMR characterization of the ubiquitin homodimer (Fig. 5B).

DISCUSSION

CXMS has been increasingly used to characterize protein-protein interactions and to model protein complex structures (Walzthoeni et al. 2013; Schmidt and Robinson 2014). However, when experimental cross-links cannot be accounted for with a unique structure, previous CXMS applications generally ignored "incompatible" ones or relaxed the C_α-C_α distance restraints (Herzog et al. 2012; Politis et al. 2014). Here we show that CXMS is exquisitely sensitive to encounter

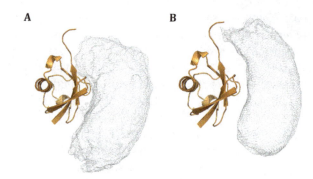

Fig. 5 Ensemble structure for the ubiquitin homodimer. With one ubiquitin subunit fixed (*orange cartoon*), the probabilistic distribution of the other ubiquitin subunit in the dimer is plotted at 20% threshold (*gray meshes*). The ensemble structures of ubiquitin homodimer were calculated by refining against **A** intermolecular CXMS restraints or **B** intermolecular NMR restraints

and fleeting protein-protein interactions that have apparent K_D values in mmol/L, and those seemingly incompatible cross-links contain the information about the dynamics of protein-protein interactions.

To account for the intermolecular cross-links identified with high confidence, we established a rigid-body refinement protocol. The protocol enabled the depiction of the relative subunit distributions in a complex. We first show that the refinement protocol can model the structures of stable complexes to high precision and accuracy. For transient and fleeting ones, however, when a single conformation failed to satisfy all the intermolecular cross-links, we invoked ambiguous distance restraints, in which a distance restraint was accounted for by any one of the conformers in the ensemble (Fig. S10). Demonstrated with EIN/HPr and ubiquitin homodimeric complexes, we showed that the resulting structures satisfied the experimental intermolecular cross-links and recapitulated alternative modes of protein-protein interactions. Moreover, the lysine- and carboxylate-specific cross-links for the EIN/HPr complex corroborate each other, which attests the power of CXMS in revealing the dynamics in protein interactions. Nevertheless, it should be noted that, though a qualitative validation of the ensemble structure can be readily performed, a complete cross-validation may not be feasible owing to the sparsity of the CXMS restraints.

Protein interaction dynamics have been mostly characterized using NMR spectroscopy. Though NMR afforded more structural details than CXMS does, it only works for relatively small protein complexes and requires a large amount of isotopically labeled proteins. In contrast, CXMS is not limited by the size of the proteins, and can be performed on μg or ng of proteins of natural isotope abundance. CXMS is often used conjunction with other techniques like electron microscopy

(EM; Rappsilber 2011; Thalassinos et al. 2013). Nevertheless, the data from other technique are sometimes at odds with the CXMS data (Plaschka et al. 2015). Since proteins dynamically interact with each other, we envision that the ensemble refinement protocol presented herein will allow the reconciliation of different types of data and enable the characterization of subunit rearrangement in these large complexes. The method described herein does not take into account the flexibility of each subunit. Yet we anticipate that CXMS would allow the visualization of the dynamics for each individual protein, providing that a large number of intra-molecular cross-links of high confidence are identified using cross-linking reagents of different lengths and chemical properties.

MATERIALS AND METHODS

Cross-linking reaction and analysis

CDK9, Cyclin-T1, EIN, HPr, and ubiquitin proteins were purified as previously described (Garrett et al. 1999; Baumli et al. 2008; Liu et al. 2012). To prepare ^{15}N-labeled protein, bacterial cells expressing ubiquitin were grown in M9 minimum medium with U-^{15}NH$_4$Cl as the sole nitrogen source. The two subunits in each complex were mixed at a 1:1 ratio—0.6 μmol/L for CDK9/Cyclin-T1, 16 μmol/L for EIN/HPr and 70 μmol/L for the ubiquitin homodimer. Cross-linking reactions were performed at room temperature in 20 mmol/L HEPES buffer (pH 8.0, 7.2 and 7.5 for CDK9/Cyclin-T1, EIN/HPr and ubiquitin, respectively) containing 150 mmol/L NaCl and 0.5 mmol/L BS^3 (Thermo Scientific) or BS^{2}G (Thermo Scientific) for 1 h, and were quenched with 20 mmol/L NH$_4$HCO$_3$. Cross-linking reactions using PDH for EIN/HPr complex were performed at 37 °C in 20 mmol/L HEPES buffer pH 7.2 containing 150 mmol/L NaCl and 11 mmol/L 4-(4,6-dimethoxy-1,3,5-triazin-2-yl)-4-methylmorpholinium chloride for 1 h, and were quenched with 20 mmol/L NH$_4$HCO$_3$. The proteins were subsequently precipitated with ice-cold acetone, air dried, and resuspended in 8 mol/L urea, 100 mmol/L Tris pH 8.5. The cross-linked samples were assessed with SDS-PAGE; about 30%–50% of the protein remains monomeric, whereas the remaining proteins correspond to the singly cross-linked form.

After trypsin (Promega) digestion, LC-MS/MS analysis was performed on an Easy-nLC 1000 UPLC (Thermo Fisher Scientific) coupled with a Q Exactive Orbitrap mass spectrometer (Thermo Fisher Scientific). The top ten most intense precursor ions from each full scan (resolution 70,000) were isolated for MS2 analysis. The pLink (Yang et al. 2012) program was used to search a database containing the sequences of the proteins in question and the cross-linked peptides were identified with the following criteria: false discovery rate smaller than 0.05 followed by an E-value cutoff of 10^{-3} at the spectral level; at the peptide level, spectral count ≥ 2 and the best E-value $<10^{-8}$ for each identification. The lower the E-value, the less likely the putative identification is a false discovery (Yang et al. 2012). For each complex, the cross-linking reaction was repeated twice on different samples, which afforded almost identical cross-links.

To identify the intermolecular cross-links between two ubiquitin molecules, we mixed the ^{15}N- and ^{14}N-labeled (natural isotope abundance) ubiquitin at a 1:1 ratio. The ^{14}N-/^{14}N-labeled and ^{15}N-/^{15}N-labeled cross-linked peptide pairs were identified using pLink (Yang et al. 2012). Based on a strategy previously described (Taverner et al. 2002; Petrotchenko et al. 2014), we assigned cross-links between the ^{15}N and the ^{14}N-labeled peptides as intermolecular if the ratio in mass intensity in liquid chromatography of ^{15}N-/^{14}N-labeled (or ^{14}N-/^{15}N-labeled) cross-linked peptide relative to the corresponding ^{14}N-/^{14}N-labeled (or ^{15}N-/^{15}N-labeled) cross-linked peptide in the extracted ion chromatogram is >0.14. At this ratio, the intermolecular contribution is >25%.

Refinement of protein complex structures

The starting structures for the specific complexes and for constituting proteins were retrieved from the PDB. The accession codes for trypsin, BPTI, and trypsin/BPTI complex are 4GUX, 1JV8, and 2PTC, respectively. The accession codes for PP2Ac and PP2Ac/IGBP1 complex are 2NYL and 4IYP (Jiang et al. 2013), respectively. Only the coordinates for the catalytic core domain were extracted from the PDB structure 2NYL. The coordinates for IGBP1 in the complex were obtained from the PDB structure 3QC1 (free) and 4IYP (bound to PP2Ac). Since many residues in free IGBP1 structure are missing (residues V122–M144), the free structure was spliced with the bound structure, and the resulting structure was solvated in a cubic box containing the TIP3P water molecules with a 10 Å padding in all directions. The structure was subjected 10 ns MD simulation in Amber 14 (Case et al. 2012) to relax the conformation, to generate the initial coordinates for the unbound IGBP1. The accession code for the CDK9/Cyclin-T1 complex was 3BLH. The accession codes for EIN, HPr, and EIN/HPr complexes were 1ZYM, 1POH, and 3EZA (Garrett et al. 1999), respectively. The PDB accession code for ubiquitin monomer is 1UBQ (Vijay-Kumar et al. 1987). The theoretical CXMS distance restraints for trypsin/BPTI were calculated using Xwalk (Kahraman et al. 2011) with 24 Å cutoff. The intermolecular cross-links for PP2Ac/IGBP1 complex were taken from a previous study (Herzog et al. 2012). In that report, the authors identified seven cross-links, one of which involves IGBP1 Lys[306]; since the known structure for IGBP1 encompasses residues 1–221, this cross-link is not used for the structural refinement.

Structural refinement against the CXMS restraints was performed using Xplor-NIH (Schwieters et al. 2006). The refinement started from the coordinates for the free proteins. Each protein subunit was treated as a rigid body, and only CXMS and van der Waals repulsive terms between the subunits are considered. In the refinement, one subunit was fixed, and the other subunit was manipulated with a random rotation and translation, away from the fixed subunit. For each intermolecular cross-link, a square-well energy function was used to enforce the C_α-C_α distance of the cross-linked lysine residues less than 24 and 20 Å for the BS^3 and BS^2G cross-links, respectively (Lee 2009; Kahraman et al. 2011). The upper limits of the distance restraints for cross-linking involving a protein N-terminus were 19 and 15 Å for the BS^3 and BS^2G cross-linkers, respectively. The lengths correspond to a fully extended cross-linker and side chains of two cross-linked residues; no energy penalty was applied when the back-calculated C_α-C_α distance was within the maximally allowed lengths. The penalty for a distance violation was defined as $k\Delta^2$, as the force constant k was gradually ramped from 1 to 30 kcal/(mol · Å^2), as the bath temperature cooled from 3000 K to room temperature in the simulated annealing protocol. Upper limits for BS^2G were used when intermolecular cross-links were observed with both BS^2G and BS^3; upper limits for BS^3 were used for intermolecular cross-links were observed with only BS^3. In addition to the distance restraint derived from CXMS, the restraints also included covalent terms, and van der Waals repulsive energy term. For the ensemble refinement of ubiquitin homodimer, a C_2 non-crystallographic symmetry term was applied for each pair of interacting proteins.

For a protein complex, the structural refinement against CXMS restraints was first performed with a single-conformer ($N = 1$) representation for the complex. All the CXMS restraints could be satisfied for trypsin/BPTI and PP2Ac/IGBP1 complex. For EIN/HPr or ubiquitin/ubiquitin complexes, however, not all the cross-links could be accounted for. Thus we replicate the moving subunit to generate an $N = 2, 3, 4,$ or 5 ensemble to represent the complex, and different conformers in the ensemble can overlap. Ambiguous distance restraints were employed: each restraint was applied to the C_α atom of Lys(i) of the fixed subunit and to the C_α atom of Lys(j) of any conformer of the moving subunit, in which i and j are the residue numbers of cross-linked lysine residues in Table 1. We defined the CXMS energy to be related to inverse sixth power of the distance between the C_α atoms of two cross-linked residues, and to be averaged over all conformers in the ensemble. As a result, the CXMS term has a steep dependence on distance and is biased towards the conformer with the shortest C_α-C_α distance, which can be satisfied providing that one of the conformers in the ensemble has shorter-than-maximum lysine C_α-C_α atom distance. The calculation was repeated 512 times starting from different random positions for each conformer of the moving subunit, and each calculation afforded a slightly different quaternary arrangement of the complex. Structures with no violations against CXMS restraints and no steric clashes were selected for further analysis. The flowchart for the ensemble refinement protocol against CXMS data was illustrated in Fig. S10.

The center-of-mass for one subunit with respect to the other subunit in the each CXMS model was calculated using an in-house Python script. The map projection with spherical coordinates was plotted using Gnuplot. The intermolecular NMR paramagnetic relaxation data were taken from previously published studies for EIN/HPr complex (Tang et al. 2006; Fawzi et al. 2010) and for ubiquitin homodimer (Liu et al. 2012), and ensemble refinement against the NMR data was performed as previously described. Reweighted atomic probability maps depicting the distribution of one subunit relative to another were calculated in Xplor-NIH (Schwieters et al. 2006) and were plotted at respective thresholds (Schwieters and Clore 2002). Structural figures were prepared with PyMOL (the PyMOL molecular graphics system).

Abbreviations

CXMS	Chemical cross-linking of proteins coupled with mass spectrometry analysis
NMR	Nuclear magnetic resonance
EM	Electron microscopy
BS^3	Bis-sulfosuccinimidyl suberate
BS^2G	Bis-sulfosuccinimidyl glutarate
PDH	Pimelic acid dihydrazide
BPTI	Bovine pancreatic trypsin inhibitor
PP2Ac	Phosphatase 2A catalytic subunit

IGBP1   Immunoglobulin binding protein 1
RMSD   Root-mean-square deviation

Acknowledgments This work has been supported by grants from the Chinese Ministry of Science and Technology (2013CB910200), and the National Natural Science Foundation of China (31225007, 31400735, 31400644 and 21375010). The research of C.T. was supported in part by an International Early Career Scientist Grant from the Howard Hughes Medical Institute.

Compliance with Ethical Standards

Conflict of Interest Zhou Gong, Yue-He Ding, Xu Dong, Na Liu, E. Erquan Zhang, Meng-Qiu Dong, and Chun Tang declare that they have no conflict of interest.

Human and Animal Rights and Informed Consent This article does not contain any studies with human or animal subjects performed by the any of the authors.

References

Baumli S, Lolli G, Lowe ED, Troiani S, Rusconi L, Bullock AN, Debreczeni JE, Knapp S, Johnson LN (2008) The structure of P-TEFb (CDK9/cyclin T1), its complex with flavopiridol and regulation by phosphorylation. EMBO J 27:1907–1918

Berg OG, Winter RB, Von Hippel PH (1981) Diffusion-driven mechanisms of protein translocation on nucleic acids. 1. Models and theory. Biochemistry (Mosc) 20:6929–6948

Case DA, Darden TA, Cheatham TEI, Simmerling CL, Wang J, Duke RE, Luo R, Walker RC, Zhang W, Merz KM, Roberts B, Hayik S, Roitberg A, Seabra G, Swails J, Goetz AW, Kolossváry I, Wong KF, Paesani F, Vanicek J, Wolf RM, Liu J, Wu X, Brozell SR, Steinbrecher T, Gohlke H, Cai Q, Ye X, Wang J, Hsieh MJ, Cui G, Roe DR, Mathews DH, Seetin MG, Salomon-Ferrer R, Sagui C, Babin V, Luchko T, Gusarov S, Kovalenko A, Kollman PA (2012) AMBER 12. University of California, San Francisco

Fawzi NL, Doucleff M, Suh JY, Clore GM (2010) Mechanistic details of a protein–protein association pathway revealed by paramagnetic relaxation enhancement titration measurements. Proc Natl Acad Sci USA 107:1379–1384

Gabdoulline RR, Wade RC (2002) Biomolecular diffusional association. Curr Opin Struct Biol 12:204–213

Garrett DS, Seok YJ, Peterkofsky A, Gronenborn AM, Clore GM (1999) Solution structure of the 40,000 Mr phosphoryl transfer complex between the N-terminal domain of enzyme I and HPr. Nat Struct Biol 6:166–173

Herzog F, Kahraman A, Boehringer D, Mak R, Bracher A, Walzthoeni T, Leitner A, Beck M, Hartl FU, Ban N, Malmstrom L, Aebersold R (2012) Structural probing of a protein phosphatase 2A network by chemical cross-linking and mass spectrometry. Science 337:1348–1352

Jiang L, Stanevich V, Satyshur KA, Kong M, Watkins GR, Wadzinski BE, Sengupta R, Xing Y (2013) Structural basis of protein phosphatase 2A stable latency. Nat Commun 4:1699

Jones S, Thornton JM (1996) Principles of protein–protein interactions. Proc Natl Acad Sci USA 93:13–20

Kahraman A, Malmstrom L, Aebersold R (2011) Xwalk: computing and visualizing distances in cross-linking experiments. Bioinformatics 27:2163–2164

Kahraman A, Herzog F, Leitner A, Rosenberger G, Aebersold R, Malmstrom L (2013) Cross-link guided molecular modeling with ROSETTA. PLoS One 8:e73411

Kalisman N, Adams CM, Levitt M (2012) Subunit order of eukaryotic TRiC/CCT chaperonin by cross-linking, mass spectrometry, and combinatorial homology modeling. Proc Natl Acad Sci USA 109:2884–2889

Kastritis PL, Moal IH, Hwang H, Weng Z, Bates PA, Bonvin AM, Janin J (2011) A structure-based benchmark for protein–protein binding affinity. Protein Sci 20:482–491

Lasker K, Forster F, Bohn S, Walzthoeni T, Villa E, Unverdorben P, Beck F, Aebersold R, Sali A, Baumeister W (2012) Molecular architecture of the 26S proteasome holocomplex determined by an integrative approach. Proc Natl Acad Sci USA 109:1380–1387

Lee YJ (2009) Probability-based shotgun cross-linking sites analysis. J Am Soc Mass Spectrom 20:1896–1899

Leitner A, Joachimiak LA, Unverdorben P, Walzthoeni T, Frydman J, Forster F, Aebersold R (2014) Chemical cross-linking/mass spectrometry targeting acidic residues in proteins and protein complexes. Proc Natl Acad Sci USA 111:9455–9460

Liu Z, Zhang WP, Xing Q, Ren X, Liu M, Tang C (2012) Noncovalent dimerization of ubiquitin. Angew Chem Int Ed Engl 51:469–472

Liu Z, Gong Z, Dong X, Tang C (2016) Transient protein–protein interactions visualized by solution NMR. Biochim Biophys Acta 1864(1):115–122

Lossl P, Kolbel K, Tanzler D, Nannemann D, Ihling CH, Keller MV, Schneider M, Zaucke F, Meiler J, Sinz A (2014) Analysis of nidogen-1/laminin gamma1 interaction by cross-linking, mass spectrometry, and computational modeling reveals multiple binding modes. PLoS One 9:e112886

Marquart M, Walter J, Deisenhofer J, Bode W, Huber R (1983) The geometry of the reactive site and of the peptide groups in trypsin, trypsinogen and its complexes with inhibitors. Acta Crystallogr B 39:480–490

Merkley ED, Rysavy S, Kahraman A, Hafen RP, Daggett V, Adkins JN (2014) Distance restraints from crosslinking mass spectrometry: mining a molecular dynamics simulation database to evaluate lysine–lysine distances. Protein Sci 23:747–759

Nilges M (1995) Calculation of protein structures with ambiguous distance restraints. Automated assignment of ambiguous NOE crosspeaks and disulphide connectivities. J Mol Biol 245:645–660

Nooren IM, Thornton JM (2003) Diversity of protein–protein interactions. EMBO J 22:3486–3492

Petrotchenko EV, Serpa JJ, Makepeace KA, Brodie NI, Borchers CH (2014) (14)N(15)N DXMSMS Match program for the automated analysis of LC/ESI-MS/MS crosslinking data from experiments using (15)N metabolically labeled proteins. J Proteomics 109:104–110

Plaschka C, Lariviere L, Wenzeck L, Seizl M, Hemann M, Tegunov D, Petrotchenko EV, Borchers CH, Baumeister W, Herzog F, Villa E, Cramer P (2015) Architecture of the RNA polymerase II-Mediator core initiation complex. Nature 518:376–380

Politis A, Stengel F, Hall Z, Hernandez H, Leitner A, Walzthoeni T, Robinson CV, Aebersold R (2014) A mass spectrometry-based hybrid method for structural modeling of protein complexes. Nat Methods 11:403–406

Rappsilber J (2011) The beginning of a beautiful friendship: cross-linking/mass spectrometry and modelling of proteins and multi-protein complexes. J Struct Biol 173:530–540

Rinner O, Seebacher J, Walzthoeni T, Mueller LN, Beck M, Schmidt A, Mueller M, Aebersold R (2008) Identification of cross-linked peptides from large sequence databases. Nat Methods 5:315–318

Schilder J, Ubbink M (2013) Formation of transient protein complexes. Curr Opin Struct Biol 23:911–918

Schmidt C, Robinson CV (2014) Dynamic protein ligand interactions—insights from MS. FEBS J 281:1950–1964

Schreiber G, Fersht AR (1996) Rapid, electrostatically assisted association of proteins. Nat Struct Biol 3:427–431

Schwieters CD, Clore GM (2002) Reweighted atomic densities to represent ensembles of NMR structures. J Biomol NMR 23:221–225

Schwieters CD, Kuszewski JJ, Clore GM (2006) Using Xplor-NIH for NMR molecular structure determination. Prog Nucl Magn Reson Spectrosc 48:47–62

Suh JY, Tang C, Clore GM (2007) Role of electrostatic interactions in transient encounter complexes in protein–protein association investigated by paramagnetic relaxation enhancement. J Am Chem Soc 129:12954–12955

Tang C, Iwahara J, Clore GM (2006) Visualization of transient encounter complexes in protein–protein association. Nature 444:383–386

Tang C, Louis JM, Aniana A, Suh JY, Clore GM (2008) Visualizing transient events in amino-terminal autoprocessing of HIV-1 protease. Nature 455:U692–U693

Taverner T, Hall NE, O'Hair RA, Simpson RJ (2002) Characterization of an antagonist interleukin-6 dimer by stable isotope labeling, cross-linking, and mass spectrometry. J Biol Chem 277:46487–46492

Thalassinos K, Pandurangan AP, Xu M, Alber F, Topf M (2013) Conformational states of macromolecular assemblies explored by integrative structure calculation. Structure 21:1500–1508

The PyMOL molecular graphics system, Version 1.7.4 Schrödinger, LLC

Vijay-Kumar S, Bugg CE, Cook WJ (1987) Structure of ubiquitin refined at 1.8 A resolution. J Mol Biol 194:531–544

Vinogradova O, Qin J (2012) NMR as a unique tool in assessment and complex determination of weak protein–protein interactions. Top Curr Chem 326:35–45

Walzthoeni T, Leitner A, Stengel F, Aebersold R (2013) Mass spectrometry supported determination of protein complex structure. Curr Opin Struct Biol 23:252–260

Xing Q, Huang P, Yang J, Sun JQ, Gong Z, Dong X, Guo DC, Chen SM, Yang YH, Wang Y, Yang MH, Yi M, Ding YM, Liu ML, Zhang WP, Tang C (2014) Visualizing an ultra-weak protein–protein interaction in phosphorylation signaling. Angew Chem Int Ed Engl 53:11501–11505

Yang B, Wu YJ, Zhu M, Fan SB, Lin J, Zhang K, Li S, Chi H, Li YX, Chen HF, Luo SK, Ding YH, Wang LH, Hao Z, Xiu LY, Chen S, Ye K, He SM, Dong MQ (2012) Identification of cross-linked peptides from complex samples. Nat Methods 9:904–906

Zheng C, Yang L, Hoopmann MR, Eng JK, Tang X, Weisbrod CR, Bruce JE (2011) Cross-linking measurements of in vivo protein complex topologies. Mol Cell Proteomics 10:M110 006841

Roles of H3K36-specific histone methyltransferases in transcription: antagonizing silencing and safeguarding transcription fidelity

Chang Huang[1], Bing Zhu[1,2]

[1] National Laboratory of Biomacromolecules, CAS Center for Excellence in Biomacromolecules, Institute of Biophysics, Chinese Academy of Sciences, Beijing 100101, China
[2] College of Life Sciences, University of Chinese Academy of Sciences, Beijing 100049, China

Abstract Histone H3K36 methylation is well-known for its role in active transcription. In *Saccharomyces cerevisiae*, H3K36 methylation is mediated solely by SET2 during transcription elongation. In metazoans, multiple H3K36-specific methyltransferases exist and contribute to distinct biochemical activities and subsequent functions. In this review, we focus on the H3K36-specific histone methyltransferases in metazoans, and discuss their enzymatic activity regulation and their roles in antagonizing Polycomb silencing and safeguarding transcription fidelity.

Keywords H3K36 methylation, Histone methyltransferase, SETD2, Ash1L, NSD

INTRODUCTION

Chromatin features, including DNA modifications, histone modifications, histone variants, nucleosome occupation, and chromatin organization, regulate the regional accessibility of chromatin and thus modulate various chromatin-based biological processes, including replication, transcription, and repair. Histone acetylation and methylation at lysine residues are two of the most studied histone modifications, and they have interesting differences. Histone acetylation generally promotes active transcription by altering the positive charge at the lysine residues and the interactions between DNA and histone tails. Therefore, with few exceptions, the majority of histone acetyltransferases and deacetylases display broad substrate specificity and function for multiple lysine residues on various histones (Kouzarides 2007b). In contrast, histone lysine methylation does not change the charge at the histone tails and these methylated lysine moieties function by recruiting downstream reader proteins that are involved in gene activation or repression. The reader proteins generally display site-specificity due to the recognition of neighboring residues. This is probably the reason that histone methyltransferases and demethylases have co-evolved to have specific lysine site preferences (Kouzarides 2007a).

Histone H3K36 methylation is a hallmark of active transcription. Pioneer studies on budding yeast SET2, the first H3K36 methyltransferase (Strahl et al. 2002), have established a paradigm for the recruitment of SET2 and the function of H3K36 methylation: SET2 is recruited by Ser-2-phosphorylated Pol II during elongation to deposit H3K36me3, which functions as a docking site for the Rpd3S histone deacetylase complex to suppress cryptic transcription initiation (Venkatesh and Workman 2013). In metazoans, the characterization of multiple H3K36-specific methyltransferases has expanded the function of H3K36 methylation from transcription elongation to developmental gene regulation (Wagner and Carpenter 2012). In metazoans, SETD2 (also known as HYPB) is the sole enzyme responsible for H3K36me3 (Edmunds et al. 2008; Yuan

✉ Correspondence: zhubing@ibp.ac.cn (B. Zhu)

et al. 2009); MES-4 (maternal-effect sterile 4) in *C. elegans* and *Drosophila* and its mammalian homologs, the NSD (nuclear receptor-binding SET domain) family proteins (including NSD1, NSD2, and NSD3), are the main contributors of global H3K36me2 (Bell et al. 2007; Bender et al. 2006; Kuo et al. 2011; Li et al. 2009b); Ash1 (absent, small and homeotic-1) in *Drosophila* and its mammalian homolog Ash1L (Ash1-like) can also produce H3K36me2, but they are limited more to the active Hox genes (An et al. 2011; Huang et al. 2017; Miyazaki et al. 2013; Schmahling et al. 2018; Tanaka et al. 2007; Yuan et al. 2011). MES-4 and Ash1 play critical roles in maintaining developmental gene expression, and the mutations and deregulation of human NSD family proteins and Ash1L are linked to various developmental diseases (Bennett et al. 2017; Rogawski et al. 2016; Wagner and Carpenter 2012). Moreover, the recent discoveries that "oncohistones" with H3K36M/I mutations can drive chondroblastoma tumorigenesis by inhibiting H3K36 methyltransferases and reprogramming the H3K36 methylation landscape further underscore the functional significance of H3K36 methylation (Fang et al. 2016; Lu et al. 2016). *In vitro*, H3K36 methylation directly inhibits PRC2, which catalyzes repressive H3K27 methylation (Schmitges et al. 2011; Yuan et al. 2011). *In vivo*, the genomic landscape of H3K36 and H3K27 methylation are anti-correlated (Gaydos et al. 2012; Lu et al. 2016; Papp and Muller 2006; Popovic et al. 2014; Yuan et al. 2011). These findings clearly underscore the role of H3K36 methylation in antagonizing Polycomb silencing. On the other hand, the distinct regulation of each H3K36-specific methyltransferase remains unclear. Here, we summarize the recent progress regarding the regulation of H3K36 methyltransferase activity and their roles in transcription. Notably, H3K36 methylation also participates in other aspects of chromatin events such as DNA repair and mRNA splicing, discussion of which is beyond the scope of this review, but has been reviewed by others (Fahey and Davis 2017; Li 2013; McDaniel and Strahl 2017; Wagner and Carpenter 2012).

AUTO-INHIBITION IS A CONSERVED REGULATORY MECHANISM OF H3K36 METHYLTRANSFERASES

Studies reporting the characterization of the substrate specificity of H3K36-specific methyltransferases, especially dimethylases, displayed quite a number of disputes. After years of study, chromatin researchers have adopted the common belief that SETD2 is the sole enzyme responsible for H3K36me3, NSD family enzymes are the main contributors of H3K36me2, and Ash1/Ash1L is an enzyme governing H3K36me2 at specific regions (An et al. 2011; Dorighi and Tamkun 2013; Edmunds et al. 2008; Huang et al. 2017; Kuo et al. 2011; Li et al. 2009b; Miyazaki et al. 2013; Qiao et al. 2011; Streubel et al. 2018; Tanaka et al. 2007; Yuan et al. 2009, 2011). One likely explanation for the initial conflicting observations is that most H3K36-specific methyltransferases are nucleosome-specific enzymes; these methyltransferases are highly specific for H3K36 methylation at nucleosome substrates, but they display weak non-specific activities for non-nucleosomal histones (An et al. 2011; Byrd and Shearn 2003; Gregory et al. 2007; Li et al. 2009b; Tanaka et al. 2007; Yuan et al. 2009). The structural basis for the nucleosome-specific activities of H3K36-specific methyltransferases has not yet been resolved, and the exact molecular mechanisms of how nucleosomes confine the specificity and stimulate the catalytic activity of H3K36-specific methyltransferases remain unclear. Nevertheless, the structures of the catalytic domains of all three sub-types of H3K36 methyltransferases have been resolved, and interestingly, all of them share a conserved auto-inhibitory mechanism (Fig. 1) (An et al.

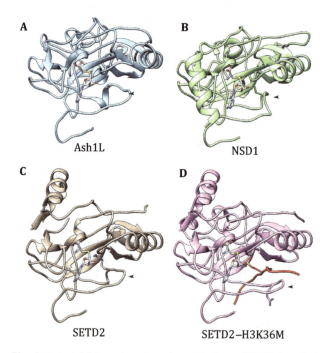

Fig. 1 Auto-inhibitory loop is a shared feature of H3K36 methyltransferases that undergoes dynamic changes during catalysis. Structures of the catalytic domain of Ash1L (PDB code: 3OPE) (**A**), NSD1 (PDB code: 3OOI) (**B**), SETD2 (PDB code: 4H12) (**C**), and SETD2 bound with the H3K36 M peptide (PDB code: 5JJY) (**D**) are shown with arrowheads indicating the auto-inhibitory loop. Note that the side chain of S2259(Ash1L)/C211(NSD1)/R1670(SETD2) within the inhibitory loop occupies the positioning pocket of H3K36. In the SETD2–H3K36 M complex, R1670 flips out and allows the catalytic center to accommodate substrate binding

2011; Qiao et al. 2011; Zheng et al. 2012). A loop at the post-SET region occupies the binding channel for the histone H3 tail, thus blocking the access of lysine 36 to the catalytic center. Obviously, this loop must undergo a conformational change to remove this steric hindrance upon activation. Indeed, a half-opened conformation of the inhibitory loop was observed in SETD2, indicating the dynamic nature of this loop (Yang et al. 2016). Furthermore, when engaged with a K36M-mutated H3 peptide, which mimics the methylated product but cannot be released from the catalytic center, a fully opened state was observed (Fig. 1D) (Yang et al. 2016). In addition, biochemical studies suggest that nucleosomal DNA may act as an allosteric effector for NSD proteins (Li et al. 2009b), and computational docking and simulation suggest that the inhibitory loop of NSD1 may come into contact with the DNA and lead to lysine binding channel widening (Qiao et al. 2011). Therefore, it is reasonable to speculate that the engagement of nucleosome substrates with H3K36-specific methyltransferases may alter the auto-inhibitory loop to a conformation that favors H3K36 methylation catalysis.

Auto-inhibition is also an intrinsic characteristic of E(z), the catalytic subunit of the H3K27 methyltransferase PRC2. While E(z) alone is inactive (Antonysamy et al. 2013; Wu et al. 2013), engagement with two other core subunits of PRC2—EED and SUZ12—alters the configuration of its catalytic center and transforms it into an active conformation (Broaun et al. 2016; Jiao and Liu 2015; Justin et al. 2016). Therefore, another interesting speculation is that certain interaction partner(s) may induce a conformational change and activate H3K36-specific methyltransferases. SETD2 and Ash1/Ash1L have stable interaction partners (Huang et al. 2017; Schmahling et al. 2018; Yuan et al. 2009). In human cells, SETD2 interacts stably with HnRNP-L, which facilitates H3K36me3 deposition in vivo. However, HnRNP-L does not stimulate the enzymatic activity of SETD2 in vitro, which suggests that HnRNP-L does not function as a catalytic activator of SETD2 (Yuan et al. 2009). Recently, we and others demonstrated that Drosophila Ash1 and human Ash1L form stable complexes with Mrg15 (or human Morf4L1/2) and Nurf55 (or human RbAp46/48) (Huang et al. 2017; Schmahling et al. 2018). Interestingly, MRG domain-containing proteins, including Mrg15 and human Morf4L1/2, stimulate the catalytic activity of Ash1/Ash1L significantly. It will be interesting to determine whether Mrg15 activates Ash1 by inducing a conformational change that eliminates the blockage of the catalytic center by the auto-inhibitory loop of Ash1. Although the exact mechanism of induction remains to be determined, this is the first case of allosteric activation directed by an interaction partner among H3K36-specific methyltransferases.

In addition to the positive regulation of catalytic activities, all three sub-types of mammalian H3K36-specific methyltransferases are inhibited directly by H2A ubiquitination (Yuan et al. 2013). Whether such negative regulation involves the stabilization of the auto-inhibitory loop is an interesting question for future exploration.

The physiological roles of this auto-inhibition remain unclear, but this process has been proposed to protect enzymes from hyperactivation (Wang et al. 2015). An intriguing hypothesis is that auto-inhibition and its bypass are ideal interfaces for additional regulation, such as signaling events. Thus, the catalytic process of H3K36-specific methyltransferases is subject to intricate regulations directed by both intrinsic and external mechanisms.

DIVERSIFIED CHROMATIN RECRUITMENT OF H3K36-SPECIFIC METHYLTRANSFERASES ASSIGNS DISTINCT PHYSIOLOGICAL FUNCTION TO H3K36 METHYLATION

In Saccharomyces cerevisiae, SET2 generates all forms of H3K36 methylation, with trimethylation being the primary effective mark; however, H3K36me2 can work as efficiently as H3K36me3 in recruiting Rpd3S and suppressing cryptic initiation (Li et al. 2009a), suggesting indiscriminate roles of di- and trimethylation under this context. In metazoans, ChIP-sequencing showed that H3K36me2 demarcates chromatin differently from H3K36me3: within genic regions, H3K36me2 is preferentially enriched proximal to the transcription start sites and gradually decays downstream into the H3K36me3-enriched 3' region; in addition, the large amount of H3K36me2 spread across intergenic regions implies a role at distal regulatory elements (Kuo et al. 2011). Although SETD2 preserves the capacity to generate all three states of methylation, SETD2 depletion affects only H3K36me3, not H3K36me1/2, at the bulk level (Edmunds et al. 2008; Yuan et al. 2009), indicating the specific assignment of trimethylation for transcription elongation. Moreover, the evolvement of several methyltransferases specific for H3K36me2 and the fact that mutations and deregulation of these enzymes cause varied developmental diseases further suggest that H3K36me2 may have a function distinct from that of H3K36me3. In convergence, these di-methyltransferases are recruited to chromatin differently from SETD2 (see discussion below), which expands the H3K36 methylation territory and diversifies the function of H3K36 methylation. H3K36 also exists in mono-methylated state. However, as an intermediate product of methylation reactions catalyzed by the aforesaid di- and tri-methyltransferases, the chromatin distribution and functional role of H3K36me1 remain poorly characterized.

SET2/SETD2 and H3K36me3: recruited by elongating Pol II to safeguard transcription fidelity

As mentioned, in *Saccharomyces cerevisiae*, SET2 is recruited by the Ser-2-phosphorylated C-terminal domain (CTD) of elongating RNA polymerase II, and it deposits H3K36 methylation at the gene bodies of active genes (Krogan et al. 2003; Li et al. 2002, 2003; Xiao et al. 2003). Successive transcription may cause histone hyperacetylation in gene bodies, which allow cryptic transcription initiation. To prevent such deleterious events, transcription elongation-coupled H3K36 methylation serves as a docking site for the histone deacetylase complex Rpd3S, which restores the repressive chromatin environment following Pol II passage to prevent cryptic transcription initiation (Carrozza et al. 2005; Joshi and Struhl 2005; Keogh et al. 2005; Li et al. 2007).

Transcription elongation-coupled SETD2 recruitment and H3K36me3 deposition are conserved in mammals, as well as the role of H3K36me3 to prevent aberrant transcription initiation; however, whether the repressive environment promoted by H3K36me3 depends on histone deacetylases remains unexplored. Intriguingly, in mammals, the PWWP domain containing *de novo* DNA methyltransferases DNMT3A/B is recruited by H3K36me3 to methylate intragenic DNA, which may, in turn, recruit methylated DNA-binding proteins and histone deacetylases (Jones 2012; Neri et al. 2017), to safeguard transcription initiation. Indeed, mouse ES cells lacking DNA methylation exhibited intragenic transcription initiation (Neri et al. 2017).

Overall, SETD2 deposits H3K36me3 in gene bodies to recruit downstream machineries to restore the non-permissive chromatin state following Pol II passage and to maintain transcription fidelity at the genome level.

H3K36me2 methyltransferases and H3K36me2: demarcating active chromatin by antagonizing silencing

Loss-of-function studies in multiple species, including *C. elegans*, *Drosophila*, and mammalian cells, indicate that NSD family proteins are responsible for bulk chromatin H3K36me2 levels (Bell et al. 2007; Bender et al. 2006; Kuo et al. 2011), implying the widespread distribution of these enzymes.

In *C. elegans*, MES-4, a homolog of mammalian NSD proteins, is vital for germ cell viability (Bender et al. 2006) and is highly abundant in H3K36me2-enriched autosomes, but not in the H3K27me3-enriched X chromosome in germline cells (Bender et al. 2006; Fong et al. 2002). Moreover, an MES-4 ChIP-chip analysis of early embryos revealed that MES-4 is distributed around the gene bodies of approximately 20% of genes, among which germline-specific genes are highly enriched (Rechtsteiner et al. 2010). MES-4 signals arise near TSS regions, peak proximally, and gradually decrease towards the 3′ end of gene bodies, correlating well with the pattern of H3K36me2 at genic regions (Rechtsteiner et al. 2010).

In *Drosophila*, MES-4 is also enriched at the 5′ end of target genes (Bell et al. 2007). Importantly, dMES-4 is recruited by an insulator-binding protein to promote the transcription of flanking genes by antagonizing the spread of H3K27 methylation from nearby regions (Lhoumaud et al. 2014); this finding revealed the functional role of the MES-4/NSD family of enzymes at the intergenic cis-regulatory regions.

In mammals, there are three NSD family proteins, and their chromatin localization has not been thoroughly analyzed. Knocking down NSD1 in ESC cells reduces H3K36me2 levels throughout the genome—at gene promoters, gene bodies, and intergenic regions (Streubel et al. 2018). NSD2 localizes to active transcripts, with a greater preference for elongating regions and distal regulatory regions (Ram et al. 2011). Consistently, in multiple myeloma, t(4;14) chromosomal translocation resulting in NSD2 overexpression led to the aberrant accumulation of H3K36me2 at both the intragenic and intergenic regions, supporting the widespread targeting of NSD2 (Kuo et al. 2011; Popovic et al. 2014). The distribution of full-length NSD3 has not been reported, but a short isoform of NSD3 possessing the PWWP domain localizes preferentially to enhancers and promoters (Shen et al. 2015). Importantly, in chondroblastomas, recurrent H3K36M mutations reprogrammed the transcriptome through inhibiting and sequestering H3K36 methyltransferases, resulting in a global reduction in H3K36 methylation; in this process, intragenic and intergenic H3K36me2 were mediated largely by NSD proteins (Fang et al. 2016; Lu et al. 2016). These findings further underscore the physiological significance of the broad targeting of NSD proteins.

NSD proteins can interact with nuclear receptors, suggesting their recruitment by transcription factors (Huang et al. 1998); however, the general targeting mechanism remains unknown, especially for the intergenic regions. Notably, NSD proteins contain multiple chromatin reader modules, including the PWWP and PHD domains, which may contribute to the spread of NSD proteins (He et al. 2013; Sankaran et al. 2016). Overall, the NSD family of proteins targets numerous genes involved in many development pathways and extensive intergenic regions.

Different from the MES-4/NSD family of enzymes, the other H3K36me2-specific methyltransferase Ash1 functions as a trithorax protein in *Drosophila* to maintain the expression of a small collection of developmental genes, the HOX genes. ChIP analysis showed that Ash1 is distributed throughout its target genes (Huang et al. 2017; Schwartz et al. 2010). A subset of cis-regulatory elements in the *Drosophila* genome that can recruit Trithorax/Polycomb group proteins was identified and defined as Trithorax/Polycomb response elements (TRE/PRE) (Ringrose and Paro 2007). Transgenic TRE/PREs can recruit Trithorax/Polycomb proteins ectopically to maintain the active/repressive states of reporter genes, underscoring their significance in Trithorax/Polycomb recruitment. Importantly, the sequences of TRE/PREs are the same in different cell types, but Trithorax and Polycomb proteins, including Ash1, have different locations in each cell type, suggesting additional regulators beyond DNA sequences. Moreover, Ash1 cooperates with other Trithorax group members to maintain Hox gene expression, among which Trithorax and Kismet may directly promote the chromatin recruitment of Ash1: Ash1 and the N-terminus of Trithorax display an interdependency on chromatin localization (Schwartz et al. 2010); knocking down kismet, a CHD family chromatin remodeler, greatly reduces the chromatin retention of Ash1 (Srinivasan et al. 2008), suggesting that chromatin accessibility directed by a chromatin remodeler also affects Ash1 recruitment. Given that the Ash1 protein also contains multiple chromatin binding domains, including Bromo, BAH, and PHD, and that its partner protein Mrg15 also contains a Chromo domain that recognizes H3K36 methylation, it is natural to expect that the recognition of a combination of histone modifications will contribute another regulatory layer to Ash1 recruitment. Taken together, DNA elements, transcription factors, and the chromatin environment may function coordinately to shape the binding profile of Ash1. The recruitment of Ash1L in mammalian systems is not well studied, and PREs/TREs are not defined in mammals. Despite this, Ash1L also regulates the HOX genes in mammals, indicating the conservation of the recruitment and function of Ash1L (Miyazaki et al. 2013).

Overall, NSD proteins and Ash1L are associated with active transcription, the malfunction of which leads to gene inactivation. Mechanistically, H3K36me2/3 inhibits the catalytic activity of PRC2 (Schmitges et al. 2011; Yuan et al. 2011), and the mutually exclusive distribution of H3K36 methylation and H3K27 methylation along chromatin has been observed in many biological systems (Gaydos et al. 2012; Lu et al. 2016; Papp and Muller 2006; Popovic et al. 2014; Yuan et al. 2011). While the anti-silencing mechanisms of NSD proteins and Ash1L are similar, it would be of great interest to uncover their distinct recruitment mechanisms, which will help in the understanding of the biological impact of these distinct enzymes.

SUMMARY AND PERSPECTIVES

In higher eukaryotes, in addition to its conservative role in transcription elongation, H3K36 methylation has an additional function: anti-silencing. In addition to targeting certain developmental genes, PRC2-mediated transcription silencing through H3K27 methylation seems to establish and maintain a default repressive state for most of the inert genome. PRC2 may be recruited initially by cis-elements and then propagate along chromatin through a reinforced spreading mechanism (Wang et al. 2018). On the other hand, active genes and regulatory regions have adapted multiple mechanisms, including installation of H3K4 and H3K36 methylation, as well as an open chromatin status, which are all repulsive substrates for PRC2 catalysis, to overcome the silencing effect (Schmitges et al. 2011; Yuan et al. 2011, 2012). Both H3K36me2 and H3K36me3 can inhibit PRC2 activity efficiently *in vitro* (Schmitges et al. 2011; Yuan et al. 2011). However, they may antagonize PRC2 in different ways *in vivo*: PRC2 generally targets promoters and enhancers, but not gene bodies, for the initiation of silencing, so protecting these cis-elements with H3K36me2 via NSD/Ash1L may abolish the initial recruitment of PRC2; in addition, the general H3K36me2 enrichment and the gene body-enriched H3K36me3 at active genes may inhibit the spread of H3K27 methylation from adjacent regions (Fig. 2). Taken together, we propose that NSD/Ash1L-mediated H3K36me2 may act as the primary effector to actively repel PRC2 silencing at developmental genes, thus maintaining their expression.

Although we attempt to dissect the specific roles of each H3K36 methyltransferase and the different forms of H3K36 methylation to clarify their intrinsic biological functions, they often work in concert and affect each other in many cases. For instance, the transcription of

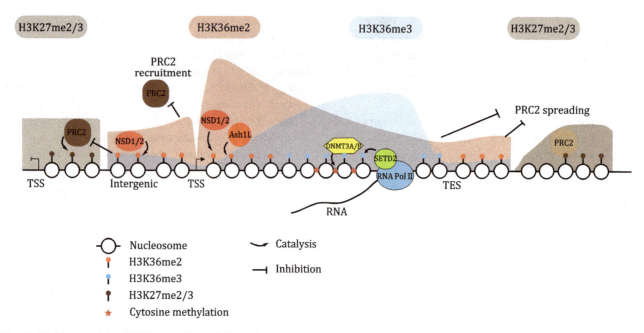

Fig. 2 Working model of H3K36-specific methyltransferases and H3K36 methylation in mammalian gene transcription regulation

genes maintained by the NSD family proteins/Ash1L will certainly upregulate SETD2 and H3K36me3, and the loss of the NSD family proteins/Ash1L, resulting in transcription inactivation, will surely cause the loss of SETD2 and H3K36me3.

Now two decades old, our knowledge of H3K36-specific methyltransferases and H3K36 methylation is still expanding. The links between the deregulation of H3K36-specific methyltransferases and various biological outcomes in diseases have not been fully established. Hopefully, a thorough understanding of the mechanism and function of H3K36-specific methyltransferases will help to pave the way for designing specific and rational targeting strategies for these diseases in the future.

Acknowledgements The authors are supported by grants from the Natural Science Foundation of China (31530037, 31701101), the Chinese Ministry of Science and Technology (2017YFA0504100), and the Youth Innovation Promotion Association of the Chinese Academy of Sciences (2018127).

Compliance with Ethical Standards

Conflict of interest Chang Huang and Bing Zhu declare that they have no conflict of interest.

Human and animal rights and informed consent This article does not contain any studies with human or animal subjects performed by any of the authors.

References

An S, Yeo KJ, Jeon YH, Song JJ (2011) Crystal structure of the human histone methyltransferase ASH1L catalytic domain and its implications for the regulatory mechanism. J Biol Chem 286:8369–8374

Antonysamy S, Condon B, Druzina Z, Bonanno JB, Gheyi T, Zhang F, MacEwan I, Zhang A, Ashok S, Rodgers L, Russell M, Gately LuzJ (2013) Structural context of disease-associated mutations and putative mechanism of autoinhibition revealed by X-ray crystallographic analysis of the EZH2-SET domain. PLoS ONE 8:e84147

Bell O, Wirbelauer C, Hild M, Scharf AN, Schwaiger M, MacAlpine DM, Zilbermann F, van Leeuwen F, Bell SP, Imhof A, Garza D, Peters AH, Schübeler D (2007) Localized H3K36 methylation states define histone H4K16 acetylation during transcriptional elongation in *Drosophila*. EMBO J 26:4974–4984

Bender LB, Suh J, Carroll CR, Fong Y, Fingerman IM, Briggs SD, Cao R, Zhang Y, Reinke V, Strome S (2006) MES-4: an autosome-associated histone methyltransferase that participates in silencing the X chromosomes in the *C. elegans* germ line. Development 133:3907–3917

Bennett RL, Swaroop A, Troche C, Licht JD (2017) The role of nuclear receptor-binding SET domain family histone lysine methyltransferases in cancer. Cold Spring Harb Perspect Med. https://doi.org/10.1101/cshperspect.a026708

Brooun A, Gajiwala KS, Deng YL, Liu W, Bolanos B, Bingham P, He YA, Diehl W, Grable N, Kung PP, Sutton S, Maegley KA, Yu X, Stewart AE (2016) Polycomb repressive complex 2 structure with inhibitor reveals a mechanism of activation and drug resistance. Nat Commun 7:11384

Byrd KN, Shearn A (2003) ASH1, a *Drosophila* trithorax group protein, is required for methylation of lysine 4 residues on histone H3. Proc Natl Acad Sci USA 100:11535–11540

Carrozza MJ, Li B, Florens L, Suganuma T, Swanson SK, Lee KK, Shia WJ, Anderson S, Yates J, Washburn MP, Workman JL (2005) Histone H3 methylation by Set2 directs deacetylation of coding regions by Rpd3S to suppress spurious intragenic transcription. Cell 123:581–592

Dorighi KM, Tamkun JW (2013) The trithorax group proteins Kismet and ASH1 promote H3K36 dimethylation to counteract Polycomb group repression in *Drosophila*. Development 140:4182–4192

Edmunds JW, Mahadevan LC, Clayton AL (2008) Dynamic histone H3 methylation during gene induction: HYPB/Setd2 mediates all H3K36 trimethylation. EMBO J 27:406–420

Fahey CC, Davis IJ (2017) SETting the stage for cancer development: SETD2 and the consequences of lost methylation. Cold Spring Harb Perspect Med. https://doi.org/10.1101/cshperspect.a026468

Fang D, Gan H, Lee JH, Han J, Wang Z, Riester SM, Jin L, Chen J, Zhou H, Wang J, Zhang H, Yang N, Bradley EW, Ho TH, Rubin BP, Bridge JA, Thibodeau SN, Ordog T, Chen Y, van Wijnen AJ, Oliveira AM, Xu RM, Westendorf JJ, Zhang Z (2016) The histone H3.3K36M mutation reprograms the epigenome of chondroblastomas. Science 352:1344–1348

Fong Y, Bender L, Wang W, Strome S (2002) Regulation of the different chromatin states of autosomes and X chromosomes in the germ line of *C. elegans*. Science 296:2235–2238

Gaydos LJ, Rechtsteiner A, Egelhofer TA, Carroll CR, Strome S (2012) Antagonism between MES-4 and Polycomb repressive complex 2 promotes appropriate gene expression in *C. elegans* germ cells. Cell Rep 2:1169–1177

Gregory GD, Vakoc CR, Rozovskaia T, Zheng X, Patel S, Nakamura T, Canaani E, Blobel GA (2007) Mammalian ASH1L is a histone methyltransferase that occupies the transcribed region of active genes. Mol Cell Biol 27:8466–8479

He C, Li F, Zhang J, Wu J, Shi Y (2013) The methyltransferase NSD3 has chromatin-binding motifs, PHD5-C5HCH, that are distinct from other NSD (nuclear receptor SET domain) family members in their histone H3 recognition. J Biol Chem 288:4692–4703

Huang N, vom Baur E, Garnier JM, Lerouge T, Vonesch JL, Lutz Y, Chambon P, Losson R (1998) Two distinct nuclear receptor interaction domains in NSD1, a novel SET protein that exhibits characteristics of both corepressors and coactivators. EMBO J 17:3398–3412

Huang C, Yang F, Zhang Z, Zhang J, Cai G, Li L, Zheng Y, Chen S, Xi R, Zhu B (2017) Mrg15 stimulates Ash1 H3K36 methyltransferase activity and facilitates Ash1 Trithorax group protein function in *Drosophila*. Nat Commun 8:1649

Jiao L, Liu X (2015) Structural basis of histone H3K27 trimethylation by an active polycomb repressive complex 2. Science 350:aac4383

Jones PA (2012) Functions of DNA methylation: islands, start sites, gene bodies and beyond. Nat Rev Genet 13:484–492

Joshi AA, Struhl K (2005) Eaf3 chromodomain interaction with methylated H3-K36 links histone deacetylation to Pol II elongation. Mol Cell 20:971–978

Justin N, Zhang Y, Tarricone C, Martin SR, Chen S, Underwood E, De Marco V, Haire LF, Walker PA, Reinberg D, Wilson JR, Gamblin SJ (2016) Structural basis of oncogenic histone H3K27M inhibition of human polycomb repressive complex 2. Nat Commun 7:11316

Keogh MC, Kurdistani SK, Morris SA, Ahn SH, Podolny V, Collins SR, Schuldiner M, Chin K, Punna T, Thompson NJ, Boone C, Emili A, Weissman JS, Hughes TR, Strahl BD, Grunstein M, Greenblatt JF, Buratowski S, Krogan NJ (2005) Cotranscriptional set2 methylation of histone H3 lysine 36 recruits a repressive Rpd3 complex. Cell 123:593–605

Kouzarides T (2007a) Chromatin modifications and their function. Cell 128:693–705

Kouzarides T (2007b) SnapShot: histone-modifying enzymes. Cell 128:802

Krogan NJ, Kim M, Tong A, Golshani A, Cagney G, Canadien V, Richards DP, Beattie BK, Emili A, Boone C, Shilatifard A, Buratowski S, Greenblatt J (2003) Methylation of histone H3 by Set2 in *Saccharomyces cerevisiae* is linked to transcriptional elongation by RNA polymerase II. Mol Cell Biol 23:4207–4218

Kuo AJ, Cheung P, Chen K, Zee BM, Kioi M, Lauring J, Xi Y, Park BH, Shi X, Garcia BA, Li W, Gozani O (2011) NSD2 links dimethylation of histone H3 at lysine 36 to oncogenic programming. Mol Cell 44:609–620

Lhoumaud P, Hennion M, Gamot A, Cuddapah S, Queille S, Liang J, Micas G, Morillon P, Urbach S, Bouchez O, Severac D, Emberly E, Zhao K, Cuvier O (2014) Insulators recruit histone methyltransferase dMes4 to regulate chromatin of flanking genes. EMBO J 33:1599–1613

Li GM (2013) Decoding the histone code: role of H3K36me3 in mismatch repair and implications for cancer susceptibility and therapy. Cancer Res 73:6379–6383

Li J, Moazed D, Gygi SP (2002) Association of the histone methyltransferase Set2 with RNA polymerase II plays a role in transcription elongation. J Biol Chem 277:49383–49388

Li B, Howe L, Anderson S, Yates JR 3rd, Workman JL (2003) The Set2 histone methyltransferase functions through the phosphorylated carboxyl-terminal domain of RNA polymerase II. J Biol Chem 278:8897–8903

Li B, Gogol M, Carey M, Lee D, Seidel C, Workman JL (2007) Combined action of PHD and chromo domains directs the Rpd3S HDAC to transcribed chromatin. Science 316:1050–1054

Li B, Jackson J, Simon MD, Fleharty B, Gogol M, Seidel C, Workman JL, Shilatifard A (2009a) Histone H3 lysine 36 dimethylation (H3K36me2) is sufficient to recruit the Rpd3s histone deacetylase complex and to repress spurious transcription. J Biol Chem 284:7970–7976

Li Y, Trojer P, Xu CF, Cheung P, Kuo A, Drury WJ 3rd, Qiao Q, Neubert TA, Xu RM, Gozani O, Reinberg D (2009b) The target of the NSD family of histone lysine methyltransferases depends on the nature of the substrate. J Biol Chem 284:34283–34295

Lu C, Jain SU, Hoelper D, Bechet D, Molden RC, Ran L, Murphy D, Venneti S, Hameed M, Pawel BR, Wunder JS, Dickson BC, Lundgren SM, Jani KS, De Jay N, Papillon-Cavanagh S, Andrulis IL, Sawyer SL, Grynspan D, Turcotte RE, Nadaf J, Fahiminiyah S, Muir TW, Majewski J, Thompson CB, Chi P, Garcia BA, Allis CD, Jabado N, Lewis PW (2016) Histone H3K36 mutations promote sarcomagenesis through altered histone methylation landscape. Science 352:844–849

McDaniel SL, Strahl BD (2017) Shaping the cellular landscape with Set2/SETD2 methylation. Cell Mol Life Sci 74:3317–3334

Miyazaki H, Higashimoto K, Yada Y, Endo TA, Sharif J, Komori T, Matsuda M, Koseki Y, Nakayama M, Soejima H, Handa H, Koseki H, Hirose S, Nishioka K (2013) Ash1 l methylates Lys36 of histone H3 independently of transcriptional elongation to counteract polycomb silencing. PLoS Genet 9:e1003897

Neri F, Rapelli S, Krepelova A, Incarnato D, Parlato C, Basile G, Maldotti M, Anselmi F, Oliviero S (2017) Intragenic DNA methylation prevents spurious transcription initiation. Nature 543:72–77

Papp B, Muller J (2006) Histone trimethylation and the maintenance of transcriptional ON and OFF states by trxG and PcG proteins. Genes Dev 20:2041–2054

Popovic R, Martinez-Garcia E, Giannopoulou EG, Zhang Q, Zhang Q, Ezponda T, Shah MY, Zheng Y, Will CM, Small EC, Hua Y, Bulic M, Jiang Y, Carrara M, Calogero RA, Kath WL, Kelleher NL, Wang JP, Elemento O, Licht JD (2014) Histone methyltransferase MMSET/NSD2 alters EZH2 binding and reprograms the myeloma epigenome through global and focal changes in H3K36 and H3K27 methylation. PLoS Genet 10:e1004566

Qiao Q, Li Y, Chen Z, Wang M, Reinberg D, Xu RM (2011) The structure of NSD1 reveals an autoregulatory mechanism underlying histone H3K36 methylation. J Biol Chem 286:8361-8368

Ram O, Goren A, Amit I, Shoresh N, Yosef N, Ernst J, Kellis M, Gymrek M, Issner R, Coyne M, Durham T, Zhang X, Donaghey J, Epstein CB, Regev A, Bernstein BE (2011) Combinatorial patterning of chromatin regulators uncovered by genome-wide location analysis in human cells. Cell 147:1628-1639

Rechtsteiner A, Ercan S, Takasaki T, Phippen TM, Egelhofer TA, Wang W, Kimura H, Lieb JD, Strome S (2010) The histone H3K36 methyltransferase MES-4 acts epigenetically to transmit the memory of germline gene expression to progeny. PLoS Genet 6:e1001091

Ringrose L, Paro R (2007) Polycomb/Trithorax response elements and epigenetic memory of cell identity. Development 134:223-232

Rogawski DS, Grembecka J, Cierpicki T (2016) H3K36 methyltransferases as cancer drug targets: rationale and perspectives for inhibitor development. Fut Med Chem 8:1589-1607

Sankaran SM, Wilkinson AW, Elias JE, Gozani O (2016) A PWWP domain of histone-lysine N-methyltransferase NSD2 binds to dimethylated Lys-36 of histone H3 and regulates NSD2 function at chromatin. J Biol Chem 291:8465-8474

Schmahling S, Meiler A, Lee Y, Mohammed A, Finkl K, Tauscher K, Israel L, Wirth M, Philippou-Massier J, Blum H, Habermann B, Imhof A, Song JJ, Müller J (2018) Regulation and function of H3K36 di-methylation by the trithorax-group protein complex AMC. Development 145

Schmitges FW, Prusty AB, Faty M, Stutzer A, Lingaraju GM, Aiwazian J, Sack R, Hess D, Li L, Zhou S, Bunker RD, Wirth U, Bouwmeester T, Bauer A, Ly-Hartig N, Zhao K, Chan H, Gu J, Gut H, Fischle W, Müller J, Thomä NH (2011) Histone methylation by PRC2 is inhibited by active chromatin marks. Mol Cell 42:330-341

Schwartz YB, Kahn TG, Stenberg P, Ohno K, Bourgon R, Pirrotta V (2010) Alternative epigenetic chromatin states of polycomb target genes. PLoS Genet 6:e1000805

Shen C, Ipsaro JJ, Shi J, Milazzo JP, Wang E, Roe JS, Suzuki Y, Pappin DJ, Joshua-Tor L, Vakoc CR (2015) NSD3-short is an adaptor protein that couples BRD4 to the CHD8 chromatin remodeler. Mol Cell 60:847-859

Srinivasan S, Dorighi KM, Tamkun JW (2008) *Drosophila* Kismet regulates histone H3 lysine 27 methylation and early elongation by RNA polymerase II. PLoS Genet 4:e1000217

Strahl BD, Grant PA, Briggs SD, Sun ZW, Bone JR, Caldwell JA, Mollah S, Cook RG, Shabanowitz J, Hunt DF, Allis CD (2002) Set2 is a nucleosomal histone H3-selective methyltransferase that mediates transcriptional repression. Mol Cell Biol 22:1298-1306

Streubel G, Watson A, Jammula SG, Scelfo A, Fitzpatrick DJ, Oliviero G, McCole R, Conway E, Glancy E, Negri GL, Dillon E, Wynne K, Pasini D, Krogan NJ, Bracken AP, Cagney G (2018) The H3K36me2 methyltransferase Nsd1 demarcates PRC2-mediated H3K27me2 and H3K27me3 domains in embryonic stem cells. Mol Cell 70:371-379

Tanaka Y, Katagiri Z, Kawahashi K, Kioussis D, Kitajima S (2007) Trithorax-group protein ASH1 methylates histone H3 lysine 36. Gene 397:161-168

Venkatesh S, Workman JL (2013) Set2 mediated H3 lysine 36 methylation: regulation of transcription elongation and implications in organismal development. Wiley Interdiscip Rev Dev Biol 2:685-700

Wagner EJ, Carpenter PB (2012) Understanding the language of Lys36 methylation at histone H3. Nat Rev Mol Cell Biol 13:115-126

Wang Y, Niu Y, Li B (2015) Balancing acts of SRI and an auto-inhibitory domain specify Set2 function at transcribed chromatin. Nucleic Acids Res 43:4881-4892

Wang C, Zhu B, Xiong J (2018) Recruitment and reinforcement: maintaining epigenetic silencing. Sci China Life Sci 61:515-522

Wu H, Zeng H, Dong A, Li F, He H, Senisterra G, Seitova A, Duan S, Brown PJ, Vedadi M, Arrowsmith CH, Schapira M (2013) Structure of the catalytic domain of EZH2 reveals conformational plasticity in cofactor and substrate binding sites and explains oncogenic mutations. PLoS ONE 8:e83737

Xiao T, Hall H, Kizer KO, Shibata Y, Hall MC, Borchers CH, Strahl BD (2003) Phosphorylation of RNA polymerase II CTD regulates H3 methylation in yeast. Genes Dev 17:654-663

Yang S, Zheng X, Lu C, Li GM, Allis CD, Li H (2016) Molecular basis for oncohistone H3 recognition by SETD2 methyltransferase. Genes Dev 30:1611-1616

Yuan W, Xie J, Long C, Erdjument-Bromage H, Ding X, Zheng Y, Tempst P, Chen S, Zhu B, Reinberg D (2009) Heterogeneous nuclear ribonucleoprotein L Is a subunit of human KMT3a/Set2 complex required for H3 Lys-36 trimethylation activity *in vivo*. J Biol Chem 284:15701-15707

Yuan W, Xu M, Huang C, Liu N, Chen S, Zhu B (2011) H3K36 methylation antagonizes PRC2-mediated H3K27 methylation. J Biol Chem 286:7983-7989

Yuan W, Wu T, Fu H, Dai C, Wu H, Liu N, Li X, Xu M, Zhang Z, Niu T, Han Z, Chai J, Zhou XJ, Gao S, Zhu B (2012) Dense chromatin activates Polycomb repressive complex 2 to regulate H3 lysine 27 methylation. Science 337:971-975

Yuan G, Ma B, Yuan W, Zhang Z, Chen P, Ding X, Feng L, Shen X, Chen S, Li G, Zhu B (2013) Histone H2A ubiquitination inhibits the enzymatic activity of H3 lysine 36 methyltransferases. J Biol Chem 288:30832-30842

Zheng W, Ibanez G, Wu H, Blum G, Zeng H, Dong A, Li F, Hajian T, Allali-Hassani A, Amaya MF, Siarheyeva A, Yu W, Brown PJ, Schapira M, Vedadi M, Min J, Luo M (2012) Sinefungin derivatives as inhibitors and structure probes of protein lysine methyltransferase SETD2. J Am Chem Soc 134:18004-18014

Raman spectra of the GFP-like fluorescent proteins

Ye Yuan[1,2], Dianbing Wang[1], Jibin Zhang[2], Ji Liu[1,3], Jian Chen[3], Xian-En Zhang[1]

[1] National Laboratory of Biomacromolecules, CAS Center for Excellence in Biomacromolecules, Institute of Biophysics, Chinese Academy of Sciences, Beijing 100101, China
[2] College of Life Science and Technology, Huazhong Agricultural University, Wuhan 430070, China
[3] College of Life Science, Hubei University, Wuhan 430070, China

Abstract The objective of the study was to elucidate optical characteristics of the chromophore structures of fluorescent proteins. Raman spectra of commonly used GFP-like fluorescent proteins (FPs) with diverse emission wavelengths (green, yellow, cyan and red), including the enhanced homogenous FPs EGFP, EYFP, and ECFP (from jellyfish) as well as mNeptune (from sea anemone) were measured. High-quality Raman spectra were obtained and many marker bands for the chromophore of the FPs were identified via assignment of Raman spectra bands. We report the presence of a positive linear correlation between the Raman band shift of $C_5=C_6$ and the excitation energy of FPs, demonstrated by plotting absorption maxima (cm^{-1}) against the position of the Raman band $C_5=C_6$ in EGFP, ECFP, EYFP, the anionic chromophore and the neutral chromophore. This study revealed new Raman features in the chromophores of the observed FPs, and may contribute to a deeper understanding of the optical properties of FPs.

Keywords Raman spectra, Fluorescent protein, Chromophore

INTRODUCTION

The Raman spectrum provides a "fingerprint" of the vibration and rotation of molecules. The conformation and structure of biomacromolecules such as DNA, protein chains, membrane proteins and lipids, as well as other structural data related to such molecules can be obtained using Raman spectroscopy (Bunaciu et al. 2015; Carey 1982; Tu 1982; Tuma 2005; Xu 2005).

In recent years, the Raman spectra of fluorescent proteins (FPs) have attracted much attention due to the unique optical properties of FPs and their wider applicability in molecular and cellular imaging. Analysis of the Raman spectra of GFP and its mutants revealed that the ground-state structure of the anionic form of the chromophore may be heavily dependent on the chromophore environment Bell et al. 2000). Femtosecond-stimulated Raman spectroscopy showed that skeletal motions are related to proton transfers which makes GFP in the fluorescent form (Fang et al. 2009). In addition, the Raman spectra of the red fluorescent protein, eqFP611, from the sea anemone, *Entacmaea quadricolor*, revealed photoinduced cis–trans isomerization of the chromophore (Davey et al. 2006). Resonance and pre-resonance Raman spectra of the photochromic fluorescent protein, Dronpa, demonstrated enhanced Raman band selectively for the chromophore, thus yielding important information on the chromophore structure (Higashino et al. 2016).

Based on the above findings, our focus was directed at the relationship between emission wavelengths of FPs and their Raman spectrum characteristics. To clarify this relationship, we measured the Raman spectra of a group of commonly used GFP-like FPs with diverse emission wavelengths (green, yellow, cyan, and red),

---

National Laboratory of Biomacromolecules, CAS Center for Excellence in Biomacromolecules, Institute of Biophysics, Chinese Academy of Sciences and College of Life Science and Technology, Huazhong Agricultural University contributed equally to this paper.

✉ Correspondence: zhangxe@ibp.ac.cn (X.-E. Zhang)

including the enhanced homogenous fluorescent proteins EGFP, EYFP, and ECFP (from jellyfish) to mNeptune (from anemone). It is felt that the results of this study may not only enrich the understanding of Raman spectra in relation to FPs, but also benefit efforts associated with the rational design and directed evolution of FPs for practical purposes.

RESULTS AND DISCUSSION

Raman spectra of fluorescent proteins

Raman spectra of EGFP, ECFP, and EYFP are shown (Fig. 1). Assignment of Raman bands for these FPs are presented (Table 1). Raman spectroscopy with 785-nm excitation was used to acquire the Raman spectra of FPs. This excitation wavelength selectively enhances the intensity of vibrational bands originating in the chromophore, and thereby avoids certain issues associated with strictly on-resonance Raman experiments such as fluorescence, photoisomerization, or sample degradation (Bell et al. 2000). As a result, most Raman spectra bands obtained in the study were produced by the chromophores of FPs, and only a few Raman spectra bands were due to the main chain groups and side chain groups on the β-barrel of fluorescent proteins. The Raman spectra bands of EGFP at 1664 cm^{-1}, ECFP at 1662 cm^{-1} and EYFP at 1659 cm^{-1} are all assigned to Amide I modes (Table 1). The Raman spectra bands of EGFP at 1447 cm^{-1}, ECFP at 1450 cm^{-1} and EYFP at 1446 cm^{-1} are assigned to side-chain CH_2 group modes. The Raman spectra band at 1004 cm^{-1} is assigned to the aromatic side-chain mode of the FPs. The Raman spectra bands among 1220–1350 cm^{-1} were from Amide III.

The chromophores of EGFP, EYFP, and mNeptune are mainly composed of the phenol group and the imidazolinone ring formed by propylene group bridging. As the phenol group consists of C, H, and O, the Raman spectra bands from the phenol group are due to the vibration of the phenol ring, C–H bonds of the phenol group, and the phenolic hydroxyl group. For example, the Raman spectra bands of EGFP at 618, 1035, 1128

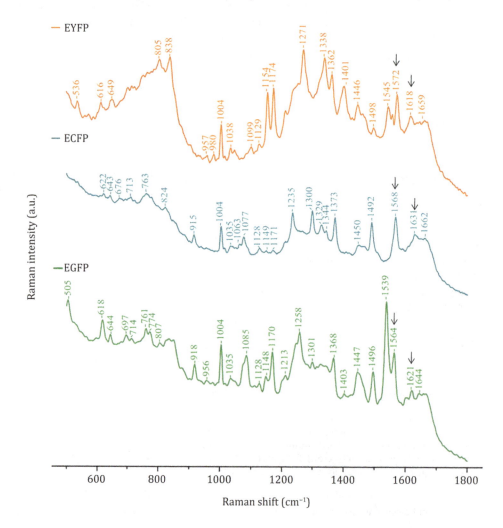

Fig. 1 Raman spectra of EGFP, ECFP and EYFP

Table 1 Raman shifts and mode assignments of EGFP, ECFP and EYFP

EGFP	ECFP	EYFP	Mode assignment
618	622	616	Ph[a] C–H in-plane H bend
644	643		Ph C–H in-plane H bend
		649	–C=C str[b]
697			–N–H def[c]
807		805	–C–H out-of-plane def
		838	–C–H out-of-plane def
		980	–C–H in-plane H bend
1004	1004	1004	Aromatic side-chain mode of protein
1035	1035	1038	–C–H in-plane H bend
	1077		–C–H in-plane H bend
		1099	–C–H in-plane H bend
1128	1128	1129	–C–H in-plane H bend, Ph ring-H scissor
		1154	Ph C–H in-plane H bend, Bridge C6–H rock, imidazolinone ring, C12OH rock
1148	1149		–C–H in-plane H bend
1170	1171	1174	Ph ring-H bend, Bridge C6–H rock, Ph C–H def, imidazolinone ring def
1213			–C–H in-plane H bend
	1235		Ring combination bend-str
1258			–C–H in-plane H bend
		1271	–C–H bend
1301	1300		Ph ring-H bend
	1329		–C–N–H str
		1338	–C6–H rock
1368		1362	C–O–H def (O–H bend and C–O str, Ph hydroxyl group)
1403		1401	Ring combination bend-str
1447	1450	1446	Side chain CH2 group
1496	1492	1498	Ph ring str
1539			Imidazolinone + C=C str
		1545	–C–N–H str
1564	1568	1572	–C3=N1 str
1621	1631	1618	C5=C6 str
1644	1662	1659	–C4=O13 str

[a]Ph: Phenol
[b]str: stretch
[c]def: deformation

and 1170 cm^{-1}, EYFP at 616, 1038, 1129 and 1174 cm^{-1} and mNeptune at 621, 1107, 1119, 1156, and 1170 cm^{-1} are likely assigned to a C–H phenol bending mode. The bands of EGFP at 1496 cm^{-1}, EYFP at 1498 cm^{-1} and mNeptune at 1480 cm^{-1} are assigned to a phenol ring stretching mode.

To configure the marker bands of the chromophore, the Raman spectra bands of ECFP were compared with those of EGFP, and EYFP. It was found that the band at 1368 cm^{-1} of EGFP and the band at 1362 cm^{-1} of EYFP were from stretching vibrations of the phenolic hydroxyl group on the side chain of 66Tyr in the chromophore. However, the bands of ECFP around 1368 cm^{-1} or 1362 cm^{-1} did not appear, due to the presence of an indolyl group, rather than a phenol group, on the side chain of 66Trp. Therefore, the band of ECFP at 1329 cm^{-1} due to the stretching vibration of indolyl C–N–H on the side chain of 66Trp, is likely a marker band of ECFP. In addition, there were other differences between the Raman spectra of EGFP, ECFP, and EYFP. For example, $C_3=N_1$ stretching of EGFP, ECFP, and EYFP each produced a Raman band at 1564, 1568 and 1572 cm^{-1}, respectively; $C_5=C_6$ stretching of EGFP, ECFP, and EYFP each produced a Raman band at 1621, 1631 and 1618 cm^{-1}, respectively; also, $C_4=O_{13}$ stretching of EGFP, ECFP, and EYFP each produced a Raman band at 1644, 1662 and 1659 cm^{-1}, respectively. These results suggested that EGFP, EYFP, and ECFP may each have their respective featured Raman bands, and therefore may be distinguished by a comparison of their Raman spectra, although these enhanced FPs are highly homogenous. These features in the Raman spectra of FPs were strongly dependent on the environment, as well as the structure of the chromophore.

We also measured the Raman spectra of mNeptune, which is a GFP-like protein, originating in a sea anemone. As expected, its Raman spectrum was significantly different from that of EGFP, as the two FPs originated in different species (Fig. 2 and Table 2). Assignment of the Raman bands indicated that the bands of mNeptune at 1170, 1156 and 1201 cm^{-1} are marker bands, by which mNeptune may be distinguished from EGFP. Besides, mNeptune produced many Raman bands at 1320–1370 cm^{-1} which arose from the imidazolinone ring-related groups, whereas EGFP produced more bands at 1400–1500 cm^{-1} due to the presence of 66Tyr in its chromophore. We, therefore, hypothesized that different features in the Raman spectra of EGFP and mNeptune were mainly due to differences in their chromophore structures which resulted in obviously different molecular vibrations.

Linear correlation between the Raman band shift of $C_5=C_6$ in the chromophores and the excitation energy for FPs

The Raman band of $C_5=C_6$ in the chromophore is at 1621 cm^{-1} for EGFP, 1631 cm^{-1} for ECFP and 1618 cm^{-1} for EYFP (Tables 1, 3). A previous study illustrated that the Raman band of $C_5=C_6$ in the anionic chromophore form is at 1628 cm^{-1} and the band in the neutral chromophore form is at 1648 cm^{-1} (Bell *et al.* 2000). There was a positive linear correlation between the absorption maxima (cm^{-1}) and the position of the Raman band of $C_5=C_6$ in EGFP, ECFP, EYFP, the anionic chromophore form, and the neutral chromophore form (Fig. 3).

Excitation energy is dependent on the chromophore structure of the FP and its surrounding microenvironment, which is related to the absorption maximum. In the chromophore of EGFP, Tyr presents a conjugated ring with π electrons because of the connection of the imidazolidone ring and the phenolic group of 66Tyr by $C_5=C_6$. In this conjugated ring, the dihedral angle of C_4–$C_5=C_6$–C_7 is 177.67° (Arpino *et al.* 2012). As a comparison, the dihedral angle in ECFP is 173.38° (Lelimousin *et al.* 2009). Therefore, the π-conjugated plane in the chromophore of EGFP is larger than that of ECFP, illustrating lower excitation energy needed for EGFP. However, EYFP requires even lower excitation energy than EGFP, as π-stacking interaction between the chromophore and 203Tyr of EYFP leads to a more stable electronic state (Wachter *et al.* 1998). For these reasons, we postulate that the presence of a linear correlation demonstrates a direct relationship between the Raman band shift of $C_5=C_6$ and the excitation energy for FPs (Fig. 3). Obviously, the lower the excitation

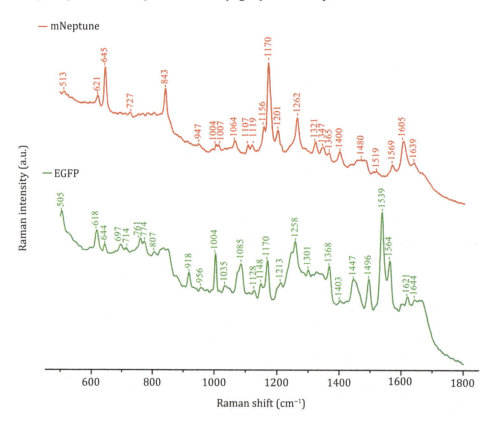

Fig. 2 Raman spectra of mNeptune and EGFP

Table 2 Raman shifts and mode assignments of EGFP and mNeptune

EGFP	mNeptune	Mode assignment
618	621	Ph[a] C–H in-plane H bend
644		Ph C–H in-plane H bend
697		–N–H def[b]
807		–C–H out-of-plane def
	645	CH=CH def cis in-phase wag
	843	–C–O skeletal str[c]
1004	1004	Aromatic side chain mode
1035		–C–H in-plane H bend
	1107	–C–H in-plane H bend
	1119	–C–H in-plane H bend
1128		–C–H in-plane H bend, Ph ring-H scissor
1148		–C–H in-plane H bend
	1156	Ph C–H in plane H bend, Bridge C_6–H rock, imidazolinone ring def, C_{12}OH rock
1170	1170	Ph ring-H bend(rock), Bridge C_6–H rock, Ph C–H def;
	1201	–C–N str
1213		–C–H in-plane H bend
1258		–C–H in-plane H bend
	1262	Phenol C–H, Ph ring C_{12}–O stretch, Ph ring-H rock
1301		Ph ring-H bend
	1321	–C–N str
	1347	–C–H rock
	1365	–C–O–H def[a]
1368		C–O–H def (O–H bend and C–O str, Ph hydroxyl group)
	1400	Ring combination bend-str;C_{12}OH rock
1403		Ring combination bend-str
1447		Side chain CH_2 group
	1480	Ph ring
1496		Ph ring str
1539		Imidazolinone + C=C str
1564		–C_3=N_1 str
	1569	Ring combination bend-str
	1605	C=C
1621		C_5=C_6 str
	1639	Amide I
1644		–C_4=O_{13} str

[a]Ph: Phenol
[b]def: deformation
[c]str: stretch

energy, the bigger the redshift of C_5=C_6 stretching mode, and vice versa.

Analysis of the correlation between the Raman band shift of C_3=N_1 in chromophore and the photostability of FPs

The protonation of N_1 in the chromophore of enhanced FPs plays an important role in chromophore stability, and exerts an effect on some optical properties of FPs (Wachter et al. 1998). We consider the interactions of N_1 with its surrounding amino-acid residues and H_2O may further stabilize the structure of the chromophore. Some evidence for this can be found in the Raman spectra of the FPs. For instance, the band at 1564 cm^{-1} in EGFP is a C_3=N_1 stretching mode (Table 3). However, the C_3=N_1 stretching mode is shifted to 1568 cm^{-1} in ECFP and to 1572 cm^{-1} in EYFP, respectively. It is evident from the chromophore hydrogen bond network of FPs that N_1 and H_2O form a hydrogen bond, through which H_2O absorbs electrons from N_1 (Fig. 4). This electron attraction effect causes a redshift of the C_3=N_1 mode, whereas such an effect is not observed in EYFP's Raman spectrum, because EYFP lacks such a hydrogen

Table 3 Comparison of characteristics of EGFP, ECFP and EYFP

FPs	Scheme of chromophore	$\lambda_{ex}/\lambda_{em}{}^a$ (nm)	Raman bands $C_3=N_1$	$C_5=C_6$	Raman spectra
EGFP		489/508	1564 cm^{-1}	1621 cm^{-1}	
ECFP		434/477	1568 cm^{-1}	1631 cm^{-1}	
EYFP		514/537	1572 cm^{-1}	1618 cm^{-1}	

[a]Excitation and emission maxima (Voityuk et al. 1998)

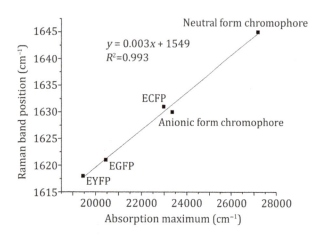

Fig. 3 Plot of absorption maxima versus $C_5=C_6$ Raman band position for neutral form of chromophore, anionic form of chromophore, EGFP, ECFP and EYFP

bond. The length of the hydrogen bond in EGFP is 3.43 Å while it is 3.52 Å in ECFP, indicating a stronger attraction effect in EGFP compared to ECFP. Therefore, the $C_3=N_1$ mode in EGFP presents a bigger redshift compared to that of ECFP. Considering the fact that the photobleaching time ratio of ECFP to EGFP is 0.85, whereas the ratio of EYFP to EGFP is only 0.35 (Patterson et al. 2001), we propose that the redshift of the $C_3=N_1$ mode in the chromophore may possibly be related to the photostability of FPs. However, this contention may require further validation via experimental data.

CONCLUSION

In summary, high-quality Raman spectra of a group of GFP-like FPs were obtained. Raman spectra of the FPs derived from GFP were evidently distinct from the RFP (mNeptune). Some marker bands were also found in the Raman spectra of GFP-derived FPs. These marker bands are mainly produced by their distinct chromophores. Among these bands, the Raman band shift of $C_5=C_6$ presents a positive linear correlation with the excitation energy for FPs. This study not only reveals new Raman features in the chromophore, but also illustrates the relationship between these features and the optical properties of FPs.

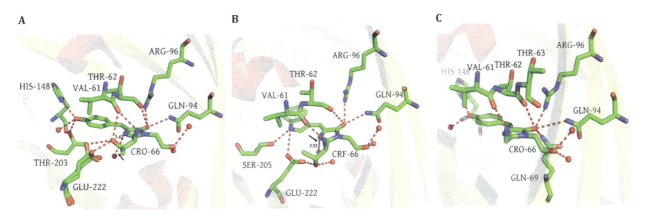

Fig. 4 Chromophore hydrogen bond network of fluorescent proteins. *Red dashed lines* represent hydrogen bond. *Red spheres* denote H_2O molecule. **A** Chromophore hydrogen bond network of EGFP (PDB Code:4EUL). **B** Chromophore hydrogen bond network of ECFP (PDB Code:2WSN). **C** Chromophore hydrogen bond network of EYFP (PDB Code:1YFP). *Black arrow* indicates the hydrogen bond between N1 and H_2O molecule

MATERIALS AND METHODS

Reagents and materials

The vector pQE30, containing the gene clone of FPs, was purchased from Qiagen (Hilden, Germany). *E. coli* TG1 was used to express FPs. The primers used in this study are listed in Table 4.

Expression and purification of fluorescent proteins

The cDNAs of FPs were cloned into the BamH I and Sac I restriction sites of the pQE30 vector, using forward primers and reverse primers, respectively (Table 4). FPs were expressed in the *E. Coli* TG1 strain. Bacterial cultures were grown overnight in LB media containing 100 μg/mL ampicillin at 180 r/min at 37 °C, and further incubated at 25 °C for 12 h. The cells were harvested via centrifugation at 4500 r/min at 4 °C for 5 min, and the cell pellets were resuspended in binding buffer (20 mmol/L Tris–HCl, 500 mmol/L NaCl, 20 mmol/L imidazole, pH 8.0). Following cell lysis by French pressure (JNBIO, JN-02C, China), the FPs were purified using Ni–NTA His-Bind resin (GE Healthcare, USA) and a Superdex-200 size exclusion column (GE Healthcare, USA) according to the manufacturer's instructions. Purified proteins were characterized using SDS–polyacrylamide electrophoresis. FPs were stored in Start Buffer (20 mmol/L Tris pH 7.9; NaCl 100 mmol/L) with concentration of 15–20 mg/mL for further analysis.

Raman spectroscopy

For Raman measurements, 45 μl of protein solution was added into a sample cell and placed on the object stage. The sample was excited with a 785-nm pulsed laser at an excitation power of 50 mW/cm^2, as 785-nm excitation allows probing of the chromophore site with minimal spectral interference from the surrounding protein environment.

The Raman spectra were acquired using Renishaw inVia Reflex confocal Raman microscope (Renishaw, UK) with a total collection time of 60 s for the recording of the spectra region from approximately 400–2400 cm^{-1}. The laser beam was focused on the 200–300 μm point under the sample surface.

Table 4 Primers used in this study

Primer name	Primer sequence(5′–3′)
EGFP-BamHI-F	ATATGGATCCATGGTGAGCAAGGGCGAGGA
EGFP-SacI-R	GAGCGAGCTCTTACTTGTACAGCTCGTCCAT
ECFP-BamHI-F	ATATGGATCCATGGTGAGCAAGGGCGAGGA
ECFP-SacI-R	GAGCGAGCTCTTACTTGTACAGCTCGTCCAT
EYFP-BamHI-F	ATATGGATCCATGGTGAGCAAGGGCGAG
EYFP-SacI-R	GCGTGAGCTCTTACTTGTACAGCTCGTCCAT
mNeptune-BamHI-F	ATAGGATCCATGGTGTCTAAGGGCGAAGAGCTGATTA
mNeptune-SacI-R	ATAGAGCTCTTACTTGTACAGCTCGTCCATGCCATTA

Raw data were processed by KnowItAll software, (Bio-Rad, USA) and Raman spectra were calculated by Savitzky–Golay smoothing. Raman spectra assignment was performed using KnowItAll software functional group database and references.

Acknowledgements This work was supported by the Strategic Priority Research Program of Chinese Academy of Sciences (XDPB0305, CAS). The authors are very grateful to Junfang Zhao from Technical Institute of Physics and Chemistry, CAS for technical support in Raman spectra experiments and Prof. Zhou Lu from Institute of Chemistry, CAS for his valuable suggestions.

Compliance with ethical standards

Conflict of interest Ye Yuan, Dianbing Wang, Jibin Zhang, Ji Liu, Jian Chen, Xian-En Zhang declare that they have no conflict of interest.

Human and animal rights and informed consent This article does not contain any studies with human or animal subjects performed by any of the authors.

References

Arpino JA, Rizkallah PJ, Jones DD (2012) Crystal structure of enhanced green fluorescent protein to 1.35 Å resolution reveals alternative conformations for Glu222. PLoS ONE 7(10):e47132

Bell AF, He X, Wachter RM, Tonge PJ (2000) Probing the ground state structure of the green fluorescent protein chromophore using Raman spectroscopy. Biochemistry 39:4423–4431

Bunaciu AA, Aboulenein HY, Hoang VD (2015) Raman spectroscopy for protein analysis. Appl Spectrosc Rev 50:377–386

Carey P (1982) Biochemical applications of raman and resonance raman spectroscopes. In: Chapter 4 protein conformation from raman and resonance raman sptecta, pp 71–98

Fang C, Frontiera RR, Tran R, Mathies RA (2009) Mapping GFP structure evolution during proton transfer with femtosecond Raman spectroscopy. Nature 462:200–204

Higashino A, Mizuno M, Mizutani Y (2016) Chromophore structure of photochromic fluorescent protein Dronpa: acid-base equilibrium of two cis configurations. J Phys Chem B 120:3353

Lelimousin M, Noirclerc-Savoye M, Lazareno-Saez C, Paetzold B, Le VS, Chazal R, Macheboeuf P, Field MJ, Bourgeois D, Royant A (2009) Intrinsic dynamics in ECFP and Cerulean control fluorescence quantum yield. Biochemistry 48:10038–10046

Loos Davey C, Habuchi Satoshi, Flors Cristina, Hotta Junichi, Jörg Wiedenmann G, Nienhaus Ulrich, Hofkens Johan (2006) Photoconversion in the red fluorescent protein from the sea anemone *Entacmaea quadricolor*: is *cis − trans* isomerization involved? J Am Chem Soc 128:6270–6271

Patterson G, Day RN, Piston D (2001) Fluorescent protein spectra. J Cell Sci 114:837–838

Tu AT (1982) Raman spectroscopy in biology: principles and applications. Wiley, New York

Tuma R (2005) Raman spectroscopy of proteins: from peptides to large assemblies. J Raman Spectrosc 36:307–319

Voityuk AA, Michel-Beyerle ME, Rösch N (1998) Quantum chemical modeling of structure and absorption spectra of the chromophore in green fluorescent proteins. Chem Phys 231(1):13–25

Wachter RM, Elsliger MA, Kallio K, Hanson GT, Remington SJ (1998) Structural basis of spectral shifts in the yellow-emission variants of green fluorescent protein. Structure 6:1267–1277

Xu Y (2005) Raman spectroscopy in application of structure biology. In: Chapter 2 proteins. Chemical Industry Press, Beijing, pp 11–26

In situ protein micro-crystal fabrication by cryo-FIB for electron diffraction

Xinmei Li[1,2], Shuangbo Zhang[1,2], Jianguo Zhang[3], Fei Sun[1,2,3]

[1] National Key Laboratory of Biomacromolecules, CAS Center for Excellence in Biomacromolecules, Institute of Biophysics, Chinese Academy of Sciences, Beijing 100101, China
[2] University of Chinese Academy of Sciences, Beijing 100049, China
[3] Center for Biological Imaging, Institute of Biophysics, Chinese Academy of Sciences, Beijing 100101, China

Abstract Micro-electron diffraction (MicroED) is an emerging technique to use cryo-electron microscope to study the crystal structures of macromolecule from its micro-/nano-crystals, which are not suitable for conventional X-ray crystallography. However, this technique has been prevented for its wide application by the limited availability of producing good micro-/nano-crystals and the inappropriate transfer of crystals. Here, we developed a complete workflow to prepare suitable crystals efficiently for MicroED experiment. This workflow includes *in situ* on-grid crystallization, single-side blotting, cryo-focus ion beam (cryo-FIB) fabrication, and cryo-electron diffraction of crystal cryo-lamella. This workflow enables us to apply MicroED to study many small macromolecular crystals with the size of 2–10 μm, which is too large for MicroED but quite small for conventional X-ray crystallography. We have applied this method to solve 2.5 Å crystal structure of lysozyme from its micro-crystal within the size of $10 \times 10 \times 10$ μm^3. Our work will greatly expand the availability space of crystals suitable for MicroED and fill up the gap between MicroED and X-ray crystallography.

Keywords Cryo-electron microscopy, Cryo focused ion beam, Electron diffraction, *In situ* crystallization, Micro-crystal

INTRODUCTION

In 1960, the crystal structures of myoglobin and hemoglobin were solved (Kendrew *et al.* 1960; Perutz *et al.* 1960) by X-ray crystallography, opening the era of structural biology. Till now, there have been hundreds of thousands of biomacromolecular structures that were determined by X-ray crystallography. The size of biomacromolecular crystal should be large enough to gain efficient signal–noise ratio (SNR) of diffractions. For the home source of X-ray generated from rotating anode, the size of crystal needs to be larger than 200 μm normally (Mizohata *et al.* 2018). While the emergence of synchrotron radiation allows a brilliant and coherent source of X-ray, which can increase SNR of diffractions especially at the high-resolution region, thus even a smaller crystal (ca. 50–200 μm) still generate significant diffractions for structure determination (Mizohata *et al.* 2018). The emerge of the micro-focus beamline based on the third-generation synchrotron source has yield micro-crystallography, which made high-resolution data collection from very small crystals (ca. 10–50 μm) possible (Smith *et al.* 2012). The widespread use of synchrotron radiation has accelerated the development of X-ray crystallography and structural biology.

Xinmei Li and Shuangbo Zhang have contributed equally to this work.

✉ Correspondence: feisun@ibp.ac.cn (F. Sun)

However, many biomacromolecules could not be crystallized into large crystals, e.g., membrane proteins and biomacromolecular complexes. When the size of crystal is further smaller than 10 μm, the current source of X-ray from synchrotron radiation could not yield high SNR diffractions. More importantly, the severe radiation damage from high flux X-ray exposure makes it impossible to collect a complete diffraction dataset from a single crystal, even under the cryogenic condition (Johansson et al. 2017). The emerge of X-ray free electron laser (XFEL) and the development of serial femtosecond X-ray crystallography (SFX) provide an alternative solution (Chapman et al. 2011). The extremely short and highly intensive X-ray pulse makes it possible to collect high SNR and "radiation damage free" diffraction data from a single micro-crystal (ca. 1–10 μm) (Mizohata et al. 2018). Each micro-crystal generates one frame of diffraction image before obliterated (Neutze et al. 2000). Tens of thousands of micro-crystals are needed to generate a complete dataset. In recent years, SFX has been successfully applied to solve many important and difficult crystal structures, including the human angiotensin II type 1 (Zhang et al. 2015) and photon-synthesis complex II (Suga et al. 2017). Furthermore, the extremely short pulse (10–100 fs) of XFEL also enables time-resolved SFX and people can investigate the transient structural changes of photon-synthesis complex II upon light stimulation (Suga et al. 2017). However, culturing tens of thousands of micro-crystals and the limited accessibility of XFEL facility have restricted the wide application of SFX technology.

Besides diffracting X-ray, biomacromolecular crystals could also diffract electron, which is called electron crystallography when 2D crystals are investigated, or called micro-electron diffraction (MicroED) for 3D crystals. Investigation on biological specimens by electron crystallography arguably began when Parsons and Martius used electron diffraction to investigate the structure of muscle fibers in 1964 (Parsons and Martius 1964). In 2005, the 2D electron crystallography has been successfully utilized to solve the high-resolution structure of water channel AQP0 in a closed conformation at 1.9 Å (Gonen et al. 2005). Considering there are only 17 plane groups allowed for 2D protein crystals while there are 65 symmetries allowed for 3D protein crystals (Nannenga Brent et al. 2013), it is of less successful rate to grow a 2D crystal rather than a 3D crystal (Martynowycz and Gonen 2018). Furthermore, the extreme difficulty of culturing high-quality 2D crystals of bio-macromolecules has limited the wide application of 2D electron crystallography.

In 2013, Shi et al. first utilized MicroED to solve the crystal structure of lysozyme at 2.9 Å resolution (Shi et al. 2013) and later improved the data quality by changing still diffraction mode to continuous rotation diffraction mode (Nannenga et al. 2014b). Compared with X-ray, electron has much larger scattering cross-section when interacting with atoms (Henderson 2009). Thus, the size of micro-crystal is large enough to generate high SNR diffractions by using MicroED. The size of crystal used in MicroED experiment is within 500 nm (Nannenga and Gonen 2014), much smaller than the one in traditional X-ray crystallography experiments.

In recent years, MicroED has been further developed and applied to solve the crystal structures of a-synuclein (Rodriguez et al. 2015), prions (Sawaya et al. 2016) and the human fused in sarcoma low-complexity domain (FUS LC) (Luo et al. 2018) in atomic resolution. The crystals of these successful examples were always thin although they were long and wide. For example, the crystals of the a-synuclein are needle-like with the thickness of 20–50 nm (Rodriguez et al. 2015). The thickness of the crystal directly determines the quality of diffraction data (Nannenga and Gonen 2014) due to the mean free path of electron. For 300 kV electron, its mean free path for the vitrified biospecimen is ~350 nm, while for 200 kV electron, it is ~300 nm (Yan et al. 2015). When the thickness of the crystal is over beyond the mean free path of electron, multi-scattering events will become significant and then the diffraction pattern will become difficult to explain. As a result, for a success of MicroED experiment, nano-crystals not micro-crystals are actually needed.

The emergence of MicroED has provided an alternative solution to X-ray crystallography and it is possible to solve the crystal structure of biomacromolecule using a single nanocrystal by MicroED. However, there are still several bottlenecks left, limiting the wide application of MicroED. Firstly, it is difficult to grow and screen nano-crystals. Nano-crystals are invisible under light microscope and their growth process is difficult to monitor. Transmission electron microscope is the only way to screen the presence of nano-crystals, which is in a low throughput. The previous trials of breaking big crystals into tiny bricks were not successful (personal communication). Secondly, the current sample freezing procedure (blotting and plunge-freezing) for single particle analysis is not optimized for MicroED sample preparation. The viscous crystallization liquid could not be easily blotted and the double-sided blotting step could damage delicate crystals easily (Shian et al. 2017). In addition, more importantly, there are many bio-macromolecules that can be crystallized into microcrystal with the size of 2–10 μm. These crystals could not be analyzed by traditional X-ray crystallography and even by

SFX easily. For MicroED, the size of these crystals is also beyond the feasible range.

The recent emergence of cryo focused ion beam (cryo-FIB) technique (Marko et al. 2006) provides a solution to prepare suitable size of crystal for MicroED experiment. This technique was first applied to prepare a cryo-lamella of bacterial cells (Marko et al. 2007) and later to eukaryotic microbial cells (Rigort et al. 2012) and mammalian cells (Strunk et al. 2012; Wang et al. 2012). We also developed this technique with the name of D-cryo-FIB (Zhang et al. 2016), which has been used to prepare cryo-lamella for subsequent cryo-electron tomography experiment (Li et al. 2015).

Here, based on our D-cryo-FIB technique, we report a workflow of in situ protein crystallization and fabrication for subsequent successful MicroED experiment. We grew protein crystals on grids directly to reduce the possibility of crystal missing during sample transfer. The grids were blotted from back side to alleviate the crystal damage due to blot force. Then the crystals frozen on the grid were selected and milled into a thin lamella by cryo-FIB. The cryo-lamella was then used to collect electron diffraction dataset for structure determination. We show here that we successfully solved the crystal structure of lysozyme at 2.5 Å resolution by using seven micro lysozyme crystals within the size of $10 \times 10 \times 10$ μm^3.

RESULTS AND DISCUSSION

To set up a workflow from in situ protein crystallization, cryo-vitrification, and cryo-FIB fabrication to the subsequent cryo-electron diffraction data collection, we selected lysozyme as the testing sample since it is well characterized and was previously used for MicroED experiments (Shi et al. 2013).

Originally, we followed the previously published protocol (Shi et al. 2013) to grow lysozyme crystals by hanging drop vapor diffusion method, and tried to add the crystallization drop from the cover slip to the grid for 1 min adsorption before washing and vitrification. However, by examining in cryo-electron microscope, we found the low successful rate of the crystal absorption on the grid and the crystal could be easily destroyed by the pipette during the drop transfer process. Thus, to overcome this difficulty, we sought to try growing crystals directly on the grid to reduce the loss of the crystal and minimize its potential damage during sample transfer. A holy carbon coated metal grid (Fig. 1A, B) was used in the present study. The sitting crystallization drop was directly added to the surface of a grow-discharge treated grid that is supported by a clean plastic micro-bridge (Fig. 1C). This experimental setup allows the crystals to grow on the carbon surface of the grid.

Different metal grids were screened to select the most suitable one for in situ crystallization. For the copper grids, we found the copper is chemically reactive in crystallization buffer and the dissolved copper ion could denature the protein and affect the quality of the crystal severely (see the green colored crystal in Fig. 2A), which was subsequently proved by X-ray diffraction experiments (data not shown here). Then the titanium (Fig. 2B), molybdenum (Fig. 2C) and non-magnetic nickel grids (Fig. 2D) were tested. All these three grids could allow crystal grow on the carbon surface. Since these grids are stainless with good chemical inertia, there were no chemical effects observed to decrease the crystal quality. Considering the cost and availability, we selected the D-shaped non-magnetic nickel grid (Fig. 2D) for the subsequent experiment.

The crystallization drop contains 15% PEG5000MME that has a high viscosity and is difficult to blot to get a thin layer during cryo-vitrification process. To overcome this difficulty, we used the washing buffer that contains 5% PEG200 instead of 15% PEG5000MME to wash the grid before blotting, which helped to get a thin ice layer on the grid after freezing. To optimize the washing buffer that does not affect the quality of the crystals, we used to pick out some crystals and transfer them to the washing buffer and observed under a light microscope. There should be no obvious dissolving phenomenon appeared within one hour. After washing, the grid was blotted for 6 s from backside using Leica EMGP and then fast frozen in liquid ethane. The backside blotting is also important to prevent potential crystal loss and damage.

The vitrified D-shaped grid was transferred into the chamber of FIB/SEM by keeping the direction of the straight edge perpendicular to FIB (Zhang et al. 2016). The crystals were first identified and visualized by SEM (Fig. 3A, B). To choose the area of interest for FIB milling, there are a few criteria to be considered. Firstly, the position of the crystal should be close to the middle of the square. Otherwise, the metal grid bar would block FIB, causing the trimming process not thorough. In addition, the grid bar could also potentially block the electron beam during diffraction data collection when the grid is tilted. Secondly, it is important to select a separated single crystal to avoid potential twin diffraction images during data collection. In addition, a proper size (5–20 μm) of the crystal was selected. Smaller size would decrease the successful rate of the cryo-lamella production and also yield a very small area for MicroED. Larger size would increase significantly the time of FIB

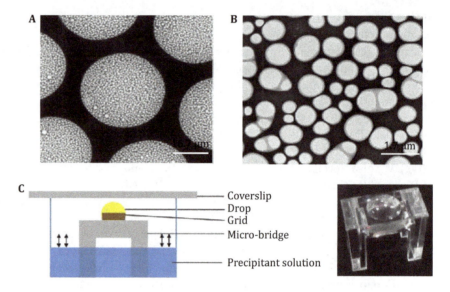

Fig. 1 *In situ* on-grid protein crystallization setup. **A** A magnified SEM image of the non-magnetic nickel grid coated with holy carbon film. Scale bar, 16.7 μm. **B** A further magnified SEM image in **A** showing the structural details of the holy carbon film. Scale bar, 1.7 μm. **C** The schematic diagram of the sitting drop vapor diffusion setup for *in situ* on-grid protein crystallization. The real photo of the micro-bridge is shown in right. Scale bar, 0.35 mm

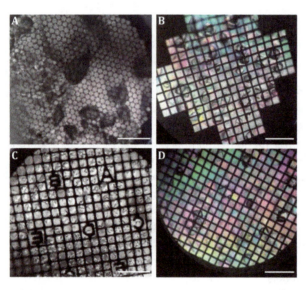

Fig. 2 Photographs of the lysozyme crystals growing on different grids. **A** Copper grid. **B** Titanium grid. **C** Molybdenum grid. **D** Non-magnetic nickel grid. Scale bar, 500 μm

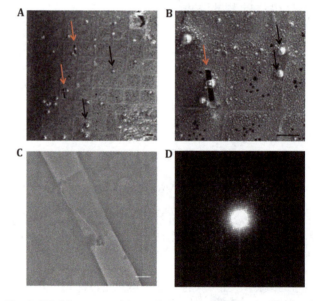

Fig. 3 FIB fabrication of frozen lysozyme crystals on grid. **A** and **B** SEM images (SE detector) of frozen lysozyme crystals on grid with low magnification (**A**) and high magnification (**B**). The areas with strong contrast indicate the positions of the crystals. The *black arrows* indicate the crystals close to the grid bars and the *red arrows* indicate the crystals trimmed by FIB. Scale bars, 50 μm. **C** TEM micrograph of the FIB fabricated crystal lamella. Scale bar, 500 nm. **D** Cryo-electron diffraction pattern of the FIB fabricated crystal lamella in **C**

trimming and thus lower the throughput. After the cryo-lamella of crystal was formed, its thickness was measured from FIB image and could be judged from the low magnification TEM image (Fig. 3C). For a good crystal lamella, its electron diffraction could reach to 2.0 Å resolution (Fig. 3D) with our current experiment hardware.

During electron diffraction data collection, crystals could be either tilted discretely or rotated continuously in the electron beam, which yield still diffraction image or continuous diffraction image. We wrote a simple script of SerialEM (see Supplementary Information) to collect tilt series of still diffraction images automatically. We were aware of that collecting continuous diffraction images could increase the accuracy of reciprocal spot intensity measurement. However, for our current camera setup, it was difficult to synchronize the stage rotation with the data recording of camera due to the imperfect mechanics of the stage and the significant lag of camera recording system. Thus, we made an appropriate approach to collect continuous diffraction images. In our approach, we utilized our script to collect tilt series of still diffraction frames with a very small angle interval. In the present work, the angle interval of 0.2° was used. The exposure time for each still diffraction frame was properly determined to balance its SNR and the total electron dose, and was set 0.2 s/frame in the present work. Then, numbers of frames were simply integrated to form a final image that approximates the continuous diffraction one. Here, every five continuous frames were merged into one image with a rotation angle width of 1° (Movie S1), which was ready for the subsequent processing in iMOSFLM (Fig. 4).

The profiles of diffraction spots suggested that the diffraction spots can be properly indexed, predicted and integrated (Fig. 4A). We found a significant jump of crystal-to-detector distance at the first few frames (Fig. 4B), which might be due to the pre-calibration error of the camera length. While the detected variations of the crystal-to-detector distance and the crystal orientation (Fig. 4B, C) suggest that the stability of the microscope mechanical stage needs to be further optimized for MicroED experiments. From the statistics of the data processing (Fig. 4D), we also observed significant shift of electron beam position, which needs to be further investigated to find the reason. The existence of electron radiation damage and decay of crystal lattice could be indicated from the reduced SNR of diffraction spots during data collection (Fig. 4E). Finally, we collected seven datasets and merged into one dataset with space group of $P4_32_12$, the resolution of 2.5 Å, and the overall cumulative completeness of 94.0% (Table 1). The relative high R_{merge} of 0.356 would be caused by the inaccuracy of reciprocal spot intensity measurement, which could be most probably due to the poor hardware of our camera.

The final merged and scaled intensity data can be directly used for molecular replacement with a single significant solution. The calculated electron potential map based on electron scattering factor and the refined structural model shows a clear envelope of the molecule in the crystal (Fig. 4A). The resolution and quality of the map can further be reflected by the unambiguity of residue assignment in Fig. 4B. The final structure was refined to 2.5 Å with R_{work} of 35.9% and R_{free} of 40.0% (Table 1). Again, we believe that the relative high R value is due to the inaccuracy of reciprocal spot intensity measurement. Considering the current resolution of 2.5 Å, we did not intend to assign water molecule in the map (Fig. 5).

Overall, in the present study, we developed a new approach to efficiently prepare suitable crystals for MicroED experiment. The *in situ* on-grid crystallization method avoids the potential loss of crystals during

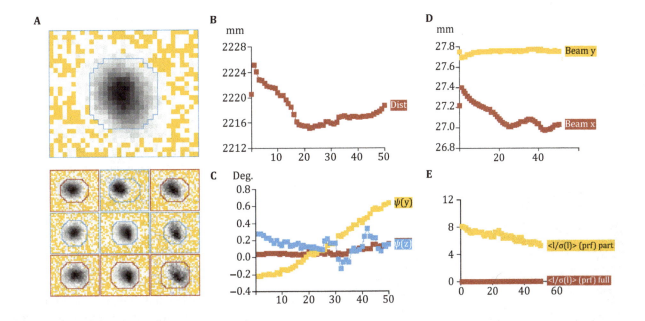

Fig. 4 Statistic parameters during electron diffraction dataset processing by iMOSFLM. **A** A representative average spot profile of one diffraction image (up) and a representative standard profile for different regions of the detector (down). The *red line* indicates the profile is poor and averaged by including reflections from inner regions. **B** The crystal-to-detector distance changes with different diffraction images. **C** The crystal orientation changes with different diffraction images. **D** The electron beam position changes with different diffraction images. **E** The averaged SNR of diffraction spots changes with different diffraction images. The *yellow curve* represents all partial spots and the *red one* for all full spots

Table 1 Statistics of data collection, processing, and structural refinement

Data collection	
Excitation voltage (kV)	200
Electron source	Field emission gun
Wavelength (Å)	0.025
Total electron dose per crystal ($e^-/Å^2$)	5–9
Number of patterns per crystal	150–400
No. of crystals used	7
Nominal camera length (m)	1.35
Real (corrected) camera length (m)	2.22
Selective area aperture (μm)	100
Rotation step (°)	0.2
Data processing	
Resolution (Å)	18.4–2.5
Space group	$P4_32_12$
Unit cell dimensions	
$a = b$ (Å)	77
c (Å)	37
$α = β = γ$ (°)	90
No. of total reflections	53,003
No. of unique reflections	4235
$CC_{1/2}$ (overall/outer shell)	0.803/0.158
$<I/σ>$	5.5
Completeness (%) (overall/outer shell)	94.0/86.8
Multiplicity (overall/outer shell)	12.5/12.7
R_{merge}	0.356
Structural refinement	
Resolution (Å)	18.4–2.5
Reflections in working set	4206
Reflections in test set	191
R_{work}/R_{free} (%)	35.9/40.0
r.m.s.d. bond length	0.005
r.m.s.d. bond angle	0.976

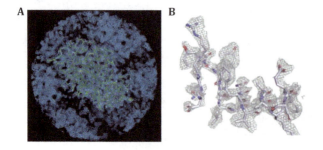

Fig. 5 Electron potential map determined by molecular replacement. **A** The overall map fitted with the whole lysozyme structural model. **B** A zoomed-in view of the electron potential map around a selected region showing the resolution and quality of the map

sample transfer, the single-side blotting method prevents the potential damage to the fragile crystals during vitrification, and the cryo-FIB fabrication method allows the crystals with the size of a few microns to be thinned for MicroED experiment. Finally, the large area of crystal cryo-lamella would yield a strong electron diffraction with good SNR. Thus a few crystals are enough to merge into a complete dataset for structural determination. Our work will greatly expand the availability space of crystals suitable for MicroED and fill up the gap between MicroED and X-ray crystallography.

In the future, we will further systematically investigate the influence of the thickness of crystal cryo-lamella for the MicroED data quality, and study whether the current and energy of focused ion beam would induce observable damage of crystal, which eventually affects the diffraction ability of the crystal.

MATERIALS AND METHODS

In situ crystallization and cryo-vitrification

Lysozyme was purchased by Sangon Biotech Company. A 200 mg/mL solution of lysozyme was prepared in 50 mmol/L sodium acetate pH 4.5. A grow-discharge treated nonmagnetic nickel grid with holy carbon film was placed facing up on a micro-bridge, and the protein was mixed 1:1 with the precipitant solution (0.35 mol/L sodium chloride; 15% PEG5000MME; 50 mmol/L sodium acetate pH 4.5) on the grid. Then the crystals were grown by the sitting drop vapor diffusion method (Fig. 1).

After the crystals formed, the grid was washed four times by washing buffer (0.35 mol/L sodium chloride; 5% PEG200; 50 mmol/L sodium acetate pH 4.5) before cryo-vitrification. Then the grid was blotted 6 s from the backside where no crystals grew on, and plunged into liquid ethane using the Leica EMGP. The frozen grids were transferred and stored in liquid nitrogen for the subsequent experiments.

Crystals fabrication by cryo-FIB

The frozed grid was loaded onto a home-made cryo-shuttle (Zhang *et al.* 2016) and then transferred into the chamber of a dual beam scanning electron microscope (FEI Helios NanoLab 600i) that is equipped with a Quorum PT3000 cryo-stage. The grid has been pre-tilted 45° on the shuttle. Then the grid was imaged and examined using electron beam and the signal from secondary electron with the following experimental parameters, an accelerating voltage of 2 kV, a beam current of 0.34 nA, and a dwell time of 3 μs. The region of interest (ROI) with crystals was identified and marked. Then focused ion gallium beam was utilized to perform fabrication of crystals. Before cryo-FIB milling, a thin layer of Pt was coated using GIS system to reduce the radiation damage during milling. The inclination angle of the beam was kept ~10° against the grid plane. The accelerating voltage of ion beam was 30 kV. Considering the block shape of the crystal, a large beam current of 0.43 nA was used to skive crystals efficiently. When a thick slice of lamella was made, the beam current was reduced to 80 pA for the fine trimming and also the reduction of potential radiation damage. The final thickness of the crystal cryo-lamella was controlled ~300 nm. Several crystal cryo-lamellas can be made in one grid. After cryo-FIB fabrication, the grid was transferred and kept in liquid nitrogen for the subsequent MicroED experiment.

Cryo-electron diffraction and data collection

Cryo-electron diffraction of cryo-FIB fabricated crystal was collected using cryo-electron microscope FEI Talos F200C equipped with a field-emission gun operated at 200 kV (λ = 0.0251 Å). And the diffraction patterns were recorded by the FEI Ceta camera with 4096 × 4096 pixels and the physical pixel size of 14 μm. We utilized SerialEM (Mastronarde 2005) to control the microscope and collect the diffraction datasets.

The frozen grid was loaded into the microscope using a Gatan cryo-transfer holder (Model 626) that was precooled in liquid nitrogen. The straight side of the D-shaped grid (Zhang et al. 2016) was kept parallel to the rotation axis of the holder. ROIs that were trimmed by cryo-FIB were located in low magnification of SA2600X (view mode in SerialEM).

Then, at the exposure mode in SerialEM, the spot size and the illumination area were adjusted to yield a very low electron dose of 0.07 $e^-/(Å^2 \cdot s)$ for the diffraction experiment. To measure the electron dose accurately, the microscope should be in image mode and high magnification so that the electron beam can spread over the entire screen. Eventually, in our experiment setup, the spot size was selected as nine and the excitation level of C2 lens was kept at 45.3000%. After the electron dose was determined, the microscope was switched to diffraction mode. The nominal camera length was set to 1.35 m that was calibrated to 2.22 m by measuring the diffraction pattern of gold crystal. The excitation level of objective lens was set and kept at 85.5148%. The diffraction lens was adjusted to focus the central beam. Then, we switched back and forth between the image and diffraction modes to make sure that the excitation levels of objective and C2 lens did not change. Finally, we switched to diffraction mode and saved all the parameters to the exposure mode in SerialEM.

When all the parameters were setup, we moved the cryo-FIB fabricated crystal to the center of the screen in the view mode of SerialEM, and then chose a selective aperture of 100 μm in diameter to just cover the area of the crystal. Then we switched to the exposure mode in SerialEM. A customized SerialEM script (see Supplementary Information) was written to collect diffraction data in an approximate continuous rotation mode, in which the crystal was initially rotated to −40° (or other degree) and then started to rotate to 40° (or other degree). For every 0.2° increment, a diffraction frame was recorded with the exposure time of 0.2 s and stored in TIFF format. For the rotation angle range from −40° to 40°, we collected 400 frames with the approximate total dose of ~6 $e^-/(Å^2 \cdot s)$. In total, we collected multiple diffraction datasets from different cryo-FIB fabricated crystals. For each crystal, the total electron dose was kept below 9 $e^-/(Å^2 \cdot s)$.

Processing of electron diffraction datasets

We first exacted the dark background from every raw diffraction image and then summed every five frames to the final one with an "expected" oscillation angle width of 1°. The program developed in Tamir Gonen's lab for converting TVIPS camera image to the SMV format (http://cryoem.janelia.org/pages/MicroED) was slightly modified based on our microscope and camera system, and then used to convert our datasets from TIFF format to SMV format.

Then the converted diffraction datasets were processed (index and integration) by iMOSFLM 7.2.1 (Battye et al. 2011). Different with processing X-ray crystallography datasets, a wide rotation angle range was used for a successful index because the Ewald sphere in electron diffraction is very flat. Secondly, due to the instability of microscope stage, the crystal orientation and the distance from the crystal to the detector would change during data collection, which should be carefully considered during data integration process. Furthermore, our camera FEI CETA was not well characterized for electron diffraction experiment, and the GAIN parameter defined in iMOSFLM is generally unknown. Thus, it is important and necessary to try different GAIN values (1.5–2.5) during data integration process.

Due to the instability of our microscope, the first several frames of each dataset were dropped. Seven datasets were merged, scaled and converted to structure factor amplitudes using AIMLESS (Evans and Murshudov 2013) to increase the completeness of diffraction data. Then molecular replacement was performed using PHASER (McCoy et al. 2007) with the starting model (PDB code, 4AXT) of lysozyme structure. Finally, REFMAC5 (Murshudov et al. 2007) was used to perform structural refinement by taking electron scattering factors into consideration.

The statistics of data collection, processing, and structural refinement are summarized in Table 1.

Acknowledgements We would like to thank Ping Shan and Ruigang Su (F.S. lab) for their assistances. We are grateful to Dr. Xiaojun Huang, Dr. Zhenxi Guo and Mr. Deyin Fan (CBI, IBP) for their support on the microscope operation and calibration. This work was supported by grants from Chinese Academy of Sciences (ZDKYYQ20170002 and XDB08030202) and the Ministry of Science and Technology of China (2017YFA0504700 and 2014CB910700). All the EM works were performed at Center for Biological Imaging (CBI, http://cbi.ibp.ac.cn), Institute of Biophysics, Chinese Academy of Sciences.

Compliance with ethical standards

Conflict of interest Xinmei Li, Shuangbo Zhang and Fei Sun declare that they have no conflict of interest.

Human and animal rights and informed consent This article does not contain any studies with human or animal subjects performed by any of the authors.

References

Battye TGG, Kontogiannis L, Johnson O, Powell HR, Leslie AGW (2011) iMOSFLM: a new graphical interface for diffraction-image processing with MOSFLM. Acta Crystallogr Sect D 67:271–281

Chapman HN, Fromme P, Barty A, White TA, Kirian RA, Aquila A, Hunter MS, Schulz J, DePonte DP, Weierstall U, Doak RB, Maia FR, Martin AV, Schlichting I, Lomb L, Coppola N, Shoeman RL, Epp SW, Hartmann R, Rolles D, Rudenko A, Foucar L, Kimmel N, Weidenspointner G, Holl P, Liang M, Barthelmess M, Caleman C, Boutet S, Bogan MJ, Krzywinski J, Bostedt C, Bajt S, Gumprecht L, Rudek B, Erk B, Schmidt C, Hömke A, Reich C, Pietschner D, Strüder L, Hauser G, Gorke H, Ullrich J, Herrmann S, Schaller G, Schopper F, Soltau H, Kühnel KU, Messerschmidt M, Bozek JD, Hau-Riege SP, Frank M, Hampton CY, Sierra RG, Starodub D, Williams GJ, Hajdu J, Timneanu N, Seibert MM, Andreasson J, Rocker A, Jönsson O, Svenda M, Stern S, Nass K, Andritschke R, Schröter CD, Krasniqi F, Bott M, Schmidt KE, Wang X, Grotjohann I, Holton JM, Barends TR, Neutze R, Marchesini S, Fromme R, Schorb S, Rupp D, Adolph M, Gorkhover T, Andersson I, Hirsemann H, Potdevin G, Graafsma H, Nilsson B, Spence JC (2011) Femtosecond X-ray protein nanocrystallography. Nature 470:73–77

Evans PR, Murshudov GN (2013) How good are my data and what is the resolution? Acta Crystallogr Sect D 69:1204–1214

Gonen T, Cheng Y, Sliz P, Hiroaki Y, Fujiyoshi Y, Harrison SC, Walz T (2005) Lipid–protein interactions in double-layered two-dimensional AQP0 crystals. Nature 438:633

Henderson R (2009) The potential and limitations of neutrons, electrons and X-rays for atomic resolution microscopy of unstained biological molecules. Q Rev Biophys 28:171–193

Iadanza MG, Gonen T (2014) A suite of software for processing MicroED data of extremely small protein crystals. J Appl Crystallogr 47:1140–1145

Johansson LC, Stauch B, Ishchenko A, Cherezov V (2017) A bright future for serial femtosecond crystallography with XFELs. Trends Biochem Sci 42:749–762

Kendrew JC, Dickerson RE, Strandberg BE, Hart RG, Davies DR, Phillips DC, Shore VC (1960) Structure of myoglobin: A three-dimensional Fourier synthesis at 2 Å. Resolution. Nature 185:422–427

Li X, Feng H, Zhang J, Sun L, Zhu P (2015) Analysis of chromatin fibers in Hela cells with electron tomography. Biophys Rep 1:51–60

Luo F, Gui X, Zhou H, Gu J, Li Y, Liu X, Zhao M, Li D, Li X, Liu C (2018) Atomic structures of FUS LC domain segments reveal bases for reversible amyloid fibril formation. Nat Struct Mol Biol 25:341–346

Marko M, Hsieh C, Moberlychan W, Mannella CA, Frank J (2006) Focused ion beam milling of vitreous water: prospects for an alternative to cryo-ultramicrotomy of frozen-hydrated biological samples. J Microsc 222:42–47

Marko M, Hsieh C, Schalek R, Frank J, Mannella C (2007) Focused-ion-beam thinning of frozen-hydrated biological specimens for cryo-electron microscopy. Nat Methods 4:215

Martynowycz MW, Gonen T (2018) From electron crystallography of 2D crystals to MicroED of 3D crystals. Curr Opin Colloid Interface Sci 34:9–16

Mastronarde DN (2005) Automated electron microscope tomography using robust prediction of specimen movements. J Struct Biol 152:36–51

Mayer J, Giannuzzi LA, Kamino T, Michael J (2011) TEM sample preparation and FIB-induced damage. MRS Bull 32:400–407

McCoy AJ, Grosse-Kunstleve RW, Adams PD, Winn MD, Storoni LC, Read RJ (2007) Phaser crystallographic software. J Appl Crystallogr 40:658–674

Mizohata E, Nakane T, Fukuda Y, Nango E, Iwata S (2018) Serial femtosecond crystallography at the SACLA: breakthrough to dynamic structural biology. Biophys Rev 10:209–218

Murshudov GN, Vagin AA, Dodson EJ (2007) Refinement of macromolecular structures by the maximum-likelihood method. Acta Crystallogr Sect D 53:240–255

Nannenga BL, Gonen T (2014) Protein structure determination by MicroED. Curr Opin Struct Biol 27:24–31

Nannenga BL, Shi D, Hattne J, Reyes FE, Gonen T (2014a) Structure of catalase determined by MicroED. eLife 3:e03600

Nannenga BL, Shi D, Leslie AGW, Gonen T (2014b) High-resolution structure determination by continuous-rotation data collection in MicroED. Nat Methods 11:927

Nannenga-Brent L, Iadanza-Matthew G, Vollmar-Breanna S, Gonen T (2013) Overview of electron crystallography of membrane proteins: crystallization and screening strategies using negative stain electron microscopy. Curr Protoc Protein Sci 72:17.15.1–17.15.11

Neutze R, Wouts R, van der Spoel D, Weckert E, Hajdu J (2000) Potential for biomolecular imaging with femtosecond X-ray pulses. Nature 406:752

Parsons DF, Martius U (1964) Determination of the α-helix configuration of poly-γ-benzyl-l-glutamate by electron diffraction. J Mol Biol 10:530–534

Perutz MF, Rossmann MG, Cullis AF, Muirhead H, Will G, North AC (1960) Structure of haemoglobin: a three-dimensional Fourier synthesis at 5.5-Å resolution, obtained by X-ray analysis. Nature 185:416–422

Rigort A, Bäuerlein FJB, Villa E, Eibauer M, Laugks T, Baumeister W, Plitzko JM (2012) Focused ion beam micromachining of eukaryotic cells for cryoelectron tomography. Proc Natl Acad Sci 109:4449

Rodriguez JA, Ivanova MI, Sawaya MR, Cascio D, Reyes FE, Shi D, Sangwan S, Guenther EL, Johnson LM, Zhang M, Jiang L, Arbing MA, Nannenga BL, Hattne J, Whitelegge J, Brewster AS, Messerschmidt M, Boutet S, Sauter NK, Gonen T, Eisenberg DS (2015) Structure of the toxic core of α-synuclein from invisible crystals. Nature 525:486

Sawaya MR, Rodriguez J, Cascio D, Collazo MJ, Shi D, Reyes FE, Hattne J, Gonen T, Eisenberg DS (2016) Ab initio structure determination from prion nanocrystals at atomic resolution by MicroED. Proc Natl Acad Sci 113:11232

Shi D, Nannenga BL, Iadanza MG, Gonen T (2013) Three-dimensional electron crystallography of protein microcrystals. eLife 2:e01345

Shian L, Hattne J, Reyes FE, Silvia SM, Jason de la Cruz MJ, Dan S, Tamir G (2017) Atomic resolution structure determination by the cryo-EM method MicroED. Protein Sci 26:8–15

Smith JL, Fischetti RF, Yamamoto M (2012) Micro-crystallography comes of age. Curr Opin Struct Biol 22:602–612

Strunk KM, Wang K, Ke D, Gray JL, Zhang P (2012) Thinning of large mammalian cells for cryo-TEM characterization by cryo-FIB milling. J Microsc 247:220–227

Suga M, Akita F, Sugahara M, Kubo M, Nakajima Y, Nakane T, Yamashita K, Umena Y, Nakabayashi M, Yamane T, Nakano T, Suzuki M, Masuda T, Inoue S, Kimura T, Nomura T, Yonekura S, Yu L-J, Sakamoto T, Motomura T, Chen J-H, Kato Y, Noguchi T, Tono K, Joti Y, Kameshima T, Hatsui T, Nango E, Tanaka R, Naitow H, Matsuura Y, Yamashita A, Yamamoto M, Nureki O, Yabashi M, Ishikawa T, Iwata S, Shen J-R (2017) Light-induced structural changes and the site of O=O bond formation in PSII caught by XFEL. Nature 543:131

Wang K, Strunk K, Zhao G, Gray JL, Zhang P (2012) 3D structure determination of native mammalian cells using cryo-FIB and cryo-electron tomography. J Struct Biol 180:318–326

Yan R, Edwards TJ, Pankratz LM, Kuhn RJ, Lanman JK, Liu J, Jiang W (2015) Simultaneous determination of sample thickness, tilt, and electron mean free path using tomographic tilt images based on Beer-Lambert law. J Struct Biol 192:287–296

Yonekura K, Kato K, Ogasawara M, Tomita M, Toyoshima C (2015) Electron crystallography of ultrathin 3D protein crystals: atomic model with charges. Proc Natl Acad Sci 112:3368

Zhang H, Unal H, Gati C, Han GW, Liu W, Zatsepin NA, James D, Wang D, Nelson G, Weierstall U, Sawaya MR, Xu Q, Messerschmidt M, Williams GJ, Boutet S, Yefanov OM, White TA, Wang C, Ishchenko A, Tirupula KC, Desnoyer R, Coe J, Conrad CE, Fromme P, Stevens RC, Katritch V, Karnik SS, Cherezov V (2015) Structure of the angiotensin receptor revealed by serial femtosecond crystallography. Cell 161:833–844

Zhang J, Ji G, Huang X, Xu W, Sun F (2016) An improved cryo-FIB method for fabrication of frozen hydrated lamella. J Struct Biol 194:218–223

Mapping disulfide bonds from sub-micrograms of purified proteins or micrograms of complex protein mixtures

Shan Lu[1], Yong Cao[1], Sheng-Bo Fan[2,3], Zhen-Lin Chen[2,3], Run-Qian Fang[2,3], Si-Min He[2,3 ✉], Meng-Qiu Dong[1 ✉]

[1] National Institute of Biological Sciences, Beijing, Beijing 102206, China
[2] Key Lab of Intelligent Information Processing of Chinese Academy of Sciences (CAS), University of CAS, Institute of Computing Technology, CAS, Beijing 100190, China
[3] University of Chinese Academy of Sciences, Beijing 100049, China

Abstract Disulfide bonds are vital for protein functions, but locating the linkage sites has been a challenge in protein chemistry, especially when the quantity of a sample is small or the complexity is high. In 2015, our laboratory developed a sensitive and efficient method for mapping protein disulfide bonds from simple or complex samples (Lu *et al.* in Nat Methods 12:329, 2015). This method is based on liquid chromatography–mass spectrometry (LC–MS) and a powerful data analysis software tool named pLink. To facilitate application of this method, we present step-by-step disulfide mapping protocols for three types of samples—purified proteins in solution, proteins in SDS-PAGE gels, and complex protein mixtures in solution. The minimum amount of protein required for this method can be as low as several hundred nanograms for purified proteins, or tens of micrograms for a mixture of hundreds of proteins. The entire workflow—from sample preparation to LC–MS and data analysis—is described in great detail. We believe that this protocol can be easily implemented in any laboratory with access to a fast-scanning, high-resolution, and accurate-mass LC–MS system.

Keywords Disulfide, Identification of disulfide bonds, Cross-linking, Mass spectrometry, PLink, PLink-SS

INTRODUCTION

Functions of disulfide bonds

Formation of disulfide bonds is a common post-translational modification that has important biological functions. Many secreted proteins such as antibodies, growth factors, extracellular matrix proteins, and cell surface receptors or transporters, which happen to be of great therapeutic interest, are rich in disulfide bonds. As a structural building block, a disulfide bond covalently links two cysteine residues in the same protein or in different proteins to strengthen the correct conformation of a protein or protein complex, thereby improving stability. Further, the reversible nature of a disulfide bond enables it to act as a molecular switch to regulate the activity of enzymes or transcription factors in response to the redox state of the environment (Hogg 2003). To fully understand the biological function of a disulfide-containing protein and its regulation, it is necessary to map precisely the position of each disulfide bond and to determine the redox state of the two cysteine residues involved—in the disulfide form or as free thiols, or else—under conditions studied.

Methods for disulfide bond analysis

In the past, a variety of methods had been used for the analysis of protein disulfide bonds including Edman

---

Shan Lu and Yong Cao have contributed equally to this work.

✉ Correspondence: smhe@ict.ac.cn (S.-M. He), dongmengqiu@nibs.ac.cn (M.-Q. Dong)

degradation (Haniu et al. 1994), diagonal electrophoresis (McDonagh 2009), mutagenesis of cysteine residues coupled with reducing and non-reducing SDS-PAGE (Itakura et al. 1994), X-ray crystallography (McCarthy et al. 2000), and nuclear magnetic resonance spectroscopy (Sharma and Rajarathnam 2000). However, none of these methods are ideal; the ones that can provide precise linkage information of disulfide bonds demand highly specialized skills and devoted efforts of structural biologists, and the ones that can be executed in an average biology lab do not afford linkage information directly. Further, they usually require milligrams of purified proteins, and none of them work on complex samples.

In recent years, rapid technological development in liquid chromatography–mass spectrometry (LC–MS) has made it possible to map disulfide bonds from as little as micrograms of proteins in a relatively high throughput way (Choi et al. 2009; Götze et al. 2012; Huang et al. 2014; Liu et al. 2014, 2017a; Murad and Singh 2013; Wang et al. 2014; Wefing et al. 2006; Wu et al. 2008; Xu et al. 2007). Among the methods in this category, the more straightforward ones do not reduce disulfide bonds prior to LC–MS analysis, so the linkage information can be extracted from the fragmentation spectra of peptides containing disulfide bonds. Different fragmentation methods including collision-induced dissociation (CID), higher-energy collisional dissociation (HCD), electron-transfer dissociation (ETD), and electron-transfer/higher-energy collision dissociation (EThcD) (Liu et al. 2014) have been used to varying degrees of success. Table 1 summarizes the data analysis tools that have been developed for LC–MS analysis of disulfide bonds. Most of them are designed to identify disulfide bonds directly from fragmentation spectra. From these endeavors, two challenges have become apparent, one is to identify all the disulfide bonds in a protein and the other is to identify disulfide bonds at a proteome scale, that is, from highly complex samples such as cell lysates, isolated mitochondria, or secretomes. Another constant problem is false identification of disulfide bonds, the source of which may be faulty data analysis or disulfide bond scrambling during sample preparation.

Table 1 Software tools for MS-based disulfide bond identification

Software	MS function required	Type of data analyzed	Advantages (A) and limitations (L)	References
SearchXLinks	MALDI source	MS1	L: Works only for very simple samples	Wefing et al. (2006)
MassMatrix	CID or HCD	CID or HCD MS2	A: Complex forms of disulfide bonds are taken into consideration; L: Only for low-complexity samples digested using specific proteases	Xu et al. (2007)
DBond	CID or HCD	CID or HCD MS2	A: Disulfide-specific fragment ions are considered; L: No automatic FDR control, only for low-complexity samples	Choi et al. (2009)
MS2DB+	CID or HCD	CID or HCD MS2	L: No FDR control	Murad and Singh (2013)
MixDB	CID or HCD	CID or HCD MS2	A: Automatic FDR control, can handle large protein databases; L: Not easy to use, not tested with real-world samples	Wang et al. (2014)
RADAR	HCD	HCD MS2	A: Specific dimethyl labeling at peptide N-terminus improves accuracy of identification; L: Requires labeling after digestion; No FDR control	Huang et al. (2014)
PepFinder	EThcD or high-resolution ETD	HCD MS2	L: One reduced and one non-reduced samples need to be analyzed side by side, works only for low-complexity samples	From Thermo Scientific
SlinkS	EThcD and high-resolution ETD	ETD or EThcD MS2	A: ETD and EThcD complement each other; L: Requires a unique ion pattern, efficiency of ETD varies depending on the peptides	Liu et al. (2014)
pLink-SS	HCD or high-resolution ETD	HCD MS2	A: Automatic workflow with FDR estimation, disulfide-specific ions and internal ions are considered, works for complex samples; L: highly complex forms such as three peptides inter-linked through two disulfide bonds cannot be identified	Lu et al. (2015)

About the method used in this protocol

With these problems in mind, we developed a method that enabled us to prevent most, if not all disulfide scrambling events and to identify all the native disulfide bonds of a single protein or a mixture of ten proteins from micrograms or even a few hundred nanograms of samples. It also enabled us to map native disulfide bonds at a proteome scale, for instance, 199 disulfide bonds were identified from a periplasmic fraction of *Escherichia coli* cells and 568 disulfide bonds were identified from proteins secreted by human umbilical vein endothelial cells. This method was published in 2015 (Lu *et al.* 2015). Since then, to our knowledge, it has been used successfully in dozens of studies and seven of them have been published (Hartman *et al.* 2016; Hung *et al.* 2016; Liu *et al.* 2017b; Mauney *et al.* 2017; Wang *et al.* 2016; Wu *et al.* 2016, 2017). The three key components of this method are as follows. First, disulfide bond scrambling is prevented by blocking free thiols with *N*-ethylmaleimide (NEM) and by maintaining an acidic pH throughout the sample preparation process, the latter of which includes precipitating freshly prepared protein samples with trichloroacetic acid (TCA) as early as possible and carrying out all protease digestions at pH 6.5 (Fig. 1). Second, to identify all the disulfide bonds of a protein, multiple proteases are utilized, even the non-specific ones such as proteinase K. This is because some disulfide bonds may be present in a complex form that is difficult to identify if the sample is digested only with Lys-C and trypsin (Fig. 2). Last and the most important, a data analysis program called pLink-SS has been developed and carefully tuned to identify disulfide-bonded peptides from HCD spectra. The types of disulfide-bonded peptides that can be identified using pLink-SS are shown in Fig. 3. Presently, pLink-SS has been incorporated into pLink 2, which is an upgraded version of pLink and remains free for academic users. pLink 2 is ~40 times faster than pLink, with a friendly graphical interface and some further improvements in accuracy. pLink 2 was officially released on January 1, 2018 and can be downloaded at http://pfind.ict.ac.cn/software/pLink/.

In this paper, we present a step-by-step disulfide mapping protocol using the method we developed in 2015. As shown in Fig. 4, this protocol contains three alternative sub-protocols that are each optimized for low-complexity samples in solution, protein gel bands, or high-complexity samples.

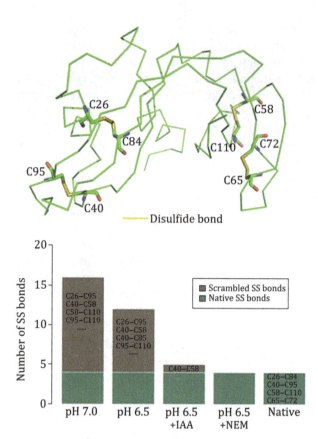

Fig. 1 Non-native disulfide bonds of RNase A are abolished by blocking free thiols with NEM and carrying out protease digestions at pH 6.5

REAGENTS

Chemicals

- Acetic acid (J. T. Baker, cat. no. 9508)
- Acetone (J. T. Baker, cat. no. 9002-02)
- Acetonitrile (ACN), HPLC grade (Fisher Scientific, cat. no. A998)
- Ammonium acetate, 7.5 mol/L solution (Sigma-Aldrich, cat. no. A2706)
- BCA protein assay kit (Pierce, cat. no. 23228)
- Formic acid (FA) (J. T. Baker, cat. no. 0129)
- Frit kit (including formamide, Kasil®-1 and Kasil®-1624 potassium silicate solution and a ceramic tubing cutter, from Next Advance)
- Guanidine hydrochloride (GndCl) (Sigma-Aldrich, cat. no. G3272)
- N-ethylmaleimide (NEM) (Pierce, cat. no. 23030)
- Trichloroacetic acid (TCA), 6.1N or 100% (*w/v*) solution (Sigma-Aldrich, cat. no. T0699)
- Tris (Amresco, cat. no. 0497)

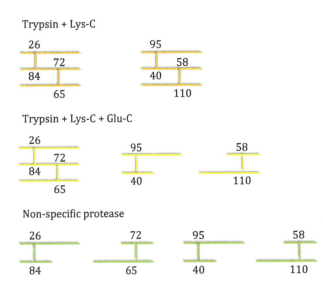

Fig. 2 Digestion of RNase A with Lys-C and trypsin results in a complex form comprising three peptides linked together through two disulfide bonds, which cannot be identified using existing software tools. Digestion with additional proteases generates simpler forms that can be identified

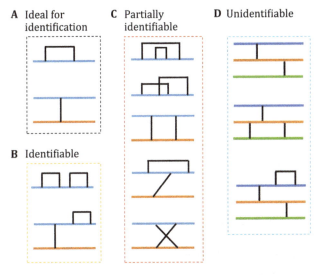

Fig. 3 Disulfide bonds that can or cannot be identified using pLink. A Peptides containing a single disulfide bond can be identified with precise linkage information. B Disulfide bonds can be identified with precise linkage information in some cases. C The presence of disulfide bonds may be identified but without linkage information. D Disulfide bonds cannot be identified

- Urea (Sigma-Aldrich, cat. no. U0631)
- Water, HPLC grade (Sigma-Aldrich, cat. no. V270733)

Enzymes

- Asp-N (Promega, cat. no. V162A)
- Elastase (Sigma-Aldrich, cat. no. E7885)
- Glu-C (Promega, cat. no. V1651)
- Lys-C (Wako, cat. no. 125-05061)
- PNGase F (NEB, cat. no. P0704)
- Proteinase K (Sigma-Aldrich, cat. no. P3910)
- Subtilisin (Sigma-Aldrich, cat. no. 43538)
- Trypsin (Promega, cat. no. V5117)

Supplies

- Fused silica capillary tubing with a polyimide coating for reverse-phase (RP) liquid chromatography–mass spectrometry (LC–MS), 75-μm inner diameter (ID), 360-μm outer diameter (OD) (Polymicro Technologies, Part No. 1068150019)
- Fused silica capillary tubing for off-line SCX fractionation, 200-μm ID, 360-μm OD (Polymicro Technologies, Part No. 1068150204)
- MicroTight Union for 360-μm OD tubing (Upchurch, Part No. P-772)
- Luna C18 resin, 3 μm particle size, 100 Å pore size (Phenomenex, cat. no. 04A-4251)
- Luna SCX resin, 5 μm particle size, 100 Å pore size, (Phenomenex, cat. no. 04A-4398)
- Sample vials for LC–MS (Thermo La-pha-pak, cat. no. 11190933)
- Welch Ultimate UHPLC-XB-C18 resin, C18, 1.8 μm particle size, 120 Å pore size (Welch Materials, Shanghai, cat. no. UHP101.02)
- YMC*GEL, C18, 10 μm particle size, pore size 120 Å (YMC, cat no. AQ12S11)

BUFFERS

- 0.5 mol/L Tris buffer, pH 6.5
- Strong cation exchange (SCX) elution buffers: 25, 50, 100, 250, 500 mmol/L, and 1 mol/L ammonium acetate in 5% ACN, 0.1% FA
- For SCX fractionation: 80% ACN, 0.1% FA
- For in-gel digestion, peptide extraction (Buffer I): 0.5% FA, 50% ACN
- For in-gel digestion, peptide extraction (Buffer II): 1% FA, 75% ACN
- LC mobile phase A (Buffer A): H_2O/FA (100/0.1, v/v), good at room temperature (RT) for several weeks
- LC mobile phase B (Buffer B): ACN/FA (100/0.1, v/v), good at RT for several months

EQUIPMENT

Small equipment

- ThermoMixer, a bench top temperature-controlled tube mixer (Eppendorf)

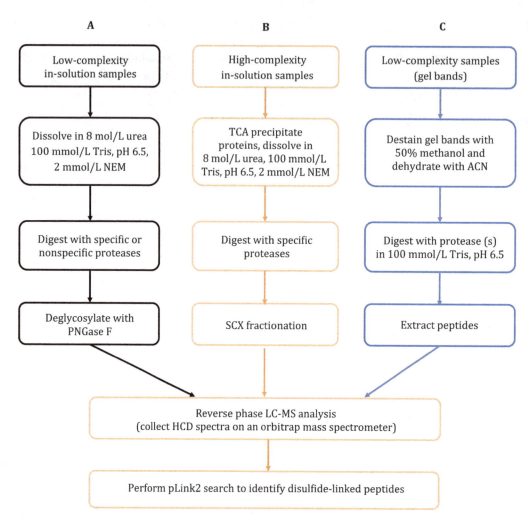

Fig. 4 Flowchart of this protocol

- Refrigerated bench top centrifuge (Eppendorf)
- SpeedVac™ concentrator (Fisher Scientific)
- A common laboratory oven or dryer
- Laser-based micropipette puller, model P-2000 (Sutter Instrument)
- Column packing setup, consisting of a pressure injection cell (also known as pressure loading cell or bomb loader) model PC77-MAG (Next Advance), a high-purity nitrogen gas canister (from a local supplier) fitted with a high-pressure regulator, and a stainless steel 1/8-inch diameter tubing that connects the regulator to the pressure injection cell. The last two items are parts of a column packing kit (Next Advance)

LC–MS system

- Easy-nLC1000 liquid chromatography system (Thermo Fisher Scientific) or a similar nano-flow (200–600 nl/min) HPLC system with an autosampler
- Q-Exactive™ Q-Orbitrap mass spectrometer (Thermo Fisher Scientific) or a similar fast-scanning, high-resolution, accurate-mass MS instrument that can collect ten or more high-resolution ($R > 7000$) MS2 spectra per second

Do-it-yourself capillary chromatography columns

Needless to say, skip this section if one chooses to purchase pre-packed columns of equivalent properties.

RP analytical column with a spray tip

i. Cut a 50-cm-long 75-μm ID fused silica tubing, burn the center segment (2–3 cm) over an ethanol burner, and wipe the blackened coating off the tubing with a sheet of kimwipe moistened with methanol.

ii. Mount the tubing into the Sutter laser puller with the clear segment in the path of the laser beam and pull a 5-μm tip.
   Note: The exact setup of the pulling program varies between instruments and over time. The example below serves as a starting point for optimization.

   cycle 1   HEAT = 270, FIL = 0, VEL = 35, DEL = 128, PUL = 0
   cycle 2   HEAT = 260, FIL = 0, VEL = 30, DEL = 128, PUL = 0
   cycle 3   HEAT = 250, FIL = 0, VEL = 25, DEL = 128, PUL = 0
   cycle 4   HEAT = 240, FIL = 0, VEL = 25, DEL = 128, PUL = 0

iii. Dismount the two empty columns each with a pulled tip.
iv. In a tube or a glass vial that will fit inside the pressure injection cell, add a small amount (roughly the size of a grain of millet) of Welch UHPLC-XB-C18 resin into methanol and make a slurry.
v. Using the column packing setup, pack the 3-μm Luna C18 resin into a 75-μm ID analytical column with a pulled tip for a length of ∼2 cm, then switch the resin to 1.8-μm Welch UHPLC-XB-C18, and pack another 10–12 cm. The total length of the reverse phase is 13 ± 1 cm.
vi. Condition the column by running a RP gradient through it and ending with Buffer A wash.

RP trap column

i. Take from the Frit kit 60 μl Kasil-1624 and 20 μl Kasil-1, mix well in a small tube, then add 20 μl formamide, mix well.
ii. Cut a 20-cm-long 75-μm ID fused silica tubing, dip one end into the mixture just made, and then pull out immediately. Inspect the column for the appearance of a segment of liquid inside.
iii. Place the column in an oven of 100 °C for 4–12 h to obtain a porous frit at one end of the column. Before polymerization, handle the column with great care to avoid displacing the liquid away from the end. Keep fritted columns at RT for long-term storage.
iv. Before packing a column, cut the frit end with a tubing cutter so only 1–2 mm frit is left.
v. In a tube or a glass vial that will fit inside the pressure injection cell, add a small amount (roughly the size of a grain of millet) of Luna C18 resin into methanol and make a slurry.
vi. Using the column packing setup, pack the 10-μm YMC*GEL C18 resin into an empty, 75-μm ID column against the frit for a length of 7 ± 1 cm.
vii. Condition the trap column by passing Buffer A through it.

SCX fractionation column

i. Take from the Frit kit 60 μl Kasil-1624 and 20 μl Kasil-1, mix well in a small tube, then add 20 μl formamide, mix well.
ii. Cut a 25-cm-long 200-μm ID fused silica tubing, dip one end into the mixture just made, and then pull out immediately. Inspect the column for the appearance of a segment of liquid inside.
iii. Place the column in an oven of 100 °C for 4–12 h to obtain a porous frit at one end of the column. Before polymerization, handle the column with great care to avoid displacing the liquid away from the end. Keep fritted columns at RT for long-term storage.
iv. Before packing a column, cut the frit end with a tubing cutter so only 1–2 mm frit is left.
v. In a tube or a glass vial that will fit inside the pressure injection cell, add a small amount (roughly the size of a grain of millet) of Luna SCX resins into methanol and make a slurry.
vi. Using the column packing setup, pack the SCX resins into an empty column against the frit for a length of 2–3 cm.
vii. Pass Buffer A through the column to pack the SCX segment more tightly.
viii. Change the packing material to 3-μm Luna C18 resin and pack a RP segment of 2–3 cm.
ix. Condition the SCX column by passing Buffer A through it.

SOFTWARE

- Computer workstation with Microsoft Windows 7 or a newer operating system
- NET framework 4.5
- MSFileReader, both 32 and 64-bit version, for pLink 2 to access the raw file
- pLink 2, version 2.3.0
- Python 3.6

REAGENT SETUP

1 mol/L NEM stock solution

Dissolve 125.1 mg of solid N-ethylmaleimide in 1 ml of 100% ACN, dispense into 4-μl aliquots, store in a desiccator at −20 °C, and use within two months. Each aliquot is for a single use.

8 mol/L urea in 100 mmol/L Tris pH 6.5

Dissolve 240 mg of solid urea in 100 μl 0.5 mol/L Tris at pH 6.5 and 220 μl H$_2$O. Prepare the fresh solution each time to minimize urea degradation and subsequent carbamylation of proteins or peptides.

Stock solutions of proteases

Prepare 0.5 μg/μL trypsin or a different protease in H$_2$O or a stock buffer specified by the vendor, dispense into aliquots of 10 μl or less, and store at −80 °C. Ideally, each aliquot is for a single use.

EQUIPMENT SETUP

LC–MS

On Q-Exactive, generate a MS method as follows: spray voltage 2.0–2.3 kV, data-dependent mode, full scan resolution 140,000, MS2 scan resolution 17,500, isolation window 2.0 *m/z*, AGC target at 1e6 for FTMS full scan and 5e4 for MS2, minimal signal threshold for MS2 at 4e4; normalized collision energy at 27%; peptide match preferred, and HCD spectra were collected for the ten most intense precursors carrying +3, +4,..., or +7 positive charges, dynamic exclusion 60 s. To increase the identification of loop-linked disulfide bonds, a technical repeat run is recommended in which +2 precursors were also included. So, generate another MS method that is the same as above except that +2 precursors are not excluded.

On Easy-nLC 1000 UHPLC, set the sample loading and RP gradient method as follows. For each sample, 0.5 μg of digested peptides are loaded onto the trap column at 1 μl/min and desalted with 10 μl of Buffer A. The peptides are separated through a 100 min linear gradient from 100% Buffer A to 30% Buffer B and then going up to 100% Buffer B in 1 min, followed by a 3-min 100% Buffer B wash before returning to 100% Buffer A in 2 min and maintaining at 100% Buffer A for 4 min. Set the flow rate at 250 nl/min.

SOFTWARE SETUP

(1) MSFileReader: Instructions are available at https://github.com/pFindStudio/pLink2/wiki/FAQ#how-to-install-msfilereader.
(2) pLink 2: Free download at http://pfind.ict.ac.cn/software/pLink/index.html#Downloads. Double click the installation package, choose the language and directory. pLink 2 will then finish the installation automatically.
[CRITICAL] License is required to run pLink 2 the first time. To receive the license file, send an e-mail to pLink@ict.ac.cn. Follow instructions during installation. The pLink 2 installation package includes pParse, a MS data conversion tool, and pLabel—a very convenient and powerful tool for annotating peaks in a MS2 spectrum. Once pLink 2 is installed, pParse and pLabel are ready.
(3) Python 3.6: Installation instructions are available at https://www.python.org/.
(4) Python package "openpyxl" and "xlrd": Enter in the following commands in windows "Command Prompt":

– pip3 install openpyxl,
– pip3 install xlrd.

[CRITICAL] Here we provide a python script to organize the pLink 2 search results and generate a simple and informative report file. To run this script, Python 3.6 and packages openpyxl and xlrd are required.

(5) Python script SS_sim.py: Download from https://github.com/daheitu/pLink2_ss_results_analysis.github.io.git.

SAMPLE PREPARATION

In-solution digestion (low-complexity samples) [TIMING ~ 1 day]

Note: Low-complexity samples refer to a purified protein or a protein complex of no more than 50 subunits.

(1) Determine the concentration of a freshly purified protein sample using a BCA protein assay kit. Do not rely on OD_{280} measurements.
(2) Based on the sequences of the proteins in question and the amount of proteins available, decide which proteases to use and how many digestions to carry out. Common choices are Lys-C/trypsin, Lys-C/trypsin/Glu-C, Lys-C/Asp-N, and Lys-C/trypsin/Asp-N;

and additional options include Lys-C/elastase, subtilisin, and proteinase K.

(3) For each digestion, take 4 μg of a freshly prepared protein sample and precipitate with 25% TCA. Specifically, add 1/3 volume of 100% TCA, mix well, and leave on ice for 30 min to overnight. Spin at 4 °C in a bench top centrifuge at top speed for 30 min to pellet proteins, wash with 0.5 ml cold acetone twice, and air dry the pellet.

(4) Dissolve 4 μg of the freshly precipitated protein sample in 10 μl of 8 mol/L urea, 100 mmol/L Tris, pH 6.5; add NEM to a final concentration 2 mmol/L; and incubate at 37 °C for 2 h.

[CRITICAL] NEM can gradually hydrolyze in water and lose activity, so 15 min before use, transfer a frozen aliquot of 1 mol/L NEM to a desiccator at RT.

[? TROUBLESHOOTING]

(5) Digest the protein(s) with one or more proteases of choice. The digestion conditions of various proteases are listed in Table 2. If two or more proteases are to be combined in one digestion, perform the digestion sequentially to reduce mutual digestion between proteases. For example, if a sample is to be digested with Lys-C/trypsin/Glu-C, digest with Lys-C first in 8 mol/L urea at 37 °C for 2 h; then dilute to 2 mol/L urea with 30 μl of 100 mmol/L Tris, pH 6.5, add trypsin, and incubate at 37 °C for 12 h; and lastly, dilute to 1 mol/L urea with 40 μl of 100 mmol/L Tris, pH 6.5, add Glu-C, and incubate at 37 °C for 12 h.

[CRITICAL] As disulfide-linked proteins are often resistant to proteases, digestion in the presence of a high concentration of denaturant is necessary. Therefore, Lys-C digestion in 8 mol/L urea usually precedes other proteases except for subtilisin and proteinase K, both of which have activity high enough for digesting any protein, even at pH 6.5. Avoid under- or over-digestion.

[? TROUBLESHOOTING]

(6) To remove glycosylation, which may interfere with data analysis, add PNGase F (112 NEB units per 6 μg of proteins) to the digest and incubate at 37 °C for 2 h.

[CRITICAL] Many secreted proteins are glycosylated and disulfide-linked. Unexpected glycosylation prevents identification of disulfide bonds. PNGase F remains its activity in 2 mol/L urea or 1 mol/L GndCl but loses most of its activity in 2 mol/L GndCl, so dilute the proteinase K digest with an equal volume of 100 mmol/L Tris, pH 6.5 before adding PNGase F.

(7) Quench the reaction by adding 90% FA to a 5% final concentration.

For LC–MS analysis of a low-complexity sample, load 0.2–0.5 μg of protein for a single run.

[PAUSE POINT] The samples can be stored up to several weeks at −20 °C or −80 °C before LC–MS analysis.

In-solution digestion (high-complexity samples) [TIMING 1–2 days]

Note: High-complexity samples refer to whole-cell lysates, subcellular fractions, or crude immunoprecipitated proteins that contain hundreds or thousands of proteins.

(1) Precipitate 30 μg of a freshly prepared protein sample with 25% TCA and wash with cold acetone as described above. For details, see "**In-solution digestion (low-complexity samples)**".

(2) Dissolve precipitated proteins in 25 μl of 8 mol/L urea, 100 mmol/L Tris, pH 6.5; add NEM to a final concentration of 2 mmol/L; and incubate at 37 °C for 2 h.

[CRITICAL] Sonication is recommended to help dissolve proteins in 8 mol/L urea. 20–40 μg proteins are acceptable.

(3) Digestion with Lys-C/trypsin/Glu-C. For details, see "**In-solution digestion (low-complexity samples)**".

[CRITICAL] Avoid using proteases of poor specificity to digest high-complexity samples as it will lead to search space explosion in data analysis, reducing the speed and sensitivity of identification.

Table 2 Proteases digestion conditions

	Lys-C	Trypsin	Glu-C	Asp-N	Elastase	Subtilisin	Proteinase K
Denaturant	8 mol/L urea	2 mol/L urea	1 mol/L urea	2 mol/L urea	2 mol/L urea	2 mol/L GndCl	2 mol/L GndCl
Digestion time	4 h	12 h	12 h	12 h	8 h	4 h	4 h
Enzyme: Protein (w:w)	1:100	1:20	1:40	1:50	1:20	1:20	1:20

Digestion temperature is 37 °C for all

In our experience, adding Glu-C on top of Lys-C/trypsin digestion significantly increases the number of disulfide bond identifications. If desired, Lys-C/trypsin/Asp-N is another option.

(4) Quench the digestion by adding FA to a final concentration of 5%.
(5) Using a pressure injection cell, load digested peptides onto a SCX fractionation column. For better flow-rate control, connect the fritted end of the SCX column to an empty 75-μm ID, 360-μm OD fused silica tubing (with a pulled tip if necessary) with a MicroTight union. Wash the column with 15 μl of 0.1% FA, followed by 15 μl 80% ACN, 0.1% FA, then 10 μl of Buffer A, at a flow rate of 1 μl/min. Now the peptides are bound to SCX resin and ready to be fractionated.
(6) Elute sequentially with 20 μl of 5% Buffer B (5% ACN, 0.1% FA) containing 25, 50, 75, 100, 500, or 1000 mmol/L ammonium acetate, pH 2–3, at a flow rate of 1.0–2.0 μl/min. Collect each of the six fractions into an Eppendorf tube. Load one-fifth of each fraction for a subsequent reverse-phase LC–MS run.

[PAUSE POINT] The samples can be stored up to several weeks at $-20\ °C$ or $-80\ °C$ before LC–MS analysis.

In-gel digestion (protein bands of interest) [TIMING 1–2 days]

[CRITICAL] To maintain protein disulfide bonds during SDS-PAGE analysis, reducing reagents are forbidden and 20 mmol/L NEM must be present in the sample loading buffer.

(1) Excise the gel band of interest and dice into 1 mm^3 pieces.
(2) Destain the gel with 50% methanol and wash with ddH$_2$O twice. Then dehydrate with 100% acetonitrile.
(3) Rehydrate into 100 mmol/L Tris, pH 6.5 containing 0.5 mmol/L NEM, 5 ng/μL Lys-C, and 10 ng/μL of another protease of choice—trypsin, Glu-C, or Asp-N.
(4) Digest for 12 h at 37 °C.
(5) Extract peptides with 50–100 μl of extraction Buffer I (50% ACN, 0.5% FA) and then with 50–100 μl of extraction Buffer II (75% ACN, 1% FA).
(6) Concentrate the sample to 4–8 μl in a SpeedVac^{TM} concentrator at 2.5 Torr, RT. If the sample dries out accidently, reconstitute the sample with 6 μl of 0.1% FA, 1% ACN. Calculate or estimate the amount of protein in the gel band and prepare to load about 0.2 μg for LC–MS analysis. If this is not practical, load 1/5 of the sample.

[PAUSE POINT] The samples can be stored up to several weeks at $-20\ °C$ or $-80\ °C$ before LC–MS analysis.

LC–MS ANALYSIS [TIMING ~ 1–30 h, DEPENDING ON SAMPLE COMPLEXITY]

[CRITICAL] Each sample is to be analyzed twice; reject 2+ precursor ions in one (to increase identification of inter-linked peptides) but not in the other (to increase identification of loop-linked peptides).

(1) Connect the trap column and the analytical column to an Easy-nLC 1000 UHPLC according to the 2-column setup scheme. Cut the tail end of each column to reduce dead volumes.
(2) Connect the 2-column setup to the nano-ESI source of the Q-Exactive^{TM} mass spectrometer and position the tip of the analytical column ~0.5 cm away from the opening of the heated capillary. Adjust the tip to ensure a steady spray.
(3) Pre-equilibrate the columns with Buffer A and make sure there is no air bubble.
(4) Transfer samples to be analyzed into sample vials.
(5) Place the sample vials into the autosampler of an Easy-nLC 1000 UHPLC.
(6) Set up a method for each sample. Specify the names of the data files to be generated and the directory in which they will be stored. Load the LC and the MS methods written above (see **"Equipment setup"** for details).
(7) Start the LC–MS analysis and monitor it from time to time till it finishes.

DATA ANALYSIS [TIMING ~ 0.2–30 h]

Start a task

(8) Double click the pLink 2 icon on desktop or the pLink installation folder to start pLink 2. Click "New…" to start a new task, change the task "Name" and choose "Location", then click "OK".

Import .raw files and set up the extraction parameters (Fig. 5)

(9) In the "MS data" panel, choose the "MS Data Format". "RAW" is the default data format.

Fig. 5 Import raw file and set extraction parameters

[CRITICAL] Although pLink 2 can import ".mgf" files, we recommend the use of ".raw" files as input. The ".mgf" files extracted using other software tools may not be supported by pLink 2, and ".ms2" files are not allowed.

(10) Click the "Add" button, choose the input file(s).
(11) Make sure to uncheck "Mixture Spectra". The default setting is on, but for disulfide bond identification it is better to turn it off.

Set up the identification parameters (Fig. 6)

(12) Click to switch to the "Identification" panel.
(13) For "Flow Type": choose "Disulfide bond (HCD-SS)".
(14) In the "Set linkers" box, choose "SS" on the right side and click the arrow pointing to the left to transfer "SS" to the left side.
(15) Choose the database to use; you can add a new "fasta" file through the "Customize Database" option.
[CRITICAL] we recommend that you append the database of common contaminant proteins stored in pLink 2 to your "fasta" database file.
(16) Select the enzymes that had been used to digest the samples, for example, "Trypsin" for Lys-C/trypsin digestion, "Glu-C.Trypsin" for Lys-C/trypsin/Glu-C digestion, and "non-specific" for Lys-C/elastase or protease K digestion.
(17) Change the "mini peptide mass" to 400.
(18) Change the "mini peptide length" to 4.

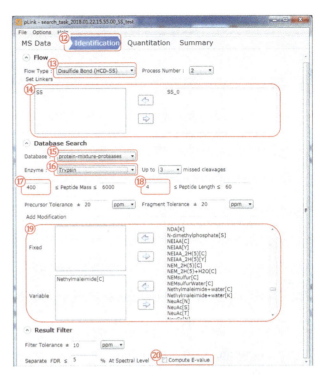

Fig. 6 Set up the identification parameters

(19) Select "N-Ethylmaleimide[C]" on the right side and transfer it to the "Variable" modification box on the left.
(20) Check "Compute E value" to output E value for each PSM.
[CRITICAL] pLink 2 uses a machine learning algorithm (SVM) to classify the target and decoy MS/MS spectra, and it only provides SVM-scores by default. Computing an E value for each PSM is highly desirable.

Check all the parameter settings and start search (Fig. 7)

(21) Once the search parameters are complete, switch to the "Summary" panel.
(22) Double check the parameters. Go back and reset if anything is wrong.
(23) After verifying the parameters, click "Save" which will activate the "Start" button.
(24) Click "Start" and now the search begins.
(25) View results. When the "Output" panel displays "[pLink] Complete report", the html results will be shown in your default browser (IE or chrome) automatically. You can also view the "csv" format results in the "reports" folder of pLink 2.

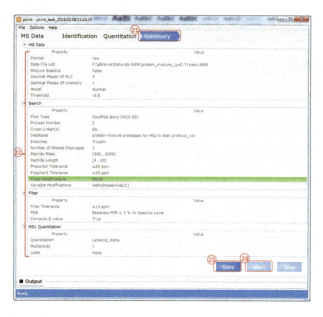

Fig. 7 Summary and starting to search

(26) To generate a concise summary of the disulfide-linked sites identified by pLink 2, copy the python script (attached to the end of this protocol) to your task folder, then open "Windows Explorer", type "cmd" in the location bar, and press "Enter" to open "Command Prompt". Lastly, enter the following in the command line:

– python SS_sim.py.

VIEW MS/MS SPECTRA USING PLABEL

Load pLabel file (Fig. 8)

(1) Double click the "pLabel" icon on desktop to open the window shown in Fig. 8. Click "FILE".
(2) Choose "Load pLabel File".
(3) Choose the "cross-linked.SS.plabel" file generated after a pLink 2 search.
[? TROUBLESHOOTING]

Find the spectra of interest (Fig. 9)

(5) Click "All".
(6) In the pop-up window, type in the scan number of interest, for example "3290", in the "Key Word" field.
(7) Click "Search".
(8) Choose the spectrum shown at the top.
(9) Click "OK".

View matched ion of MS/MS spectra (Fig. 10)

(10) Inspect the spectrum to see how well ion peaks match with calculated peptide fragments.
(11) Select the tab "Ion M/Z Deviation" to see mass deviations of matched ion peaks.

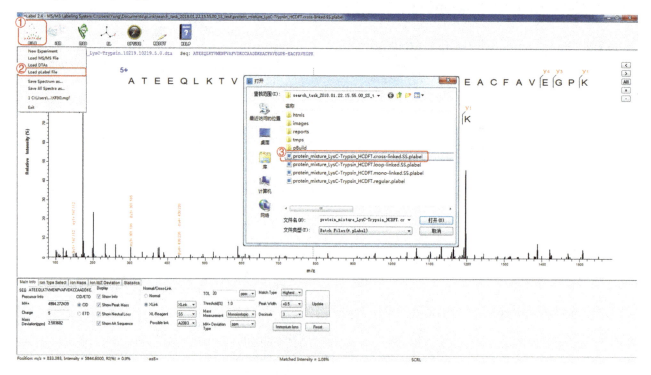

Fig. 8 Load pLabel file

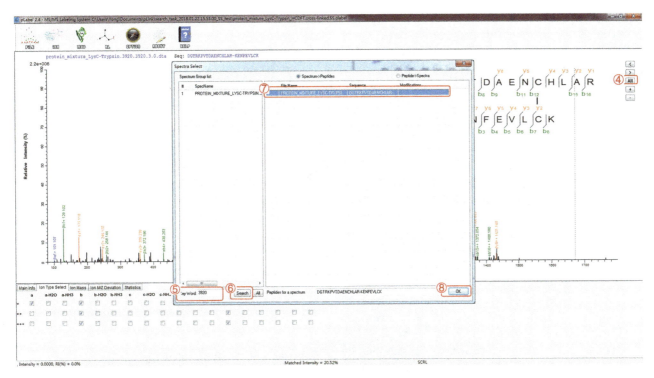

Fig. 9 Find a representative spectrum of a disulfide-linked peptide or peptide pair for manual inspection

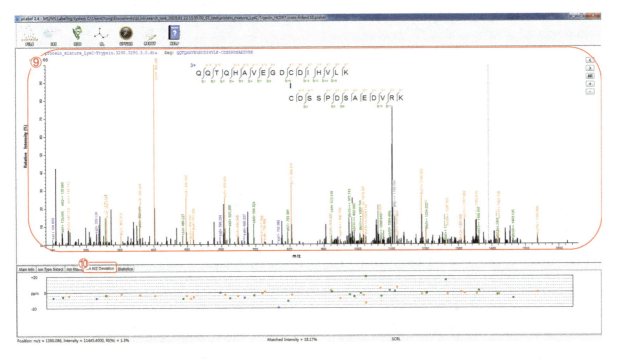

Fig. 10 View matched ion peaks and mass deviations

RESULT ANALYSIS

(1) Rank the identification results by the cysteine sites. If a cysteine residue is found to form disulfide bonds with more than one cysteine residue, it could be a result of disulfide scrambling or false identification. Compare the E values, spectral counts, and intensities of these potentially conflicting identification results to try to distinguish a native disulfide bond from scrambled ones. This is

based on the assumption that the native disulfide bond is the major form, so it should have higher signal intensity, higher spectral counts, and a smaller E value, which indicates a higher confidence in identification.

(2) For purified proteins, such as pharmaceutical protein drugs, it is often required to map all the disulfide bonds in a protein. When some cysteines are missing from the disulfide identification results, one possibility is that they exist as free cysteines and the other is that they form disulfide bonds but these disulfide bonds have escaped identification. For the first possibility, one can expect free cysteines to be modified by NEM, so the corresponding modified linear peptides should be identifiable using conventional database search engines such as pFind. To find out whether the second possibility is true, we recommend reduction of disulfide bonds followed by alkylation and conventional database search to identify linear peptides; the cysteine-containing linear peptides can then be compared with those identified from the matched, non-reduced sample. In the case that a disulfide bond has escaped identification, first consider that the disulfide-containing peptide(s) may be too long or too short. Acting accordingly, choose proteases that will generate—concerning the cysteine residues in question—peptides of 6–20 (or better, 8–16) amino acids. Also, there may be an unknown modification near the missing disulfide bond. In this case, guess what it might be by sequence analysis with the help of modification prediction software, or use open search tools such as pFind 3.0 or Peaks to find possible modifications. Then, add the variable modification in pLink 2 search and see if it helps. Lastly, manual spectrum interpretation may help, but it requires a lot of time and experience.

[? TROUBLESHOOTING]
Troubleshooting advice can be found in Table 3.

Acknowledgements The authors would like to thank the Beijing Municipal Science and Technology Commission, the Ministry of Science and Technology of China, and the Natural Science Foundation of China (21475141) for research funding.

Compliance with ethical standards

Conflict of interest Shan Lu, Yong Cao, Sheng-Bo Fan, Zhen-Lin Chen, Run-Qian Fang, Si-Min He, and Meng-Qiu Dong declare that they have no conflict of interest.

Human and animal rights and informed consent This article does not contain any studies with human or animal subjects performed by any of the authors.

Table 3 Troubleshooting table

Problem	Possible reason	Solution
Many scrambled disulfide bonds are identified	pH is off;	Make sure that the pH of the urea buffer is 6.5;
	NEM gone bad after storage	Make a fresh solution of NEM
The number of spectra of disulfide-linked peptides are too few	Insufficient digestion, masses of disulfide-linked peptides are too high;	Increase the amount of protease or the digestion time, or add another protease;
	Over-digestion by non-specific proteases	Shorten the digestion time, try different time points
Certain disulfide bonds are not identified	Peptides are too long or too short, or too complex, e.g., containing three peptides;	Digest the samples with different proteases;
	Unexpected modifications on the disulfide-linked peptides	Try to identify the modification or treat with glycosylase and see what happens
pLink 2 report "MS1 or MS2 not completely extracted"	No MSFileReader installed	Install MSFileReader https://github.com/pFindStudio/pLink2/wiki/FAQ#how-to-install-msfilereader
Error in "loading label file"	"mgf" file is missing from the directory of the raw file	Don't delete or move the "mgf" file. If it must be moved, change the path in ".plabel" file

References

Choi S, Jeong J, Na S, Lee HS, Kim H-Y, Lee K-J, Paek E (2009) New algorithm for the identification of intact disulfide linkages based on fragmentation characteristics in tandem mass spectra. J Proteome Res 9:626–635

Götze M, Pettelkau J, Schaks S, Bosse K, Ihling CH, Krauth F, Fritzsche R, Kühn U, Sinz A (2012) StavroX—a software for analyzing crosslinked products in protein interaction studies. J Am Soc Mass Spectrom 23:76–87

Haniu M, Acklin C, Kenney WC, Rohde MF (1994) Direct assignment of disulfide bonds by Edman degradation of selected peptide fragments. Chem Biol Drug Des 43:81–86

Hartman MD, Figueroa CM, Arias DG, Iglesias AA (2016) Inhibition of recombinant aldose-6-phosphate reductase from peach leaves by hexose-phosphates, inorganic phosphate and oxidants. Plant Cell Physiol 58:145–155

Hogg PJ (2003) Disulfide bonds as switches for protein function. Trends Biochem Sci 28:210–214

Huang SY, Chen SF, Chen CH, Huang HW, Wu WG, Sung WC (2014) Global disulfide bond profiling for crude snake venom using dimethyl labeling coupled with mass spectrometry and RADAR algorithm. Anal Chem 86:8742–8750

Hung C-W, Koudelka T, Anastasi C, Becker A, Moali C, Tholey A (2016) Characterization of post-translational modifications in full-length human BMP-1 confirms the presence of a rare vicinal disulfide linkage in the catalytic domain and highlights novel features of the EGF domain. J Proteomics 138:136–145

Itakura M, Iwashina M, Mizuno T, Ito T, Hagiwara H, Hirose S (1994) Mutational analysis of disulfide bridges in the type C atrial natriuretic peptide receptor. J Biol Chem 269:8314–8318

Liu F, van Breukelen B, Heck AJ (2014) Facilitating protein disulfide mapping by a combination of pepsin digestion, electron transfer higher energy dissociation (EThcD), and a dedicated search algorithm SlinkS. Mol Cell Proteomics 13:2776–2786

Liu Y, Sun W, Shan B, Zhang K (2017a) DISC: DISulfide linkage Characterization from tandem mass spectra. Bioinformatics 33:3861–3870

Liu Y, Xiao W, Shinde M, Field J, Templeton DM (2017b) Cadmium favors F-actin depolymerization in rat renal mesangial cells by site-specific, disulfide-based dimerization of the CAP1 protein. Arch Toxicol, 1–16

Lu S, Fan S-B, Yang B, Li Y-X, Meng J-M, Wu L, Li P, Zhang K, Zhang M-J, Fu Y (2015) Mapping native disulfide bonds at a proteome scale. Nat Methods 12:329

Mauney CH, Rogers LC, Harris RS, Daniel LW, Devarie-Baez NO, Wu H, Furdui CM, Poole LB, Perrino FW, Hollis T (2017) The SAMHD1 dNTP triphosphohydrolase is controlled by a redox switch. Antioxid Redox Signal 27:1317–1331

McCarthy AA, Haebel PW, Törrönen A, Rybin V, Baker EN, Metcalf P (2000) Crystal structure of the protein disulfide bond isomerase, DsbC, from *Escherichia coli*. Nat Struct Mol Biol 7:196

McDonagh B (2009) Diagonal electrophoresis for detection of protein disulphide bridges. In: Two-Dimensional Electrophoresis Protocols. Springer, pp. 305–310

Murad W, Singh R (2013) MS2DB + : a software for determination of disulfide bonds using multi-ion analysis. IEEE Trans Nanobiosci 12:69–71

Sharma D, Rajarathnam K (2000) 13C NMR chemical shifts can predict disulfide bond formation. J Biomol NMR 18:165–171

Wang J, Anania VG, Knott J, Rush J, Lill JR, Bourne PE, Bandeira N (2014) Combinatorial approach for large-scale identification of linked peptides from tandem mass spectrometry spectra. Mol Cell Proteomics 13:1128–1136

Wang T, Liang L, Xue Y, Jia P-F, Chen W, Zhang M-X, Wang Y-C, Li H-J, Yang W-C (2016) A receptor heteromer mediates the male perception of female attractants in plants. Nature 531:241

Wefing S, Schnaible V, Hoffmann D (2006) SearchXLinks. A program for the identification of disulfide bonds in proteins from mass spectra. Anal Chem 78:1235–1241

Wu S-L, Jiang H, Lu Q, Dai S, Hancock WS, Karger BL (2008) Mass spectrometric determination of disulfide linkages in recombinant therapeutic proteins using online LC–MS with electron-transfer dissociation. Anal Chem 81:112–122

Wu J, Yan Z, Li Z, Qian X, Lu S, Dong M, Zhou Q, Yan N (2016) Structure of the voltage-gated calcium channel Ca v 1.1 at 3.6 Å resolution. Nature 537:191

Wu Y, Zhu Y, Gao F, Jiao Y, Oladejo BO, Chai Y, Bi Y, Lu S, Dong M, Zhang C (2017) Structures of phlebovirus glycoprotein Gn and identification of a neutralizing antibody epitope. Proc Natl Acad Sci 114:E7564–E7573

Xu H, Zhang L, Freitas MA (2007) Identification and characterization of disulfide bonds in proteins and peptides from tandem MS data by use of the MassMatrix MS/MS search engine. J Proteome Res 7:138–144

Determining the target protein localization in 3D using the combination of FIB-SEM and APEX2

Yang Shi[1,3,4], Li Wang[2], Jianguo Zhang[2], Yujia Zhai[1,3], Fei Sun[1,2,3,4]

[1] National Key Laboratory of Biomacromolecules, CAS Center for Excellence in Biomacromolecules, Institute of Biophysics, Chinese Academy of Sciences, Beijing 100101, China
[2] Center for Biological Imaging, Institute of Biophysics, Chinese Academy of Sciences, Beijing 100101, China
[3] University of Chinese Academy of Sciences, Beijing 100049, China
[4] Sino-Danish Center for Education and Research, Beijing 100190, China

Abstract Determining the cellular localization of proteins of interest at nanometer resolution is necessary for elucidating their functions. Besides super-resolution fluorescence microscopy, conventional electron microscopy (EM) combined with immunolabeling or clonable EM tags provides a unique approach to correlate protein localization information and cellular ultrastructural information. However, there are still rare cases of such correlation in three-dimensional (3D) spaces. Here, we developed an approach by combining the focus ion beam scanning electron microscopy (FIB-SEM) and a promising clonable EM tag APEX2 (an enhanced ascorbate peroxidase 2) to determine the target protein localization within 3D cellular ultrastructural context. We further utilized this approach to study the 3D localization of mitochondrial dynamics-related proteins (MiD49/51, Mff, Fis1, and Mfn2) in the cells where the target proteins were overexpressed. We found that all the target proteins were located at the surface of the mitochondrial outer membrane accompanying with mitochondrial clusters. Mid49/51, Mff, and hFis1 spread widely around the mitochondrial surface while Mfn2 only exists at the contact sites.

Keywords Enhanced ascorbate peroxidase, Focus ion beam scanning electron microscopy, Mitochondrial dynamics, Protein location, Three-dimensional space

INTRODUCTION

Protein localization correlates with its particular function in cells or tissues. Mapping protein localization information onto their cellular ultrastructural context is of great importance for cell biology study and can be achieved via electron microscopy (EM). Generally, there are two ways to localize a target protein in EM: immune-localization and clonable tags localization. The EM contrast of immune-localization comes from the antibody-conjugated gold particles (De Mey et al. 1981) or quantum dots (Giepmans et al. 2005). This approach is significantly limited due to the limited efficiency of immunolabeling, the spatial hindrance of large antibodies, and the fact that the well-preserved ultrastructure and the antigen immuno-activity are always mutually exclusive.

Clonable tags localization has been achieved via metallothionein, a small cysteine-rich protein that can bind a variety of heavy metal ions with its cysteine residues (Hamer 1986; Kagi and Schaffer 1988; Mercogliano and DeRosier 2007). To avoid the possible heavy metal toxicity or its influence on the behavior of

✉ Correspondence: feisun@ibp.ac.cn (F. Sun)

target proteins, the cells were treated with the heavy metal during EM sample preparation (Morphew et al. 2015). However, the preservation of the cellular ultrastructure and the low tolerance of metallothionein to strong chemical fixation are still mutually exclusive. Another kind of clonable tag is based on the oxidization of 3,3′-diaminobenzidine (DAB). The contrast comes from the enriched osmium tetroxide that is recruited by the osmiophilic DAB polymer. According to the type of oxidation, these tags can be further classified into photo-oxidation tags, which include fluorescent protein (FP) (Grabenbauer et al. 2005; Meißlitzer-Ruppitsch et al. 2008), resorufin arsenical hairpin (ReAsH) (Gaietta et al. 2002), and mini singlet oxygen generator (miniSOG) (Shu et al. 2011), or they can be classified as enzyme-based oxidation tags, which include horseradish peroxidase (HRP) (Connolly et al. 1994; Li et al. 2010) and enhanced ascorbate peroxidase (APEX/APEX2) (Martell et al. 2012; Lam et al. 2015). Considering the low-yield efficiency of singlet oxygen by FPs as well as the relative low EM contrast, the usage of FPs is limited (Su et al. 2010). ReAsH works based on the reaction between tetracysteine tag and biarsenical compounds (Adams et al. 2002), which yields a problem of nonspecific labeling. Moreover, one general concern for the photo-oxidation-based tags is the inaccessibility of light in a deep tissue. Although the enzyme-based tags take advantage in dealing with thick tissue, the targeting capabilities of HRP in cytosol limit their usage (Hopkins et al. 2000). However, APEX/APEX2 has become a promising clonable EM tag because it can not only preserve cellular ultrastructure well but also give excellent EM contrast, regardless of the thickness of the specimen and the localization of target proteins (Martell et al. 2012). It is important to note that APEX2 is an A134P mutant of APEX with improved kinetics, thermal stability, heme binding, and resistance to high H_2O_2 concentrations (Lam et al. 2015).

Volume electron microscopy technique can study cellular ultrastructure in three dimensions (3D), which provides more information compared to the 2D projection or single slice. The transmission electron microscopy (TEM)-based volume EM methods include electron tomography (ET) (Hart 1968; Koster et al. 1992), serial sectioning followed by TEM (ssTEM) (Harris et al. 2006), and serial section ET (Ladinsky et al. 1999). The scanning electron microscopy (SEM)-based volume EM methods include serial section SEM (ssSEM) (Horstmann et al. 2012), serial block face scanning electron microscopy (SBF-SEM) (Denk and Horstmann 2004), and focus ion beam scanning electron microscopy (FIB-SEM) (Knott et al. 2008). For a small volume ($50 \times 50 \times 30$ μm^3), FIB-SEM approach provides an efficient and automatic way to get the 3D ultrastructure of specimen with a higher axial resolution in comparison to serial section-based techniques and SBF-SEM (Peddie and Collinson 2014; Li et al. 2016).

As a highly dynamic organelle in eukaryotic cells, mitochondria perpetually divide, fuse, and move in response to ever-changing physiological demand of the cells (Youle and Van Der Bliek 2012). There are many proteins involved in mitochondrial dynamics and regulation. In mammalian cells, mitochondrial dynamics proteins 49 and 51 kDa (MiD49 and MiD51, respectively) (Palmer et al. 2011), fission protein 1 (Fis1, and hFis1 for human homologues) (Yoon et al. 2003), and mitochondrial fission factor (Mff) (Otera et al. 2010) are mitochondrial outer membrane proteins and play an important role as receptors of dynamic-related protein 1 (Drp1), which drives mitochondrial fission (Smirnova et al. 2001). Mitofusin 1 and 2 (Mfn1 and Mfn2, respectively) control the fusion of mitochondrial outer membrane (Santel and Fuller 2001) while OPA1 in both short and long forms play the central role of mitochondrial inner membrane fusion (Song et al. 2007). Overexpression of these proteins in mammalian cells will induce significant clustering of mitochondria, which was revealed from fluoresce microscopy (Griparic et al. 2004; Huang et al. 2007; Otera et al. 2010; Zhao et al. 2011; Liu et al. 2013). However, there was no information available on how these proteins are localized within the cluster and how these clustered mitochondria look like in 3D space.

In the present study, we developed a 3D localization EM method (APEX2–FIB-SEM) by combining FIB-SEM and APEX2 techniques to enable the correlation of protein localization information and cellular ultrastructure in the 3D space of a single cell with a high resolution. We utilized this approach to study the 3D localization of mitochondrial fusion- and fission-related proteins, including MiD51, MiD49, hFis1, Mff, and Mfn2.

RESULTS

Workflow of APEX2–FIB-SEM

The gene of the target protein with APEX2 fused at the N or C terminus was cloned into an appropriate vector (e.g., pcDNA3 here) for transfection. The cells (e.g., HEK 293T cells here) were cultured on a sterile plastic coverslip within a culture dish (Fig. 1A). After transfection and further culturing, the cells were fixed, stained with DAB, dehydrated, and then embedded in resin as previously reported (Fig. 1B) (Martell et al. 2012). Considering the reproducibility, we recommend to use the

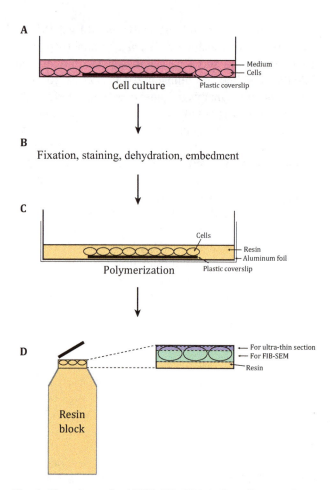

the thickness of each section was less than 80 nm to make sure there were still enough samples for subsequent FIB-SEM experiments (Fig. 1D). The details of the workflow are described in "Materials and methods".

3D Localization of Mitochondrial Dynamics-Related Proteins

We first utilized mitochondrial calcium uniporter (MCU) as a positive control that was used in the previous APEX study (Martell *et al.* 2012) to validate the performance of APEX/APEX2 staining in our system. In our TEM experiment, MCU-APEX gave a strong EM contrast within mitochondrial matrix (Fig. 2A), which was consistent with the previous report (Martell *et al.* 2012) and showed that APEX worked well in our workflow.

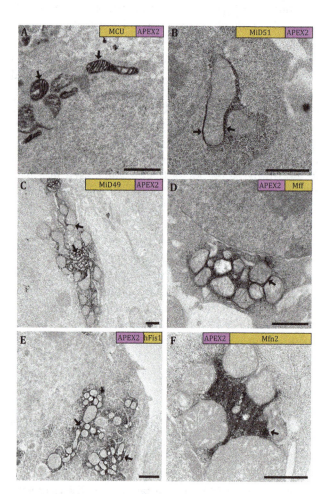

Fig. 1 The scheme for APEX2–FIB-SEM. **A** The cells are cultured on sterile plastic coverslips in dishes, and transfected with the target genes fused with APEX2. The cells on the plastic coverslip are used for further EM sample preparation. **B** The cells on the plastic coverslip undergo fixation, DAB staining, dehydration, and resin embedment. **C** Before polymerization, the plastic coverslip with cells is transferred to the dish placed with aluminum foil. The cell side of the plastic coverslip is facing up. Fresh resin with accelerator is added to cover the cells. **D** After polymerization, the block is separated from aluminum foil and plastic coverslip, and the areas with staining patterns are cut off and stuck on the tip of an empty resin block. In order to reserve enough sample for FIB-SEM (colored with green), only three to five slices (less than 80 nm) are sectioned from the block (colored with blue) for TEM screening

commercial DAB Kit (*e.g.*, CWBIO here) for DAB staining. Various embedding molds can be used for the polymerization of the resin-embedded cells with coverslip as long as they can be separated easily without damaging the cells (Fig. 1C). After polymerization, the plastic coverslip was removed carefully and a few ultrathin sections from the surface of the resin block were checked by TEM to screen for the optimally stained sample with well-preserved ultrastructure that will be further imaged by FIB-SEM. Only three to five slices were sectioned from the surface of the resin block and

Fig. 2 Visualizing the staining patterns in 2D by TEM. The micrographs show cells expressing MCU-APEX (**A**), MiD51-APEX2 (**B**), MiD49-APEX2 (**C**), APEX2-Mff (**D**), APEX2-hFis1 (**E**), and APEX2-Mfn2 (**F**). The staining areas with strong EM contrast indicate the locations of the target proteins (black arrows). The labels in the top right corner of each micrograph show the relevant position between the target protein and the tag as well as their relevant sizes. Scale bars, 1 μm

Determining the target protein localization in 3D using the combination of FIB-SEM and APEX2

In our further experiments, we selected APEX2 instead of APEX since APEX2 was more active due to its better resistance to high H_2O_2 concentration (Lam et al. 2015). By examining ultrathin sections in TEM, we found that the constructs MiD51-APEX2 (Fig. 2B), MiD49-APEX2 (Fig. 2C), APEX2-Mff (Fig. 2D), APEX2-hFis1 (Fig. 2E), and APEX2-Mfn2 (Fig. 2F) could yield a good EM contrast at the reasonable positions. These constructs were then selected for the subsequent FIB-SEM experiments. It is to be noted that, for the constructs Mff-APEX2, hFis1-APEX2, and Mfn2-APEX2, we could not find reasonable staining patterns in TEM micrographs. We speculated that it is due to the fusion of APEX2 in the proximity of the transmembrane helices of the target proteins, which induced the failure of their proper localizations in mitochondrial outer membrane (see also in "Discussions").

MiD51/49 has been found to recruit Drp1 into mitochondria, and overexpression of MiD51 or MiD49 induced mitochondrial elongation, which was thought to be the phenotype of mitochondrial fusion (Zhao et al. 2011; Liu et al. 2013). We also consistently found that mitochondria appeared as elongated tubulars or compact clusters in the cells overexpressing MiD51/49-APEX2 (Figs. 2B, C and 3A, B). However, the mitochondrial fusion phenotype induced by overexpression of MiD51/49 varied from cells to cells by the comparison of the EM images of cells expressing MiD51-APEX2 (see Figs. 2B and 3A) or those expressing MiD49-APEX2 (see Figs. 2C and 3B). We therefore speculated that the mitochondrial fusion phenotype induced by MiD51/MiD49 is highly dependent on their expression level (see also in "Discussions"). Through investigation of the 3D volume of the clustered mitochondria (Fig. 3A and B; see also Movies S1 and S2), we clearly found that the target proteins MiD51/MiD49 were largely localized at the interface among mitochondria with reduced amount of proteins surrounding the peripheral of the mitochondrial cluster. This observation implied how MiD51/MiD49 induced mitochondrial clustering by recruiting Drp1.

Mff was thought to be an essential factor for mitochondrial recruitment of Drp1 during mitochondria fission (Otera et al. 2010). In the cells with exogenous expression of APEX2-Mff, the mitochondria appeared in circular shapes and formed compact clusters (Figs. 2D and 3C) in consistency with the previous finding (Otera et al. 2010). Through investigation of the 3D volume of the clustered mitochondria, we clearly found that Mff was not only rich in the areas of mitochondrial connection, but also presented at the periphery of the mitochondrial cluster (Fig. 3C; see also Movie S3).

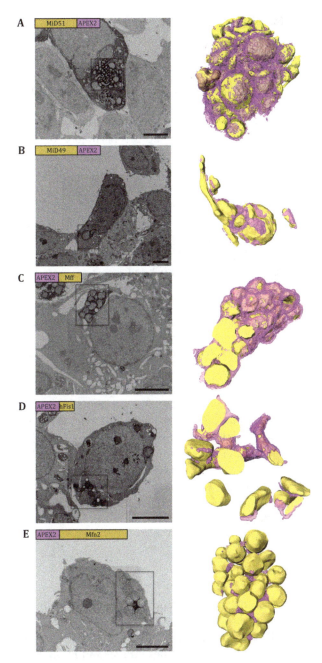

Fig. 3 Visualizing the staining pattern in 3D by FIB-SEM. The figure panels show cells expressing MiD51-APEX2 (**A**), MiD49-APEX2 (**B**), APEX2-Mff (**C**), APEX2-Fis1 (**D**), and APEX2-Mfn2 (**E**). The left columns display the representative SEM micrographs from each volume of EM data, and the right columns display the 3D rendering of boxed areas of the left with the yellow spheres for mitochondria and purple layers for APEX2-induced EM contrast. The labels in the top left corner of each micrograph show the relevant position between the target protein and the tag as well as their relevant sizes. Scale bars, 5 μm

Fis1 was the first proposed Drp1 receptor and could induce the mitochondrial fission (Mozdy et al. 2000, James et al. 2003, Yoon et al. 2003). In the APEX2-hFis1-transfected cells, the clusters of vesicular mitochondria were found (Figs. 2E and 3D) in consistency with the previous study (Otera et al. 2010). The clustering phenotype also possibly varied in an expression level-dependent manner from cells (Fig. 2E) to cells (Fig. 3D). Through investigation of the 3D volume of the clustered mitochondria, we also found that hFis1 was located in high density at the interface of clustered mitochondria with a significant amount at the periphery of mitochondrial cluster (Fig. 3D; see also Movie S4).

Mfn2 can mediate tethering and fusion of mitochondrial outer membrane during the mitochondrial fusion process (Koshiba et al. 2004; Meeusen et al. 2004), with both of its N-terminal and C-terminal domains exposed towards the cytosol (Rojo et al. 2002). In the cells overexpressing APEX2-Mfn2, the mitochondria formed clusters and most of mitochondria were in circular shapes (Figs. 2F and 3E), which was consistent with the previous description of grape-like aggregation (Huang et al. 2007). However, to our surprise, unlike the clusters induced by the four proteins above, here, the mitochondrial cluster induced by overexpressing APEX2-Mfn2 was relatively loose, especially at the inside region that was filled with strong staining signals. In addition, through investigation into the 3D volume of the clustered mitochondria, we found that Mfn2 was only located in the contact sites of clustered mitochondria, and no staining signal was found at the periphery of mitochondrial cluster (Fig. 3E; see also Movie S5). This finding was significantly different from those associated with mitochondrial fission-related proteins MiD51, MiD49, Mff, and hFis1.

DISCUSSION

In this study, we developed the APEX2–FIB-SEM approach to determine the target protein localization within 3D cellular ultrastructural context by combining two state-of-the-art techniques, APEX2 and FIB-SEM. By using APEX2–FIB-SEM, we successfully mapped the position of mitochondrial dynamics-related proteins (MiD51/49, Mff, hFis1, and Mfn2) when they were overexpressed. The clusters with tubular (MiD51/49) or circular (Mff, hFis1, and Mfn2) mitochondria were found in those cells. Through investigation of the staining patterns in the 3D volume data, we found Mff and hFis1 distributed widely around the surface of mitochondrial outer membrane, and they were consistent with MiD51/49 except that the staining signals of MiD51/49 were reduced at the periphery of mitochondrial cluster. However, the relatively loose mitochondrial clusters and extremely restrained staining patterns at the mitochondrial contact site suggested that the mechanism of the fusion phenotype induced by overexpressing Mfn2 was distinctive from that of fission factors.

For alpha-helically anchored mitochondrial outer membrane proteins, the targeting signal is confined to transmembrane domain and positive residues in its flanking regions (McBride et al. 1992; Waizenegger et al. 2003; Bruggisser et al. 2017). The failure of the constructs, such as Mff-APEX2, hFis1-APEX2, and Mfn2-APEX2, is probably due to the fusion of APEX2 to the terminal which is close to the transmembrane region, and perturbs their translocation.

Like tagged with APEX2, similar morphologies are present in cells overexpressing the above five proteins without APEX2 (Huang et al. 2007; Otera et al. 2010; Zhao et al. 2011; Liu et al. 2013), which indicates that these mitochondrial phenotypes are not tag-associated artifacts. Their ubiquitous distributions at the interface regions of clustered mitochondria suggest that these proteins participate in tethering the mitochondria together at the outer membrane surface.

The different expression levels of the exogenous genes may induce distinct phenotype of cells and different behavior of proteins. As a result, the comparison of cellular phenotypes and location of proteins should be made with approximate expression levels. With too high expression level, the EM contrast would appear at unreasonable areas, for example in Fig. 3A, the MiD51 directed staining is also found at cytosol. The dispersing signals may result from the accumulated mis-located fusion proteins in cytosol.

In addition, the endogenous oxidases can also polymerize DAB and give false-positive signals, such as the cytochrome oxidase that is located in the intermembrane space of mitochondria (Seligman et al. 1968). However, most of the endogenous oxidases should have lower tolerance to strong chemical fixative than APEX/APEX2, which is the prerequisite to ensure APEX/APEX2 can work without interruption from endogenous oxidases. As a result, the fixation level should be well regulated to ensure that the endogenous oxidases are fully inactive without disrupting the activity of APEX/APEX2. In our experiment, the detached and pelleted cells were also tested for DAB staining and EM sample preparation, and uninterpretable staining was found at the mitochondrial intermembrane space and endosome matrix (data not shown). Moreover, staining was also found in the cells without transfection. However, the false-positive staining was missing with the same procedures for the monolayer cells. So the protocol of fixation and staining in this study is suitable for monolayer cells, and slight modification should be done in the fixation step for other kinds of samples.

MATERIALS AND METHODS

Plasmids

All the genes were cloned into the pcDNA3.0 vector with five Myc tags near 3' end of the multiple cloning site, respectively. All of them were inserted before c-Myc tag with a terminator. The A134P point mutation of APEX which turned APEX to APEX2 was created using extension overlap PCR. MiD51/49 and MCU were fused at the N-terminal of the APEX/APEX2, and Mff, hFis1, and Mfn2 were fused at both N- and C-terminals of the APEX2. MiD51-APEX2, MiD49-APEX2, hFis1-APEX2, Mfn2-APEX2, and MCU-APEX were cloned between BamH I and EcoR I sites. Mff-APEX2, APEX2-Mff, APEX2-hFis1, and APEX2-Mfn2 were cloned into Kpn I and BamH I sites. These fused genes were generated by overlap extension PCR, and the linker sequence between them was 5'-CTGGACAGCACC-3'. The fused genes were inserted into vector using standard restriction cloning methods and verified by nucleotide sequencing.

Cell Culture and Transfection

HEK 293T cells were cultured in Dulbecco's modified eagle medium (DMEM, Corning) supplemented with 10% (v/v) fetal bovine serum (Gibco). The cultures were kept in the condition of 5% CO_2 and 37 °C. After culturing, the cells were placed into a 12-well plate (Corning) with each well having 200,000 cells approximately. It is to be noted that, before plating cells, a resized sterile plastic coverslip (Thermanox) was placed on the bottom of each well. After additional 24 h culturing, the cells were transfected by X-treme HP (Roche technologies), and then fixed after another 48–72 h.

Sample Preparation for Electron Microscopy

The DAB staining and EM sample preparation were as previously reported (Martell *et al.* 2012) with slight modification. The plastic coverslip with transfected cells were fixed by 2.5% glutaraldehyde (Electron Microscopy Science) for 1 h on ice. Then they were washed by phosphate-buffered solution (PBS, pH 7.4) three times. Then the cells were incubated at 20 mmol/L glycine in PBS for 5 min, and rinsed by chilled PBS 3 × 5 min. DAB Kit (CWBIO) was used to stain cells for 15 min, followed by 3 × 2 min rinses in chilled PBS. The formation of DAB polymer could be checked by transmission light microscope. The cells were further post-fixed and stained by 2% osmium tetroxide (Electron Microscopy Science) for 30 min, and rinsed by distilled water 3 × 5 min. Then 2% aqueous uranyl acetate (Electron Microscopy Sciences) was placed on cells overnight. Thereafter, cells were gradient-dehydrated in ethanol series (30%, 50%, 70%, 80%, 95%, 100%) and 100% acetone. During the dehydration process of using 100% acetone, cells were transferred from ice temperature to room temperature. Cells were embedded by diluted Epon 812 (Electron Microscopy Sciences) in acetone (25%, 50%, 75%) for 2 h each and then by 100% resin three times with each time for 2 h. Before the resin polymerization step, the plastic coverslip with cells was transferred to a six-well plate coated with aluminum foil: the aluminum foil on the bottom of plate needs to be as flat as possible. The surface of plastic coverslip with cells was upward, and fresh resin with accelerator was added into the plates to cover the cells. The polymerization was performed at 60 °C for 48 h. Then the blocks could be taken out from the six-well plate and carefully separated from the aluminum foil and plastic coverslip. The cells should be near the surface of the blocks. Before the following steps, the block could be observed under a light microscope and the areas with significant staining signals should be cut off and stuck on the tip of a blank resin block that was polymerized in standard flat embedding mold (Electron Microscopy Sciences).

Transmission Electron Microscopy

The blocks were trimmed and sectioned into 80 nm using an ultramicrotome (Leica EM UC6). Electron micrographs were recorded using FEI Tecnai spirit TEM (Thermo Fisher Scientific) operated at 120 kV.

FIB-SEM and Image Processing

The block with the target regions was glued onto a 45° inclined aluminum sample stub with conductive silver paint. To avoid the charging artifacts, the blocks were coated with a thin-layer of carbon. The data were collected by FEI Helios NanoLab 600i Dual beam SEM (Thermo Fisher Scientific). The platinum layer was deposited on the region of interest with a gas injection system to protect the specimen surface from ion beam-induced damage. The accelerating voltage of FIB was 30 kV, and the beam current was set to 0.79 nA with the milling thickness of 15 or 20 nm for different volume data. The series micrographs were recorded using an accelerating voltage of 2 kV, a current of 0.69 nA, and a dwell time of 8 μs. The FIB-SEM data were aligned, reconstructed, and segmented by the Amira software (Thermo Fisher Scientific).

Abbreviations

3D	Three dimensions
APEX	Enhanced ascorbate peroxidase
CMV	Cytomegalovirus
DAB	3,3'- diaminobenzidine
Drp1	Dynamic-related protein 1
EM	Electron microscopy
ET	Electron tomography
FIB-SEM	Focus ion beam scanning electron microscopy
Fis1	Fission protein 1
FP	Fluorescent protein
HRP	Horseradish peroxidase
MCU	Mitochondrial calcium uniporter
Mff	Mitochondrial fission factor
Mfn	Mitofusin
MiD	Mitochondrial dynamics protein
miniSOG	Mini singlet oxygen generator
ReAsH	Resorufin arsenical hairpin
ssTEM	Serial section transmission electron microscopy
ssSEM	Serial section scanning electron microscopy
SBF-SEM	Serial block face scanning electron microscopy
TEM	Transmission electron microscopy

Acknowledgements We would like to thank Prof. Quan Chen from Institute of Zoology, Chinese Academy of Sciences, for his kindness of giving the gene of Mfn2. We would like to thank Prof. Long Miao from Institute of Biophysics, Chinese Academy of Sciences, for his kindness of sharing the original plasmid of APEX. We are also grateful to Ping Shan and Ruigang Sui (F.S. lab) for their assistances. This work was supported by grants from the Strategic Priority Research Program of Chinese Academy of Sciences (XDB08030202) and the National Basic Research Program ("973 Program") of Ministry of Science and Technology of China (2014CB910700). All the EM works were performed at Center for Biological Imaging (CBI, http://cbi.ibp.ac.cn), Institute of Biophysics, Chinese Academy of Sciences.

Compliance with Ethical Standards

Conflict of interest Yang Shi, Li Wang, Jianguo Zhang, Yujia Zhai, and Fei Sun declare that they have no conflict of interest.

Human and animal rights and informed consent This article does not contain any studies with human or animal subjects performed by any of the authors.

References

Adams SR, Campbell RE, Gross LA, Martin BR, Walkup GK, Yao Y, Llopis J, Tsien RY (2002) New biarsenical ligands and tetracysteine motifs for protein labeling in vitro and in vivo: synthesis and biological applications. J Am Chem Soc 124:6063–6076

Bruggisser J, Käser S, Mani J, Schneider A (2017) Biogenesis of a mitochondrial outer membrane protein in Trypanosoma brucei: targeting signal and dependence on a unique biogenesis factor. J Biol Chem 292:3400–3410

Connolly CN, Futter CE, Gibson A, Hopkins CR, Cutler DF (1994) Transport into and out of the Golgi complex studied by transfecting cells with cDNAs encoding horseradish peroxidase. J Cell Biol 127:641

De Mey J, Moeremans M, Geuens G, Nuydens R, De Brabander M (1981) High resolution light and electron microscopic localization of tubulin with the IGS (immuno gold staining) method. Cell Biol Int Rep 5:889–899

Denk W, Horstmann H (2004) Serial block-face scanning electron microscopy to reconstruct three-dimensional tissue nanostructure. PLoS Biol 2:e329

Gaietta G, Deerinck TJ, Adams SR, Bouwer J, Tour O, Laird DW, Sosinsky GE, Tsien RY, Ellisman MH (2002) Multicolor and electron microscopic imaging of connexin trafficking. Science 296:503

Giepmans BNG, Deerinck TJ, Smarr BL, Jones YZ, Ellisman MH (2005) Correlated light and electron microscopic imaging of multiple endogenous proteins using Quantum dots. Nat Methods 2:743–749

Grabenbauer M, Geerts WJC, Fernandez-Rodriguez J, Hoenger A, Koster AJ, Nilsson T (2005) Correlative microscopy and electron tomography of GFP through photooxidation. Nat Methods 2:857–862

Griparic L, van der Wel NN, Orozco IJ, Peters PJ, van der Bliek AM (2004) Loss of the intermembrane space protein Mgm1/OPA1 induces swelling and localized constrictions along the lengths of mitochondria. J Biol Chem 279:18792–18798

Hamer DH (1986) Metallothionein. Annu Rev Biochem 55:913–951

Harris KM, Perry E, Bourne J, Feinberg M, Ostroff L, Hurlburt J (2006) Uniform serial sectioning for transmission electron microscopy. J Neurosci 26:12101

Hart RG (1968) Electron microscopy of unstained biological material: the polytropic montage. Science 159:1464

Hopkins C, Gibson A, Stinchcombe J, Futter C (2000) Chimeric molecules employing horseradish peroxidase as reporter enzyme for protein localization in the electron microscope. In: Jeremy Thorner SDE, John NA (eds) Methods in enzymology, vol 327. Academic Press, New York, pp 35–45

Horstmann H, Körber C, Sätzler K, Aydin D, Kuner T (2012) Serial section scanning electron microscopy (S3EM) on silicon wafers for ultra-structural volume imaging of cells and tissues. PLoS ONE 7:e35172

Huang P, Yu T, Yoon Y (2007) Mitochondrial clustering induced by overexpression of the mitochondrial fusion protein Mfn2 causes mitochondrial dysfunction and cell death. Eur J Cell Biol 86:289–302

James DI, Parone PA, Mattenberger Y, Martinou J-C (2003) hFis1, a novel component of the mammalian mitochondrial fission machinery. J Biol Chem 278:36373–36379

Kagi JH, Schaffer A (1988) Biochemistry of metallothionein. Biochemistry 27:8509–8515

Knott G, Marchman H, Wall D, Lich B (2008) Serial section scanning electron microscopy of adult brain tissue using focused ion beam milling. J Neurosci 28:2959

Koshiba T, Detmer SA, Kaiser JT, Chen H, McCaffery JM, Chan DC (2004) Structural basis of mitochondrial tethering by mitofusin complexes. Science 305:858

Koster AJ, Chen H, Sedat JW, Agard DA (1992) Automated microscopy for electron tomography. Ultramicroscopy 46:207–227

Ladinsky MS, Mastronarde DN, McIntosh JR, Howell KE, Staehelin LA (1999) Golgi structure in three dimensions: functional insights from the normal rat kidney cell. J Cell Biol 144:1135

Lam SS, Martell JD, Kamer KJ, Deerinck TJ, Ellisman MH, Mootha VK, Ting AY (2015) Directed evolution of APEX2 for electron microscopy and proximity labeling. Nat Methods 12:51–54

Li J, Wang Y, Chiu S-L, Cline HT (2010) Membrane targeted horseradish peroxidase as a marker for correlative fluorescence and electron microscopy studies. Front Neural Circuits 4:6

Li W, Ding W, Ji G, Wang L, Zhang J, Sun F (2016) Three-dimensional visualization of arsenic stimulated mouse liver sinusoidal by FIB-SEM approach. Protein Cell 7:227–232

Liu T, Yu R, Jin S-B, Han L, Lendahl U, Zhao J, Nistér M (2013) The mitochondrial elongation factors MIEF1 and MIEF2 exert partially distinct functions in mitochondrial dynamics. Exp Cell Res 319:2893–2904

Martell JD, Deerinck TJ, Sancak Y, Poulos TL, Mootha VK, Sosinsky GE, Ellisman MH, Ting AY (2012) Engineered ascorbate peroxidase as a genetically encoded reporter for electron microscopy. Nat Biotechnol 30:1143–1148

McBride HM, Millar DG, Li JM, Shore GC (1992) A signal-anchor sequence selective for the mitochondrial outer membrane. J Cell Biol 119:1451

Meeusen S, McCaffery JM, Nunnari J (2004) Mitochondrial fusion intermediates revealed *in vitro*. Science 305:1747

Meißlitzer-Ruppitsch C, Vetterlein M, Stangl H, Maier S, Neumüller J, Freissmuth M, Pavelka M, Ellinger A (2008) Electron microscopic visualization of fluorescent signals in cellular compartments and organelles by means of DAB-photoconversion. Histochem Cell Biol 130:407–419

Mercogliano CP, DeRosier DJ (2007) Concatenated metallothionein as a clonable gold label for electron microscopy. J Struct Biol 160:70–82

Morphew MK, O'Toole ET, Page CL, Pagratis M, Meehl J, Giddings T, Gardner JM, Ackerson C, Jaspersen SL, Winey M, Hoenger A, McIntosh JR (2015) Metallothionein as a clonable tag for protein localization by electron microscopy of cells. J Microsc 260:20–29

Mozdy AD, McCaffery JM, Shaw JM (2000) Dnm1p Gtpase-mediated mitochondrial fission is a multi-step process requiring the novel integral membrane component Fis1p. J Cell Biol 151:367–380

Otera H, Wang C, Cleland MM, Setoguchi K, Yokota S, Youle RJ, Mihara K (2010) Mff is an essential factor for mitochondrial recruitment of Drp1 during mitochondrial fission in mammalian cells. J Cell Biol 191:1141–1158

Palmer CS, Osellame LD, Laine D, Koutsopoulos OS, Frazier AE, Ryan MT (2011) MiD49 and MiD51, new components of the mitochondrial fission machinery. EMBO Rep 12:565–573

Peddie CJ, Collinson LM (2014) Exploring the third dimension: volume electron microscopy comes of age. Micron 61:9–19

Rojo M, Legros F, Chateau D, Lombès A (2002) Membrane topology and mitochondrial targeting of mitofusins, ubiquitous mammalian homologs of the transmembrane GTPase Fzo. J Cell Sci 115:1663

Santel A, Fuller MT (2001) Control of mitochondrial morphology by a human mitofusin. J Cell Sci 114:867

Seligman AM, Karnovsky MJ, Wasserkrug HL, Hanker JS (1968) Nondroplet ultrastructural demonstration of cytochrome oxidase activity with a polymerizing osmiophilic reagent, diaminobenzidine (DAB). J Cell Biol 38:1

Shu X, Lev-Ram V, Deerinck TJ, Qi Y, Ramko EB, Davidson MW, Jin Y, Ellisman MH, Tsien RY (2011) A genetically encoded tag for correlated light and electron microscopy of intact cells, tissues, and organisms. PLoS Biol 9:e1001041

Smirnova E, Griparic L, Shurland D-L, van der Bliek AM (2001) Dynamin-related protein Drp1 is required for mitochondrial division in mammalian cells. Mol Biol Cell 12:2245–2256

Song Z, Chen H, Fiket M, Alexander C, Chan DC (2007) OPA1 processing controls mitochondrial fusion and is regulated by mRNA splicing, membrane potential, and Yme1L. J Cell Biol 178:749

Su Y, Nykanen M, Jahn KA, Whan R, Cantrill L, Soon LL, Ratinac KR, Braet F (2010) Multi-dimensional correlative imaging of subcellular events: combining the strengths of light and electron microscopy. Biophys Rev 2:121–135

Waizenegger T, Stan T, Neupert W, Rapaport D (2003) Signal-anchor domains of proteins of the outer membrane of mitochondria: structural and functional characteristics. J Biol Chem 278:42064–42071

Yoon Y, Krueger EW, Oswald BJ, McNiven MA (2003) The mitochondrial protein hFis1 regulates mitochondrial fission in mammalian cells through an interaction with the dynamin-like protein DLP1. Mol Cell Biol 23:5409–5420

Youle RJ, Van Der Bliek AM (2012) Mitochondrial fission, fusion, and stress. Science 337:1062–1065

Zhao J, Liu T, Jin S, Wang X, Qu M, Uhlén P, Tomilin N, Shupliakov O, Lendahl U, Nistér M (2011) Human MIEF1 recruits Drp1 to mitochondrial outer membranes and promotes mitochondrial fusion rather than fission. EMBO J 30:2762–2778

Radioligand saturation binding for quantitative analysis of ligand-receptor interactions

Chengyan Dong[1], Zhaofei Liu[2], Fan Wang[1,2]

[1] Interdisciplinary Laboratory, Institute of Biophysics, Chinese Academy of Sciences, Beijing 100101, China
[2] Medical Isotopes Research Center and Department of Radiation Medicine, School of Basic Medical Sciences, Peking University, Beijing 100191, China

Abstract The reversible combination of a ligand with specific sites on the surface of a receptor is one of the most important processes in biochemistry. A classic equation with a useful simple graphical method was introduced to obtain the equilibrium constant, K_d, and the maximum density of receptors, B_{max}. The entire ^{125}I-labeled ligand binding experiment includes three parts: the radiolabeling, cell saturation binding assays and the data analysis. The assay format described here is quick, simple, inexpensive, and effective, and provides a gold standard for the quantification of ligand-receptor interactions. Although the binding assays and quantitative analysis have not changed dramatically compared to the original methods, we integrate all the parts to calculate the parameters in one concise protocol and adjust many details according to our experience. In every step, several optional methods are provided to accommodate different experimental conditions. All these refinements make the whole protocol more understandable and user-friendly. In general, the experiment takes one person less than 8 h to complete, and the data analysis could be accomplished within 2 h.

Keywords Equilibrium constant, Maximum density of receptors, Saturation binding assays, I-125 labeling, Radioligand

INTRODUCTION

Research on receptors has developed very quickly in the past few decades. The interactions between ligands and receptors generate and enhance signals for recognition, feedback and crosstalk in cells (Klotz 1985; Wilkinson 2004). Receptors are divided into two groups by their location: in the membrane or the nucleus. Hormones and transmitters can selectively recognize and bind to receptors to accomplish a biological process. Many drugs are designed and improved based on utilizing the ligand-receptor interaction.

Radioligand binding is widely used to define receptor function at the molecular level. The first radiolabeled binding assay was developed during the 1960s (Maguire et al. 2012). This radioligand binding assay (RBA) remains the most sensitive quantitative approach to measuring binding parameters in vitro, even in low receptor-expression cells (Rovati 1993; Keen 1995). Since the 1970s, the application of RBA has developed rapidly, with better receptor preparations, more radiolabeled ligands and higher radioactivity. Currently, appropriate ligands radiolabeled with tritium or iodine are available for the study of many receptors, including adrenergic, cholinergic, dopaminergic, serotonergic, and opiate receptors (Tallarida et al. 1988). This widespread availability has led to a rapid growth in the use of radioligand binding assays to characterize novel receptors and receptor subtypes and determine their anatomical distribution, and these assays play a vital role in the development of drugs by the pharmaceutical industry (Bylund and Toews 1993; Carpenter et al. 2002).

✉ Correspondence: wangfan@bjmu.edu.cn (F. Wang)

Unlabeled ligands require a radioactive isotope to be incorporated into the molecule. It is occasionally necessary to modify the structure of the ligands to provide a suitable site for radiolabeling. However, it is imperative that the selectivity and specificity of the ligand be retained after the modification and radiolabeling. Two basic parameters of this binding site can be studied by kinetic and saturation analysis: the affinity of the ligand for its recognition site, the K_d, and an estimate of the number of binding sites in a given tissue, the B_{max} (Williams and Jacobson 1990). The K_d is the equilibrium dissociation constant, which is the concentration of ligand that will occupy 50% of the receptors. The generally accepted standard is that a ligand-receptor binding with a K_d of 1 nmol/L or less has a high affinity, whereas ligands binding with a K_d of 1 μmol/L or more have low affinity (Davenport and Russell 1996). B_{max} signifies the maximum density of receptors. This value is unique to a particular tissue in the binding assay, and it is usually corrected using the amount of protein or cells present.

MECHANISM OF ACTION

Analysis of radioligand binding experiments is based on a simple model; the law of mass action that describes the interaction between one molecule of ligand and one receptor molecule. For example, a neurotransmitter binds to the synaptic receptor to initiate the neurobiological process, or an antibody binds to an antigen to initiate the immunological response. In the simplest and most common case, this is a bimolecular reaction between a ligand and a receptor. This model assumes that binding is reversible (Anderson 1994).

$$[R] + [L] \rightleftarrows [RL], \qquad (1)$$

where [R] is the concentration of free receptor, [L] is the concentration of free ligand, [RL] is the concentration of the complex, k_1 is the association rate constant, and k_2 is the dissociation rate constant,

$$K_d = \frac{k_2}{k_1} = \frac{[R][L]}{[RL]}. \qquad (2)$$

The density of unbound receptors [R] and ligands [L] cannot be determined but could be given:

$$[R] = [RL] - [RL]. \qquad (3)$$

Hence, Eq. 2 could be rearranged to:

$$\frac{[RL]}{[L]} = \frac{[RT]}{K_d} - \frac{[RL]}{K_d}. \qquad (4)$$

For further analysis, [SB] could represent the concentration of ligand bound to the receptor, [F] represents the concentration of unbound ligand or "free" ligand, and B_{max} represents the greatest attainable concentration of bound ligand:

$$\frac{[SB]}{[F]} = \frac{B_{max}}{K_d} - \frac{[RL]}{K_d}. \qquad (5)$$

Thus, a plot of fractional [SB]/[F] vs. [RL] will give the basic information of the ligand-receptor interactions, known as a "Scatchard plot". An alternative is the Woolf plot, a plot of fractional [F]/[SB] vs. [L]:

$$\frac{[F]}{[SB]} = \frac{K_d}{B_{max}} + \frac{[L]}{B_{max}}. \qquad (6)$$

APPLICATIONS AND LIMITATIONS

Radioligand-binding techniques are applicable to any receptor of interest, provided it has a relative ligand that could selectively bind the receptor and be labeled with radioactive isotopes. Antibodies, proteins and peptides that contain Tyr could easily be labeled with iodine. Radioligands provide precise probes to quantify the initial interaction between ligands and receptors. For example, the kinetics of the association and dissociation of radioligands can be accurately examined from a simple tissue expressing the specific target receptors. A series of concentrations of unlabeled ligands could inhibit the forces established between radioligands and receptors, which could be used to measure the equilibrium dissociation constants. Radioligand binding also permits a characterization of receptor subtypes with different affinities and provides an estimate of their relative proportions, especially in the study of central nervous system receptors, where the effects of neurotransmitters are complex, and isolated tissue preparations are unfeasible (Tallarida et al. 1988). Radioligand binding assays can also be used to monitor the changes in receptor density, perhaps resulting from the pathological conditions of pharmacological intervention. Furthermore, the Scatchard plot is a useful diagnostic tool to determine whether more than one ligand molecules bind to a single receptor (Hollemans and Bertina 1975; Rovati 1998). A concave upward plot is indicative of nonspecific binding, negative cooperativity, or multiple classes of binding sites. A concave downward plot suggests either positive cooperativity or instability of the ligand (Wilkinson 2004).

However, binding parameters could be affected by many factors including the specific radioactivity, the type and ionic strength of the buffer, the presence of divalent ions and the temperature. The results are not sufficient to reflect the real physiological response mediated by the receptor in this homogenate preparation, such as minimizing degradation of the ligand (Davenport and Russell 1996). In addition, radioligand-binding assays cannot adequately discriminate between full agonists that elicit maximal physiological responses and partial agonists that cannot elicit a maximal response (Tallarida et al. 1988).

SUMMARIZED PROCEDURE

1. Dissolve iodogen in chloroform at a concentration of 2 mg/mL, evaporate the chloroform and make the iodogen-coated tube under an N_2 stream.
2. Dissolve Protein L in 0.2 mol/L phosphate buffer (pH 7.4) at a concentration of $\sim$2 mg/mL.
3. Mix 50 μL Protein L, 50 μL PB (0.2 mol/L, pH 7.4) and 50–100 μL $Na^{125}I$ (>40 MBq) in an iodegen-coated tube. Incubate for 7–8 min at room temperature.
4. Remove the reaction mixture from the iodogen tube and purify the radiolabeled protein by size exclusion chromatography using a PD MidiTrap G-25 column.
5. Count the activity in the final recovered tube and calculate the specific activity.
6. Wash two 96-well plates with pre-cooled cell-binding buffer for three times (100 μL each well) and use the vacuum manifold to remove the buffer. Place two 96-well plates for specific binding and non-specific binding respectively.
7. From a stock of two million receptor-overexpressed cells per mL of cell-binding buffer (total volume > 1.5 mL), add 1×10^5 cells (50 μL) each well in the 96-well plate.
8. Prepare three stock solutions of different concentration ^{125}I-Protein L (e.g., 0.1, 1 and 10 μg/mL) in cell-binding buffer.
9. Add the cells and ^{125}I labeled ligand into the plates according to calculation of final concentration for each well ($N = 4$), and adjust the total volume to 200 μL per well with cell-binding buffer and incubate for 2 h at 4 °C.
10. Use the vacuum manifold to remove the incubation buffer from the plates and wash 5–10 times with cell-binding buffer (100 μL/well).
11. Heat-dry the plates in the dry bath incubator, and collect the membrane from each well into polystyrene culture test tubes.
12. Add 4–7 tubes of standard samples to measure. Then measure the radioactivity on each membrane with a γ-counter.
13. For each experimental measurement, subtract the cpm values of groups. Added activities [TA], total binding activities [TB] and non-specific binding activities [NSB] could be measured and calculated by the activities on the membranes of plates.
14. Calculate [SB], [LT], [RL], [L] and [F] using these corrected values.
15. K_d value and B_{max} could be calculated by Scatchard Plot, Woolf Plot or the software.

PROCEDURE

Radiolabeled protein preparation [TIMING] ~1 h

1. There are two options for labeling the proteins.

   (A) **Option A: Chloramine-T method** (Hunter 1970; Opresko et al. 1980).
      i. Dissolve Protein L (see "Reagent setup" section) in 0.2 mol/L phosphate buffer (pH 7.4) at a concentration of 2 mg/mL.
      ii. Mix the following reagents: 50 μL Protein L, 50–100 μL $Na^{125}I$ (>40 MBq) and 100 μg Chloramine-T (1 mg/mL in 100 μL 0.2 mol/L PB, pH 7.4). Incubate for 40 s at room temperature.
      [CRITICAL STEP] It is highly recommended to limit the reaction time in 1–3 min.
      iii. Add 100 μL $Na_2S_2O_5$ (200 μg in ddH_2O) and 100 μL 1% KI (1 mg in ddH_2O) to the mixture.

   (B) **Option B: Iodogen method** (Bailey 1996).
      i. Dissolve Protein L (see "Reagent setup" section) in 0.2 mol/L phosphate buffer (pH 7.4) at a concentration of $\sim$2 mg/mL.
      ii. Dissolve iodogen in chloroform at a concentration of 2 mg/mL. Transfer aliquots of 25 μL (50 μg) to a glass-bottomed screw cap vial. Evaporate the chloroform to dryness under an N_2 stream, leaving a thin coating of iodogen in the tube. Store the desiccated iodogen-coated tubes at −20 °C until required for iodination.
      iii. Mix 50 μL Protein L, 50 μL PB (0.2 mol/L, pH 7.4) and 50–100 μL $Na^{125}I$ (>40 MBq) in an iodogen-coated tube. Incubate for 7–8 min at room temperature.
      [CRITICAL STEP] It is highly recommended to maintain the duration of the reaction between 5–10 min.

iv. Remove the reaction mixture from the iodogen tube and apply to purifying.
[CAUTION!] It is imperative to obtain appropriate training from the institutional radiation safety office before experimenting with radioactivity. Abide by all relevant regulatory rules and use appropriate protection when handling radioactivity. Dispose of the ^{125}I-containing radioactive waste according to the institutional radioactive waste disposal guidelines.

2. Purification: Purify the radiolabeled protein by size exclusion chromatography using a PD MidiTrap G-25 column.

   (A) Preparation and equilibration: Remove the caps and the storage solution. Fill the column with PBS and discard the flow-through. Repeat this procedure twice (three times in total).

   (B) Sample application: Add a maximum of 1.0 mL of sample to the column. Apply the sample slowly in the middle of the packed bed and discard the flow through.

   (C) Elution: Place a clean tube for sample collection under the column. Elute with 1.5 mL PBS and collect the products.

3. Count the activity in the final recovered tube. The specific activity is calculated as the quotient between the recovered activity and the total amount of protein. This calculation assumes that 100% of the protein added to the iodination was recovered, which is not typical. (Analytical Techniques, T.P. Mommson)

$$\text{Specific activity} = \frac{\text{Radioactivity}}{\text{Protein mass}} (\text{Ci/g}).$$

[CRITICAL STEP] To obtain an optimal result, it is sufficient to utilize radioligands with high specific activity (>20 Ci/mmol).

Saturation binding [TIMING] 5–7 h

4. Wash two 96-well plates with pre-cooled cell-binding buffer three times (100 µL each well) and use the vacuum manifold to remove the buffer. One 96-well plate (Plate A) will be used for specific binding and the other one (Plate B) will be used for non-specific binding(Cai and Chen 2008).

5. From a stock of two million receptor-overexpressed cells per mL of cell-binding buffer (total volume > 1.5 mL), add 1×10^5 cells (50 µL) to each well in the 96-well plate.

6. Prepare three stock solutions of different concentrations of ^{125}I-Protein L (e.g., 0.1, 1 and 10 µg/mL) in cell-binding buffer. Typically a series of concentrations between 1 ng/200 µL and 1 µg/200 µL will be needed per well.
   [CAUTION!] It is imperative to obtain the appropriate preparatory training and abide by all regulatory rules when handling radioactivity.
   [? TROUBLESHOOTING]

7. Add the cells and ^{125}I-labeled ligand into Plate A following Table 1, and adjust the total volume to 200 µL per well with cell-binding buffer and incubate for 2 h at 4 °C. More than four samples are recommended for each concentration.
   [? TROUBLESHOOTING]

8. Add the cells, ^{125}I-labeled ligands and excess cold Protein L into Plate B following Table 2, and adjust the total volume to 200 µL per well with cell-binding buffer and incubate for 2 h at 4 °C as the last step. More than four samples are recommended for each concentration.
   [? TROUBLESHOOTING]

9. Use the vacuum manifold to remove the incubation buffer from the 96-well plate and wash 5–10 times with cell-binding buffer (100 µL per well).

10. Heat-dry the 96-well plates in a dry bath incubator until all filter membranes are dry. This usually takes approximately 15 min.

Table 1 Sample adding strategy in the typical 96-well plate for the specific binding assay

Specific binding group												
No.	1	2	3	4	5	6	7	8	9	10	11	12
^{125}I-Protein L (µg/mL)	0.1	0.1	0.1	0.1	1	1	1	1	10	10	10	10
^{125}I-Protein L (µL)	10	20	50	80	10	20	50	80	10	20	50	80
Binding buffer (µL)	140	130	100	70	140	130	100	70	140	130	100	70
Cell solution (µL)	50	50	50	50	50	50	50	50	50	50	50	50
Total volume (µL)	200	200	200	200	200	200	200	200	200	200	200	200

Table 2 Sample adding strategy in the typical 96-well plate for the non-specific binding assay

Non-specific binding group												
No.	1	2	3	4	5	6	7	8	9	10	11	12
^{125}I-Protein L (μg/mL)	0.1	0.1	0.1	0.1	1	1	1	1	10	10	10	10
^{125}I-Protein L (μL)	10	20	50	80	10	20	50	80	10	20	50	80
Cold Protein L (μL)	50	50	50	50	50	50	50	50	50	50	50	50
Binding buffer (μL)	90	80	50	20	90	80	50	20	90	80	50	20
Cell solution (μL)	50	50	50	50	50	50	50	50	50	50	50	50
Total volume (μL)	200	200	200	200	200	200	200	200	200	200	200	200

11. Collect the membrane from each well into polystyrene culture test tubes.
12. Add 4–7 tubes of radiolabeled samples (standard samples) to measure.
    [PAUSE POINT] The radioactivity on each membrane can be measured later, because ^{125}I has a half-life of 60 d.
13. Measure the radioactivity on each membrane with a γ-counter.

Analysis [TIMING] 1–2 h

14. For each experimental measurement, subtract the cpm values of the groups. Added activities [TA] could be measured using the standard samples. Total binding activities [TB] and non-specific binding activities [NSB] could be measured by the activities on the membranes of Plate A and B, respectively.
15. Using Eqs. (5)–(9), calculate [SB], [LT], [RL], [L] and [F] using these corrected values:

$$[SB] = [TB] - [NSB], \quad (7)$$

$$[LT] = \frac{[TA](cpm)}{E\% \times [SA](\mu Ci/nmol) \times 2.22 \times 10^6} \times \frac{10^3}{\text{Volume (L)}} \text{ (pmol/L)}, \quad (8)$$

$$[RL] = \frac{[SB](cpm)}{E\% \times [SA](\mu Ci/nmol) \times 2.22 \times 10^6} \times \frac{10^3}{\text{Volume (L)}} \text{ (pmol/L)}, \quad (9)$$

$$[L] = [LT] - [RL], \quad (10)$$

$$[F] = [TA] - [SB]. \quad (11)$$

16. K_d and B_{max} could be calculated using a Scatchard plot, Woolf plot or the software.

(A) **Option A: Scatchard plot**
   i. For each point on the concentration, enter [RL] into the X column and the value of [SB]/[F] into the Y column:

$$\frac{[SB]}{[F]} = \frac{B_{max}}{K_d} - \frac{[RL]}{K_d}. \quad (12)$$

   ii. All the points are plotted and then linear regression is used to produce the line.
   iii. Referring to the Results sheet for the regression analysis, the X- and Y-axis intercepts could be calculated. The X-intercept represents the B_{max}, and the Y-intercept represents B_{max}/K_d.

(B) **Option B: Woolf plot**
   i. For each point on the concentration, enter [L] into the X column and the value of [F]/[SB] into the Y column:

$$\frac{[F]}{[SB]} = \frac{K_d}{B_{max}} + \frac{[L]}{B_{max}}. \quad (13)$$

   ii. All the points are plotted and then linear regression is used to produce the line.
   iii. Referring to the results sheet for the regression analysis, the X- and Y-axis intercepts could be calculated. The X-intercept represents the K_d, and the Y-intercept represents K_d/B_{max}.

(C) **Option C: Saturation binding curve**
   i. Create a new project (file) on Prism(Motulsky 1996). For each point on the saturation-binding curve, enter the concentration of ligand into the X column and [SB] into the Y column.

ii. Select "One site binding" under the "Nonlinear Regression dialog" box to analyze the data and produce a binding curve.
iii. Prism displays the best-fit values (B_{max} and K_d) for the binding parameters in the results sheets.

[TIMING]

Step 1–3 Preparation of the ^{125}I-Protein L and cold ligands takes approximately 1 h.
Step 4–6 Preparation of the receptor samples takes approximately 1 h.
Step 7–13 The cell-binding assay usually takes 4–6 h, depending on how many samples are used.
Step 13–16 Activity measurements and data analyses take approximately 1–2 h.

[? TROUBLESHOOTING]

Step 6 The presence of certain metal ions (e.g., Mn^{2+} and Mg^{2+}) in the cell-binding buffer is essential for receptor binding. Binding buffer without these ions will result in low counts from the collected membrane.

Step 7 It is necessary to add the cells and buffer with multiple-channel pipettes to reduce the time of this step. It takes practice to become skilled at adding serial concentrations of radioligand and cold ligand. It is important to stay focused and patient.

Step 8 The non-specific binding assay requires a large quantity of the cold ligand. Usually the ligands are difficult to prepare or very expensive. The Scatchard plot and the Woolf plot could be completed using fewer concentrations and fewer parallel samples. In total, about 20 samples are sufficient for fitting the linear regression.

ANTICIPATED RESULTS

Figures 1 and 2 present typical representative data obtained using the method described here. EGFR overexpressing UM-SCC-22B cells were assayed against the radiolabeled antibody ^{125}I-Nimotuzumab. The labeling yield of ^{125}I-Nimotuzumab was 97.6% and the radiochemical purity was >98.5% after purification (Fig. 1). The specific activity was 24.7 Ci/g.

Figure 2A shows an example of a typical equilibrium saturation curve using the radiolabeled assay with increasing concentrations of ^{125}I-Nimotuzumab. Figure 2B shows a typical "Scatchard plot". Figure 2C

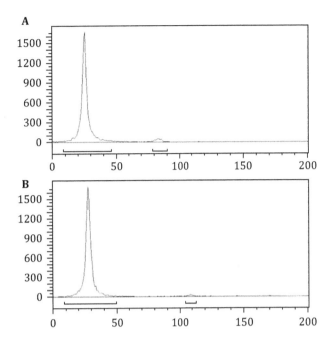

Fig. 1 ITLC analysis of ^{125}I-labeled Nimotuzumab. The labeling yield of ^{125}I-Nimotuzumab was 97.6%, and the radiochemical purity was 98.5% after purification

shows a typical "Woolf plot". In parallel comparisons from the same data obtained from the binding assays, the Scatchard plot, the Woolf plot and the saturation radioligand-binding curve gave similar estimates of the K_d (7.81, 7.935 and 8.095 pL/mol, respectively) and B_{max} (113.4, 114.5 and 115.3 pmol/L) (Table 3). The average total number of EGF receptors on each UM-SCC-22B cell could be calculated:

$$\text{EGFRS per cell} = \frac{115.3 \text{ pM} \times 200 \text{ μL} \times 6.02 \times 10^{23}}{1 \times 10^5}$$
$$= 1.4 \times 10^5.$$

MATERIALS

Reagents

- 0.2 mol/L phosphate buffer, pH 7.4 ($NaH_2PO_4 \cdot 2H_2O$ 0.4063 g, $Na_2HPO_4 \cdot 12H_2O$ 5.8077 g in 200 mL ddH_2O)
- Chloramine-T (100 μg in 0.2 mol/L PB)
- $Na_2S_2O_5$ (200 μg in 100 μL ddH_2O)
- KI (1 mg in 100 μL ddH_2O)
- Chloroform
- Iodogen (0.5 μg/μL in chloroform)

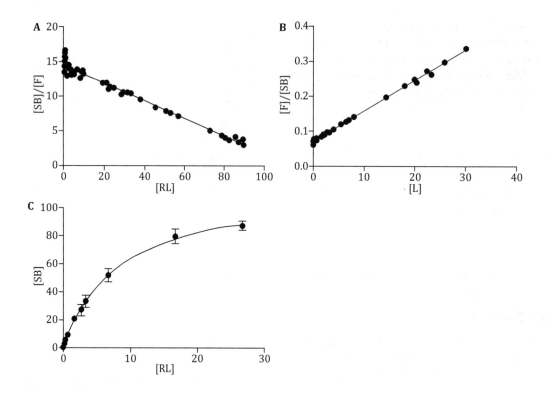

Fig. 2 The calculation of K_d and B_{max} by Scatchard plot (A), Woolf plot (B) or the software (C)

Table 3 Comparison of the results obtained by three plots

	Scatchard plot	Woolf plot	Saturation binding curve
K_d	7.81	7.935	8.095
B_{max}	113.4	114.5	115.3

- Cell-binding buffer (20 mmol/L Tris, 150 mmol/L NaCl, 2 mmol/L $CaCl_2$, 1 mmol/L $MnCl_2$, 1 mmol/L $MgCl_2$, 1% (wt/vol) bovine serum albumin; pH 7.4)
- PBS buffer ($NaH_2PO_4 \cdot 2H_2O$ 0.24 g, $Na_2HPO_4 \cdot 12H_2O$ 2.901 g and NaCl 8.5 g in 1 L ddH_2O)
- Receptor-overexpressed cells (see "Reagent setup" section)
- Medium
- Fetal bovine serum
- $Na^{125}I$ (Perkin Elmer, Waltham, MA, USA)

Equipment

- pH paper (Aladdin Inc.)
- PD MidiTrap G-25 column (GE Healthcare, cat. no. 28-9180-08)
- MultiScreen™ Vacuum Manifold 96-well plate (Millipore, cat. no. MAVM0960R)
- Vacuum pump (Zhengzhou Greatwall Inc., SHB-III)
- Dry bath incubator (Fisher Scientific, cat. no. 11-718-2)
- γ-counter (PerkinElmer, 2470 automatic gamma counter)
- Glass-bottomed screw cap vial (Agilent Technologies, cat. no. 5182-0715)
- GraphPad Prism (GraphPad Software Inc.)

Reagent setup

Cell sample preparation Culture receptor-overexpressed cells in corresponding medium under certain conditions. Collect the cells from the flasks or the plates. At least 10^5 cells are needed for each well. It takes 96 wells to test the binding of one ligand with its receptor. Wash the cell solution with 0.01 mol/L sterile PBS three times. Carefully resuspend the cells in cell binding buffer to a concentration of 2×10^6 cells/mL.

Cold protein L preparation 500 mg of protein L is dissolved in or diluted with 2.5 mL cell binding buffer. However, it takes a large amount of ligand to finish the experiment. In general, the cold ligands should be 1000 times more concentrated than the radiolabeled ligand to block the receptors. Fewer cold ligands could be used in low concentrations. Moreover, fewer concentrations (for example, eight concentrations) could conserve many ligands.

Equipment setup

γ-counter $E\%$ could be determined by comparison between the detected cpm value and the objective cpm value.

Acknowledgments This work was supported by National Natural Science Foundation of China (81125011, 81420108019 and 81427802).

Compliance with Ethical Standards

Conflict of Interest Chengyan Dong, Zhaofei Liu and Fan Wang declare that they have no conflict of interest.

Human and Animal Rights and Informed Consent This article does not contain any studies with human or animal subjects performed by any of the authors.

References

Anderson AJ (1994) A simple competitive protein-binding experiment. J Chem Educ 71:994

Bailey GS (1996) The Iodogen method for radiolabeling protein. In: Walker JM (ed) The protein protocols handbook. Humana Press, New York, pp 673–674

Bylund DB, Toews ML (1993) Radioligand binding methods: practical guide and tips. Am J Physiol 265:L421–L429

Cai W, Chen X (2008) Preparation of peptide-conjugated quantum dots for tumor vasculature-targeted imaging. Nat Protoc 3:89–96

Carpenter JW, Laethem C, Hubbard FR, Eckols TK, Baez M, McClure D, Nelson DLG, Johnston PA (2002) Configuring radioligand receptor binding assays for HTS using scintillation proximity assay technology. In: Janzen WP (ed) Methods in molecular biology, vol. 190: high throughput screening: methods and protocols. Humana Press, New York, pp 31–49

Hollemans HJ, Bertina RM (1975) Scatchard plot and heterogeneity in binding affinity of labeled and unlabeled ligand. Clin Chem 21:1769–1773

Hunter R (1970) Standardization of the chloramine-T method of protein iodination. In: Proceedings of the society for experimental biology and medicine society for experimental biology and medicine, New York, vol 133, pp 989–992

Keen M (1995) The problems and pitfalls of radioligand binding. In: Davis L (ed) Methods in molecular biology, vol 41. Humana Press Inc, Clifton, pp 1–16

Klotz IM (1985) ligand-receptor interactions: facts and fantasies. Q Rev Biophys 18:227–259

Maguire JJ, Kuc RE, Davenport AP (2012) Radioligand binding assays and their analysis. In: Davis L (ed) Methods in molecular biology, vol 897. Humana Press Inc, Clifton, pp 31–77

Williams M, Jacobson KA (1990) Radioligand binding assays for adenosine receptors. In: Williams M (ed) Adenosine and adenosine receptors, vol 2. Humana Press, New York, pp 17–55

Motulsky H (1996) The GraphPad guide to analyzing radioligand binding data. GraphPad Software, Inc., San Diego

Opresko L, Wiley HS, Wallace RA (1980) Proteins iodinated by the chloramine-T method appear to be degraded at an abnormally rapid rate after endocytosis. Proc Natl Acad Sci USA 77:1556–1560

Tallarida RJ, Raffa RB, McGonigle P (eds) (1988) Radiolabeled binding. In: Principles in general pharmacology, vol 9. Springer, New York, p 199

Rovati GE (1993) Rational experimental design and data analysis for ligand binding studies: tricks, tips and pitfalls. Pharmacol Res 28:277–299

Rovati GE (1998) Ligand-binding studies: old beliefs and new strategies. Trends Pharmacol Sci 19:365–369

Davenport AP, Russell FD (1996) Radioligand binding assays: theory and practice. In: Mather SJ (ed) Current directions in radiopharmaceutical research and development. Springer, Amsterdam, pp 169–179

Wilkinson KD (2004) Quantitative analysis of protein–protein interactions. In: Fu H (ed) Protein-protein interactions, vol 2. Humana Press, New York, pp 15–31

Using 3dRPC for RNA–protein complex structure prediction

Yangyu Huang[1], Haotian Li[1], Yi Xiao[1]

[1] Biomolecular Physics and Modeling Group, School of Physics, Huazhong University of Science and Technology, Wuhan 430074, China

Abstract 3dRPC is a computational method designed for three-dimensional RNA–protein complex structure prediction. Starting from a protein structure and a RNA structure, 3dRPC first generates presumptive complex structures by RPDOCK and then evaluates the structures by RPRANK. RPDOCK is an FFT-based docking algorithm that takes features of RNA–protein interactions into consideration, and RPRANK is a knowledge-based potential using root mean square deviation as a measure. Here we give a detailed description of the usage of 3dRPC. The source code is available at http://biophy.hust.edu.cn/3dRPC.html.

Keywords RNA–protein complex, Tertiary structure, Computational prediction, Docking, Scoring function

INTRODUCTION

RNA–protein interactions have drawn much attention recently since they might play important roles in many biological processes (Chen and Varani 2005; Glisovic et al. 2008). It was found that most of the human genome could be transcribed into RNAs but only a small fraction of these RNAs was translated into proteins (Cheng et al. 2005), i.e., most RNAs did not undergo translation. These non-coding RNAs perform their biological functions mostly through RNA–protein interactions and forming RNA–protein complexes. As the protein–protein interactions, the three-dimensional structures of RNA–protein complexes are essential to understand the mechanism of RNA–protein interactions. However, experimental determination of three-dimensional structures of RNA–protein complexes is still difficult and time-consuming at present. To solve this problem, computational methods have been proposed to predict the RNA–protein complex structures.

Most algorithms for predicting complex structure consist of two stages: sampling and scoring. The first stage is sampling conformational space and selecting candidates. Since the conformational space is very large, a fast and effective sampling method is required. The second stage is evaluation of the candidates using a ranking or scoring function. Compared to the well-developed methods for protein–protein complex structure prediction (Vakser and Aflalo 1994; Gabb et al. 1997; Chen et al. 2003; Dominguez et al. 2003; Kozakov et al. 2006), those for RNA–protein complexes remain to be developed, which mainly focus on the scoring (Chen et al. 2004; Perez-Cano et al. 2010; Tuszynska and Bujnicki 2011; Li et al. 2012; Huang and Zou 2014), while the sampling methods were borrowed from those for protein–protein complex prediction (Vakser and Aflalo 1994; Gabb et al. 1997; Chen et al. 2003). Recently, we proposed a novel protocol for predicting RNA–protein complex structures—3dRPC (Huang et al. 2013). 3dRPC originally consists of a docking procedure RPDOCK and a scoring function DECK-RP.

RPDOCK is a docking procedure specific to RNA–protein docking. Based on the fact that the atom packing at the RNA–protein interface is different from that at the protein–protein interface (Jones et al. 1999, 2001; Bahadur et al. 2008), RPDOCK applies a new set of parameters to calculate the geometric complementarity. Since the electrostatics plays an important role in RNA–protein interaction (Jones et al. 2001; Kim et al. 2006; Terribilini et al. 2006; Bahadur et al. 2008; Kumar et al. 2008; Perez-Cano et al. 2010; Perez-Cano and

✉ Correspondence: yxiao@mail.hust.edu.cn (Y. Xiao)

Fernandez-Recio 2010), RPDOCK also includes electrostatic effect. RPDOCK also accounts for the stacking interactions between aromatic side chain and bases. The scoring function DECK-RP has been replaced in the updated 3dRPC by RPRANK, a new knowledge-based potential using Root mean square deviation (RMSD) as a measure. The statistical objects of RPRANK are the conformation differences between residue-base pairs. The residue-base pairs are clustered based on the RMSD between each other. Then the energies of the residue-base pair clusters are decided by statistical method based on the number of pairs in each cluster. Different from other statistical potential, this potential does not use distance to classify the residue-base pairs directly. The RMSD-based potential RPRANK has been tested on Zou's benchmarks (Huang and Zou 2013). The success rate reaches 29.1% for top one and 41.7% for top ten. 3dRPC has been tested on two test sets(Perez-Cano et al. 2012; Huang and Zou 2013) and achieved success rates of 12.1% and 31.9% for top one prediction and 28.8% and 41.7% for top ten, respectively. In the following, we give a detailed description of the usage of 3dRPC.

3dRPC

Stage 1: rigid-body docking by RPDOCK

RPDOCK is a FFT-based, rigid-body sampling method. The overall process of RPDOCK resembles protein–protein docking algorithm FTDOCK (Gabb et al. 1997). First, the protein is discretized into three-dimensional grid and the RNA is rotated by Euler angles and then discretized into three-dimensional grid. Next, a full translation scan is performed. During the translation scan, top three poses are retained according to the RPDOCK score. Fast Fourier transform is used to accelerate the calculation. The process is repeated until full rotation scan is completed. RPDOCK score is composed of two items: geometric complementarity (GC) and electrostatics (ELEC). The electrostatics is calculated by Coulomb's formula with a distance-dependent dielectric and the charge is extracted from AMBER force field (Case et al. 2005).

Stage 2: scoring by RPRANK

Each presumptive pose generated by RPDOCK is scored by RPRANK in this stage. RPRANK extracts the residue-base pairs within 10 Å, and then the pairs from decoy complexes are compared with standard pairs that are from native structures. If the RMSD between standard pair and decoy pair is less than 6 Å, the energy of decoy pair will be recorded as same as the standard pair. Finally, the energy of the decoy complex is the sum of the energy of pairs.

PROCEDURE

3dRPC installation

1. To download 3dRPC package, visit the 3dRPC webpage (http://biophy.hust.edu.cn/3dRPC.html).

2. Set running environment for 3dRPC. Add the following lines to your "~/.bashrc":

   "export HOME_3dRPC=/home/XXX/3dRPC/",
   "export X3DNA=${HOME_3dRPC}/ext/X3DNA/",
   "export PATH=$PATH:${HOME_3dRPC}/ext/fasta/".
   Type the command in your terminal:
   "source ~/.bashrc".

3. Download and install libraries. Three external libraries are required by 3dRPC: FFTW (http://www.fftw.org/download.html), BLAS (http://www.netlib.org/blas/), and LAPACK (http://www.netlib.org/lapack/). The default path of libraries is "${HOME_3dRPC}/lib/".
   [? TROUBLESHOOTING]

4. Install FASTA. FASTA is used for sequence alignment in 3dRPC. The source code of FASTA is located on "${HOME_3dRPC}/ext/fasta/". Users can execute the following command lines to install FASTA:

   "cd ${HOME_3dRPC}/ext/fasta/",
   "make".
   After successful installation, an executable file "fasta35" can be found in "${HOME_3dRPC}/ext/fasta/".

5. Install 3dRPC program from the source code. Run the following command lines given below:

   "cd ${HOME_3dRPC}/source",
   "make".
   [? TROUBLESHOOTING]

Docking by RPDOCK

6. Prepare two PDB structures for docking, with one being protein and the other one being RNA. An example is shown in Fig. 1.

7. Prepare the parameter files for RPDOCK. The parameter files must follow the following formats:

   RPDock.receptor = 1DFU_r_u.pdb,
   RPDock.receptor.chain = V,
   RPDock.ligand = 1DFU_l_u.pdb,
   RPDock.ligand.chain = CB,
   RPDock.outfile = 1DFU.out,
   RPDock.grid_step = 1,

Fig. 1 An example of docking. The case is obtained from RNA–protein docking benchmark. The PDB code is 1DFU. Unbound protein (A) and unbound RNA (B) are shown in cartoon presentation

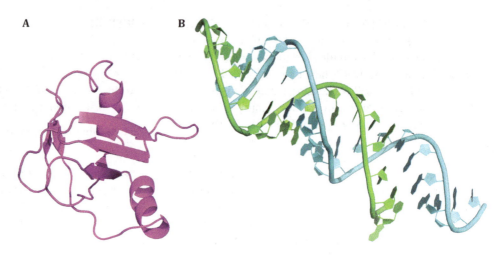

RPDock.out_pdb = 10.

The parameter files are further explained in Table 1.

8. Run RPDOCK by the following command line:

   "$HOME_3dRPC/source/3dRPC -mode 9 -system 9 -par RPDock.par".

   "RPDock.par" is the parameter file described previously. After docking is finished, RPDOCK will generate an output file "1DFU.out" and a number of docked complexes ("complex1.pdb", ..., "complex*.pdb"). An example of the output files is shown below:

G_DATA	13	0	−946.00	13	25	1	3	48.0	0.0	0.0
G_DATA	10	0	−897.00	10	25	5	2	36.0	0.0	0.0
G_DATA	14	0	−858.00	14	25	2	3	48.0	0.0	0.0

Each line represents a docked complex with related information (Table 2). RPDOCK is a rigid-body docking procedure and the docked complexes depend on the translation vector and the rotation angles (Fig. 2).

9. Generate complexes by the following command line:

   "$HOME_3dRPC/source/3dRPC -mode 9 -system 8 -par RPDock.par".

   "RPDock.par" is the same parameter file that is used for docking. Users can change the number of complexes generated.

Scoring with RPRANK

10. Prepare a list of complex structures to be scored by the following format:

| complex1.pdb | V | CB |
| complex2.pdb | V | CB |

The first column is the file name of the complex structures, the second column is the chain ID of protein and the last column is the chain ID of RNA.

11. Prepare the parameter file "scoring.par" for scoring:

    list = list,
    out = RMSD.score.

12. Run the command to score the complexes in the list:

    "${HOME_3dRPC}/source/3dRPC -mode 8 -system 9 -par scoring.par".

    According to the parameter, the output of scoring is saved in the file "RMSD.score". An example of the output is shown below:

| complex1.pdb | −93.2882 |
| complex2.pdb | −145.628 |

The first column is the name of the complex and the second column is the corresponding energy given by RMSD-based score.

Table 1 Explanation of parameter files for RPDOCK—"RPDock.par"

RPDock.receptor	File name of protein structure
RPDock.receptor.chain	Chain ID of protein
RPDock.ligand	File name of RNA structure
RPDock.ligand.chain	Chain ID of RNA
RPDock.outfile	Output file name of RPDOCK
RPDock.grid_step	Grid step of RPDOCK, 1 is recommended
RPDock.out_pdb	Number of complexes generated

Table 2 Explanation of information contained in the output files of RPDOCK

Column 4	RPDOCK score
Column 6–8	Translation vector
Column 9–11	Rotation angles

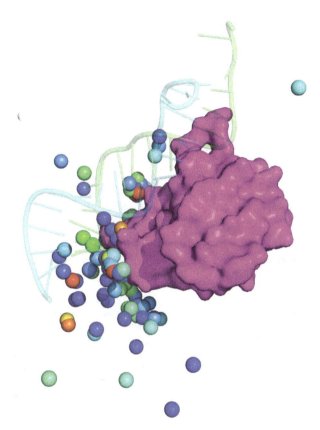

Fig. 2 An example of docking. The native complex (1DFU) is shown in cartoon. The centroids of top 100 poses according to RPDOCK score are shown in sphere with rainbow color representing RPDOCK score. The *red color* represents high RPDOCK score

Result analysis of RPDOCK decoy

13. Prepare the parameter file for analysis:

    RPDock.resfile = 1DFU.out,
    RPDock.max_matches = 10,
    native.receptor_pdb_filename = 1DFU_r_b.pdb,
    native.ligand_pdb_filename = 1DFU_l_b.pdb,
    native.receptor.chainid = P,
    native.ligand.chainid = MN,
    decoy.receptor_pdb_filename = 1DFU_r_u.pdb,
    decoy.ligand_pdb_filename = 1DFU_l_u.pdb,
    decoy.receptor.chainid = V,
    decoy.ligand.chainid = CB,
    rmsd.output = 1DFU.rmsd.dat (Table 3).

Table 3 Explanation of the parameter files

RPDock.resfile	Output of RPDOCK
RPDock.max_matches	Number of complexes
native.receptor_pdb_filename	Native protein structure
native.ligand_pdb_filename	Native RNA structure
native.receptor.chainid	Chain ID of native protein
native.ligand.chainid	Chain ID of native RNA
decoy.receptor_pdb_filename	Protein structure used for docking
decoy.ligand_pdb_filename	RNA structure used for docking
decoy.receptor.chainid	Chain ID
decoy.ligand.chainid	Chain ID
rmsd.output	Output file of result analysis

14. Run the following command:

    "${HOME_3dRPC}/source/3dRPC -mode 2 -system 0 -par rmsd.par".

    The "rmsd.par" is the parameter file described in step 15. After the calculation is finished, an outfile, named as "1DFU.rmsd.dat" according to the parameter, will be generated. The output files are formatted as following:

#Decoy	R_rmsd	L_rmsd	I_rms	fnat	fnon
1	0.744382	34.1629	14.6322	0	1
2	0.744382	32.8772	14.5631	0.0178571	0.964286

Further explanation of the files is shown in Table 4.

Table 4 Explanation of output files

#Decoy	Decoy number
R_rmsd	RMSD of receptor (protein)
L_rmsd	RMSD of ligand (RNA)
I_rms	Interface RMSD
fnat	Native contact fraction
fnon	Non-native contact fraction

[? TROUBLESHOOTING]

Step 3: How to install BLAS and LAPACK in Mac?

Open the file "BLAS/make.inc" or "LAPACK/make.inc", find the line that says: "PLAT = _LINUX" and change it to "PLAT = _MACOS". Type "make" in your terminal to install BLAS and LAPACK.

Step 5: What can I do if I get error while installing 3dRPC?

Make sure that BLAS, LAPACK and FFTW libraries are successfully installed in your system. Open the file "${HOME_3dRPC}/source/Makefile", find the line starting with 'LAPACK_LIBS' and 'BLAS_LIBS', make sure that the paths of the libraries are correctly assigned.

Acknowledgements This work is supported by the National Natural Science Foundation of China (31570722, 11374113) and the National High Technology Research and Development Program of China (2012AA020402).

Compliance with Ethical Standards

Conflict of interest Yangyu Huang, Haotian Li, and Yi Xiao declare that they have no conflict of interest.

Human and Animal Rights and Informed Consent This article does not contain any studies with human or animal subjects performed by any of the authors.

References

Bahadur RP, Zacharias M, Janin J (2008) Dissecting protein-RNA recognition sites. Nucleic Acids Res 36:2705–2716

Case DA, Cheatham TE 3rd, Darden T, Gohlke H, Luo R, Merz KM Jr, Onufriev A, Simmerling C, Wang B, Woods RJ (2005) The Amber biomolecular simulation programs. J Comput Chem 26:1668–1688

Chen Y, Varani G (2005) Protein families and RNA recognition. FEBS J 272:2088–2097

Chen R, Li L, Weng Z (2003) ZDOCK: an initial-stage protein-docking algorithm. Proteins 52:80–87

Chen Y, Kortemme T, Robertson T, Baker D, Varani G (2004) A new hydrogen-bonding potential for the design of protein-RNA interactions predicts specific contacts and discriminates decoys. Nucleic Acids Res 32:5147–5162

Cheng J, Kapranov P, Drenkow J, Dike S, Brubaker S, Patel S, Long J, Stern D, Tammana H, Helt G, Sementchenko V, Piccolboni A, Bekiranov S, Bailey DK, Ganesh M, Ghosh S, Bell I, Gerhard DS, Gingeras TR (2005) Transcriptional maps of 10 human chromosomes at 5-nucleotide resolution. Science 308:1149–1154

Dominguez C, Boelens R, Bonvin AM (2003) HADDOCK: a protein-protein docking approach based on biochemical or biophysical information. J Am Chem Soc 125:1731–1737

Gabb HA, Jackson RM, Sternberg MJ (1997) Modelling protein docking using shape complementarity, electrostatics and biochemical information. J Mol Biol 272:106–120

Glisovic T, Bachorik JL, Yong J, Dreyfuss G (2008) RNA-binding proteins and post-transcriptional gene regulation. FEBS Lett 582:1977–1986

Huang SY, Zou X (2013) A nonredundant structure dataset for benchmarking protein-RNA computational docking. J Comput Chem 34:311–318

Huang SY, Zou X (2014) A knowledge-based scoring function for protein-RNA interactions derived from a statistical mechanics-based iterative method. Nucleic Acids Res 42:e55

Huang Y, Liu S, Guo D, Li L, Xiao Y (2013) A novel protocol for three-dimensional structure prediction of RNA-protein complexes. Sci Rep 3:1887

Jones S, van Heyningen P, Berman HM, Thornton JM (1999) Protein-DNA interactions: a structural analysis. J Mol Biol 287:877–896

Jones S, Daley DTA, Luscombe NM, Berman HM, Thornton JM (2001) Protein-RNA interactions: a structural analysis. Nucleic Acids Res 29:943–954

Kim OTP, Yura K, Go N (2006) Amino acid residue doublet propensity in the protein-RNA interface and its application to RNA interface prediction. Nucleic Acids Res 34:6450–6460

Kozakov D, Brenke R, Comeau SR, Vajda S (2006) PIPER: an FFT-based protein docking program with pairwise potentials. Proteins 65:392–406

Kumar M, Gromiha AM, Raghava GPS (2008) Prediction of RNA binding sites in a protein using SVM and PSSM profile. Proteins 71:189–194

Li CH, Cao LB, Su JG, Yang YX, Wang CX (2012) A new residue-nucleotide propensity potential with structural information considered for discriminating protein-RNA docking decoys. Proteins 80:14–24

Perez-Cano L, Fernandez-Recio J (2010) Optimal protein-RNA area, OPRA: a propensity-based method to identify RNA-binding sites on proteins. Proteins 78:25–35

Perez-Cano L, Solernou A, Pons C, Fernandez-Recio J (2010) Structural prediction of protein-RNA interaction by computational docking with propensity-based statistical potentials. Pac Symp Biocomput 2010:293–301

Perez-Cano L, Jimenez-Garcia B, Fernandez-Recio J (2012) A protein-RNA docking benchmark (II): extended set from experimental and homology modeling data. Proteins 80:1872–1882

Terribilini M, Lee JH, Yan CH, Jernigan RL, Honavar V, Dobbs D (2006) Prediction of RNA binding sites in proteins from amino acid sequence. RNA 12:1450–1462

Tuszynska I, Bujnicki JM (2011) DARS-RNP and QUASI-RNP: new statistical potentials for protein-RNA docking. BMC Bioinform 12:348

Vakser IA, Aflalo C (1994) Hydrophobic docking: a proposed enhancement to molecular recognition techniques. Proteins 20:320–329

Skeletal intramyocellular lipid metabolism and insulin resistance

Yiran Li[1,2], Shimeng Xu[2,3], Xuelin Zhang[4 ✉], Zongchun Yi[1 ✉], Simon Cichello[5 ✉]

[1] Department of Biological Science and Biotechnology, School of Biological Science and Medical Engineering, Beihang University, Beijing 100191, China
[2] National Laboratory of Biomacromolecules, Institute of Biophysics, Chinese Academy of Sciences, Beijing 100101, China
[3] University of Chinese Academy of Sciences, Beijing 100049, China
[4] Capital University of Physical Education and Sport, Beijing 100191, China
[5] School of Life Sciences, La Trobe University, Melbourne, VIC 3086, Australia

Abstract Lipids stored in skeletal muscle cells are known as intramyocellular lipid (IMCL). Disorders involving IMCL and its causative factor, circulatory free fatty acids (FFAs), induce a toxic state and ultimately result in insulin resistance (IR) in muscle tissue. On the other hand, intramuscular triglyceride (IMTG), the most abundant component of IMCL and an essential energy source for active skeletal muscle, is different from other IMCLs, as it is stored in lipid droplets and plays a pivotal role in skeletal muscle energy homeostasis. This review discusses the association of FFA-induced ectopic lipid accumulation and IR, with specific emphasis on the relationship between IMCL/IMTG metabolism and IR.

Keywords Skeletal muscle, Intramyocellular lipid, Insulin resistance, Free fatty acid

INTRODUCTION

Insulin resistance (IR) is defined as the inability of target tissues to increase glucose uptake in response to insulin, which eventually leads to type II diabetes mellitus (T2DM) (Eckel et al. 2005; Samuel and Shulman 2012). IR occurs in virtually all patients with T2DM, most of whom are obese and have fat maldistribution. In addition to the association between T2DM and generalized obesity, many studies have also revealed associations between IR and body fat distribution, particularly the fat distribution in skeletal muscle, as this tissue is responsible for the majority of whole-body insulin-stimulated glucose disposal. Skeletal muscle insulin resistance is central to the pathogenesis of T2DM (Björnholm and Zierath 2005). Human skeletal muscle is a heterogeneous organ consisting of two phenotypically distinct kinds of muscle fibers. The histochemical staining for pH-sensitive myosin ATPase activity reveals two major classes of fiber types, namely type I and type II fibers. Type I (slow twitch) muscle fibers tend to be oxidative, whereas type II (fast twitch) fibers are glycolytic (Raben et al. 1998; Pette et al. 1999). Both type I and type II muscle fibers are insulin sensitive (James et al. 1985; Kern et al. 1990).

All types of lipids within myocytes are referred to as intramyocellular lipids (IMCLs), which are composed chiefly of triacylglycerol (TAG) but also include diacylglycerol (DAG), sphingolipid, and phospholipid. The accumulation of IMCL is essential for metabolism and physical exercise. Recently, the excess accumulation of IMCL has been linked to regional fat distribution, gaining considerable attention because of its association with IR. Paradoxically, both trained athletes and T2DM

✉ Correspondence: zhangxuelin@cupes.edu.cn (X. Zhang), yizc@buaa.edu.cn (Z. Yi), S.Cichello@latrobe.edu.au (S. Cichello)

patients possess higher IMCL than normal healthy individuals. However, only athletes possess a high oxidative capacity in muscle and thus enhanced insulin sensitivity. This phenomenon is called the "athlete's paradox". One type of IMCL that is stored mainly in lipid droplets (LDs), namely intramuscular triglyceride (IMTG), plays an important role in maintaining lipid homeostasis, including lipid metabolism, membrane trafficking, cell signaling, hormone production, and other molecular events. The disorder of IMCL and its derivatives leads to many metabolic diseases.

GLUCOSE-FATTY ACID CYCLE (RANDLE CYCLE)

In the early 20th century, it was recognized that both fat and carbohydrate are used as fuel during physical exercise (Krogh and Lindhard 1920). In 1926, Himwich and Rose examined muscular fuel utilization by measuring arteriovenous differences in oxygen and carbon dioxide across skeletal muscle in dogs during rest and exercise, and in fed and fasted states. The results showed that the respiratory quotient of the exercising muscle was not unity, which indicated that not only carbohydrates but also non-carbohydrates were used in muscular exercise (Himwich and Rose 1926). Later, Fritz et al. observed that fatty acid was oxidized in skeletal muscle during both rest and activity. This provided evidence that muscle oxidizes lipid to support muscle contraction (Fritz et al. 1958). Of particular note is the 1963 *Lancet* publication by Randle et al., which proposed a "glucose-fatty acid cycle", also known as the Randle cycle (Randle et al. 1963). The Randle cycle described the fuel flux between tissues as well as fuel selection by tissues. The original biochemical mechanism proposed that glucose oxidation was inhibited by fatty acids. Subsequently, lipid metabolism and glucose metabolism were linked, and researches accumulated in this field. Soon thereafter, researchers observed that IMTG could be used as fuel during exercise, and IMTG accumulation was found to be associated with IR in various studies (Watt 2009).

The Randle cycle has been contested as ignorant, because it postulates an exact correlation of metabolic fuel with competition between glucose and fatty acid during their oxidation by muscle and adipose tissue (Randle et al. 1963). Because of the prevalence of obesity and T2DM, researchers have paid increasing attention to this field. Recently, the understanding of the relationship between lipid metabolism (e.g., IMCL) and glucose metabolism (e.g., especially its related disorder, IR) has been intensely developed.

THE EFFECTS OF DIFFERENT TYPE OF LIPIDS ON INSULIN RESISTANCE

Triacylglycerol

Intramuscular lipids are stored predominantly as IMTG within LDs. The presence of IMTG was first described by Denton and Randle in 1967 (Denton and Randle 1967) and corroborated by Van Loon in 2004 (Van Loon 2004). The study by Van Loon used stable isotope methodology, ^{1}H-magnetic resonance spectroscopy, and electron and/or immunofluorescence microscopy to confirm that IMTG functions as an important substrate source during exercise. This study also found that up to 60%–70% of IMTG can be depleted in type I muscle fibers during prolonged moderate intensity exercise in trained individuals; this oxidation accounts for up to 50% of total lipid oxidized as a fuel source in the exercising muscle (Van Loon 2004). The application of these analytical techniques has facilitated the examination of IMTG and made it possible to study the function of IMTG as a metabolic fuel and its relationship with IR. Using ^{1}H-magnetic resonance spectroscopy, researchers demonstrated that IMTG could be used as a fuel source by exercising muscle (White et al. 2003) and depleted in both acute and long-term forms of exercise. Again, with the concept of the athlete's paradox, the concentration of IMTG is adaptively increased in endurance-trained individuals and in response to exercise training interventions, which paradoxically does not adversely affect insulin sensitivity and oxidative capacity (Goodpaster et al. 2001; Russell et al. 2003). In contrast, higher IMTG content is also observed in obese and T2DM individuals, or others whose insulin-sensing capability is impaired. Human studies have proposed the hypothesis that IMTG accumulation is associated with IR (Pan et al. 1995). This hypothesis has been refined over the last 15 years with more experimental and statistical results supporting the theory that accumulation of IMTG has important contributions to the development of skeletal muscle IR (Jacob et al. 1998; Bachmann et al. 2001; Goodpaster et al. 2001; Jimenez-Caballero et al. 2008; Anastasiou et al. 2009). For example, in a study of 19 non-diabetic obese and 11 diabetic obese individuals, Anasious and colleagues observed higher levels of IMTG in the diabetic obese group compared with the non-diabetic obese group, and that DAG levels were not significantly different between the study groups (Anastasiou et al. 2009). Another study from Schenk and colleagues observed that the prevention of fatty acid-induced IR following acute exercise was accompanied by enhanced skeletal muscle protein expression of key lipogenic enzymes, and further increased the rate of muscle TAG synthesis in humans (Schenk and Horowitz 2007). Furthermore, TAG-induced IR may be muscle type dependent. As previously mentioned, most IMTG is contained and used within type I muscle fibers, as IMTG content is threefold higher in type I oxidative fibers than type II glycolytic fibers (Shaw et al. 2012). This suggests that type I muscle fibers are more important in lipid toxicity (Coen and Goodpaster 2012) and abnormal lipid metabolism-induced IR. Moreover, human skeletal muscle IR is related to excess IMTG content in type I but not type II myocytes, greater ceramide content, and alterations in gene expression associated with lipid metabolism (Coen et al. 2010). Interestingly, the overexpression of acyl-CoA:diacylglycerol acyltransferase 2 (DGAT2) in type II glycolytic muscle of mice increases TAG, ceramide, and unsaturated long chain fatty acyl-CoA (LCFA-CoA) in skeletal muscle content of young adult mice, which is accompanied by impaired insulin signaling and insulin-mediated glucose uptake in glycolytic muscle and further impaired whole-body glucose and insulin tolerance (Levin et al. 2007). Moreover, diacylglycerol acyltransferase 1 (DGAT1)-deficient mice are resistant to diet-induced obesity through a mechanism involving increased energy expenditure. Chen and colleagues showed that these DGAT1-deficient mice have decreased levels of skeletal muscle TAG after induction of high-fat-diet-induced obesity, in addition to increased sensitivity to insulin and leptin (Chen et al. 2002). Their findings also demonstrated that DGAT1 deficiency in mice enhances insulin signaling in the skeletal muscle and white adipose tissue (WAT), in part through altered expression of adipocyte-derived factors that modulate insulin signaling in peripheral tissues (Chen et al. 2004). The results of these animal studies suggest that IMTG accumulation may be a causative factor of IR, a fact that is now widely accepted (Saltiel 2000).

Despite the evidence pointing to the role of IMTG in IR, it is difficult to propose that IMTG alone causes IR within the skeletal muscle tissue for several reasons. Elevation of IMTG content in T2DM is usually accompanied by higher concentrations of lipotoxic intermediates such as DAG and ceramide. Both of these metabolites inhibit insulin signaling and interfere with insulin-stimulated glucose metabolism (Samuel and Shulman 2012), and thus it is difficult to propose that IR is induced mainly by IMTG or other lipids. Furthermore, the existence of the athlete's paradox causes skepticism and prevents total acceptance of the hypothesis that IMTG causes IR within skeletal muscle. Finally, overexpression of DGAT1 in mouse skeletal muscle rescues high-fat-diet-induced IR, accompanied by high TAG levels in skeletal muscle (Liu et al. 2007) (Fig. 1).

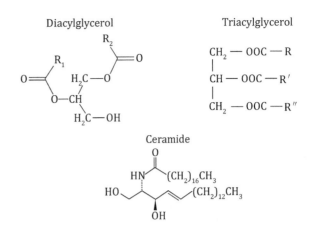

Fig. 1 Chemical structures of diacylglycerol, triacylglycerol, and ceramide. R1 and R2 in diacylglycerol, and R, R′, and R″ in triacylglycerol represent an alkyl or an alkenyl hydrocarbon chain of a fatty acid that is esterified on the glycerol, respectively

Although many studies have shown the tight connection between IMTG and IR, the mechanism linking IMTG and IR needs to be further investigated. It seems that the types of muscle fibers and lipid metabolites, such as DAG and ceramide, play important roles in the relationship between accumulation of IMTG and IR.

Diacylglycerol

It is widely accepted that sn-1,2-Diacylglycerol (DAG) derived from phospholipids by phospholipase C is an important lipotoxic mediator in IR development, although the specific mechanism remains elusive. Several clinical studies demonstrated that compared with lean controls, the intramyocellular DAG content of the vastus lateralis muscle is elevated in obese and T2DM patients (Moro et al. 2009; Bergman et al. 2012) and is also increased following acute IR induced by lipid infusion (Itani et al. 2002). Increasing DAG accumulation in skeletal muscle by altering expression of adipose triglyceride lipase (ATGL) and hormone-sensitive lipase (HSL) leads to the disruption of insulin signaling, and contributes to IR in humans (Badin et al. 2012). The relationship between DAG accumulation and IR is further confirmed in a study showing that reduced activity of diacylglycerol kinase-δ leads to high intramuscular DAG content in individuals with poorly controlled T2D (Chibalin et al. 2008). Moreover, after several weeks of endurance training and weight loss, skeletal muscle DAG content is decreased, accompanied by a parallel improvement in insulin sensitivity (Dube et al. 2011). In conclusion, these results support the theory that DAG accumulation is associated with IR.

Mechanistically, DAG is a second messenger that activates members of the protein kinase C (PKC) family.

The PKC family is divided into three isoforms: classical PKC (α, βI, βII, γ), novel PKC (δ, ε, η, θ), and atypical PKC (ζ, λ). Once activated, PKCs phosphorylate serine residues on insulin receptor substrate 1 (IRS-1), inhibiting the kinase activity and subsequently reducing activation of PI3-kinase and PKB/Akt (Timmers et al. 2008). As a result, insulin-stimulated GLUT4 translocation to the plasma membrane is impaired, therefore IR occurs. Furthermore, it is commonly accepted that in individuals with obesity and T2DM, the elevated DAG content might also increase the activity of PKCs. Itani and colleagues observed that lipid-induced IR in human muscle infused with lipids and insulin over a 6-h time period is associated with changes in DAG, PKC, and in IkB, the downstream signaling molecule of NF-κB (Itani et al. 2002). Furthermore, some researchers hypothesize that DAG mediates IR mainly through novel PKCs (Erion and Shulman 2010). PKCθ is a crucial component in skeletal muscle, and is also the most abundantly expressed PKC in this tissue. Knockout of PKCθ in skeletal muscle prevents fat-induced defects in insulin signaling and glucose transport (Kim et al. 2004). In addition, both transgenic mice with muscle-specific expression of dominant negative *PKCθ* and *PKCθ*-knockout mice exhibit age-associated or diet-associated obesity and whole-body IR (Serra et al. 2003; Gao et al. 2007). When combined, these results support the hypothesis that the accumulation of DAG in skeletal muscle leads to the activation of novel PKCs and ultimately results in IR.

On the other hand, others have shown dissociation between DAG accumulation and IR. Some studies have reported that DAG content in skeletal muscle is not elevated during obesity (Anastasiou et al. 2009), with IR (Hees et al. 2001), or in obese IR (Coen et al. 2010) compared with insulin-sensitive obese subjects. In addition, in highly trained athletes, total myocellular DAG is markedly higher, corresponding with higher insulin sensitivity (Amati et al. 2010). For animal models, overexpression of the DGAT1 enzyme in muscle results in DAG accumulation and release of IR that was induced by a high-fat diet (Timmers et al. 2010).

Researchers are still struggling to elucidate the relationship between DAG and IR in human muscle and answer the question of why DAG accumulation leads to IR. There are several possible explanations, stemming from different viewpoints, as to why DAG accumulation leads to IR. One possible explanation is the degree of FA saturation in DAG. It has been demonstrated that athletes have a lower degree of DAG saturation compared with sedentary controls (Bergman et al. 2012). Moreover, it has been observed that a higher degree of DAG saturation is associated with IR in men with metabolic syndrome (Hees et al. 2001). Conversely, others have not shown such associations (Coen et al. 2010) or even an inverse relationship (Amati et al. 2010). The subcellular location of DAG accumulation might be another possible factor that could affect the relationship of DAG and IR. DAG is present in the sarcolemma membrane, sarcoplasmic reticulum, LDs, and mitochondrial membrane. The majority of human studies only examine whole-muscle DAG concentration, which could certainly obscure the relationship between subcellular DAG concentration and IR. Moreover, it has been recently shown that membrane DAG is associated with PKC activation and insulin sensitivity in obese T2D subjects and lean athletes (Bergman et al. 2012). Finally, and perhaps most importantly, there are two distinct DAG stereoisomers, 1,3-DAG and 1,2-DAG, which may influence muscle IR to different degrees. Only 1,2-DAG has been associated with insulin signaling (Turinsky et al. 1990), and only the 1,2-DAG stereoisomer can activate PKC; 1,3-DAG lacks this ability (Boni and Rando 1985). It has been suggested that neither ATGL nor HSL has the ability to generate the 1,2-DAG stereoisomer (Zechner et al. 2012); however, this hypothesis presently lacks supporting evidence.

Ceramide

Ceramide, which belongs to the sphingolipid family, plays a role as an inert structural component of biological membranes. It also acts as an intracellular messenger in various biological mechanisms and as a lipid intermediate widely believed to be the true lipotoxic culprit behind the reported associations between IMTG and IR. In skeletal muscle, ceramide accumulation is associated with a number of cellular stresses, such as reactive oxygen species (ROS) accumulation, inflammation, hypoxia, and as part of a highly conserved stress response. All of these stresses have been identified as key mediators of IR via inhibition of the serine/threonine-specific protein kinase Akt/protein kinase B (PKB) (Chavez et al. 2003; Holland et al. 2012), and as important pathways linking insulin signaling to the translocation of GLUT4 to the sarcolemma, potentially via protein phosphatase 2 (PP2)- and protein kinase C zeta (PKCz)-dependent pathways (Stratford et al. 2004). Ceramide is also linked to mitochondrial dysfunction (Smith et al. 2013), which in turn is implicated in IMCL accumulation and IR (Coen and Goodpaster 2012).

Plasma ceramide targets skeletal muscle in T2DM (Kirwan 2013), and ceramide content is increased in skeletal muscle in obese and insulin-resistant humans. By using euglycemic-hyperinsulinemic clamps with muscle biopsies, it has been observed that muscle ceramide content is significantly correlated with the plasma FFA concentration in lean insulin-sensitive and obese insulin-resistant subjects (Adams et al. 2004). Furthermore, as previously mentioned, human skeletal muscle IR is related to greater IMTG content in type I but not type II myocytes, and it is also related to greater ceramide content, especially in type I myocytes. However, the concentration of DAG is similar in both type I and type II myocytes (Coen et al. 2010).

Similar results have been observed using the in vitro cultivation of human primary myoblast cells. By treating human vastus lateralis muscle with different kinds of FFA, Laura and colleagues observed that the application of palmitate produces more DAG and ceramide in myoblasts in addition to the induction of IR. Furthermore, oleate treatment resulted in an increase in TAG in normal insulin-sensitive muscles. These myoblasts developed IR when treated with cell-permeable analogs of ceramide, and showed normal insulin sensitivity with co-treatment of palmitate and inhibitors of de novo ceramide synthesis (Pickersgill et al. 2007). Coincidentally, inhibition of de novo ceramide synthesis reversed diet-induced IR and enhanced whole-body oxygen consumption (Ussher et al. 2010).

Several hypotheses have been proposed to explain the mechanism by which an increase in ceramide leads to IR. Increased mitochondrial oxidative stress and mitochondrial dysfunction are accepted as important causative factors for IR and T2DM (Kelley et al. 2002; Schrauwen and Hesselink 2004; Lowell and Shulman 2005; Fridlyand and Philipson 2006; Schenk and Horowitz 2007; Fleischman et al. 2009; Hernandez-Alvarez et al. 2010; Meex et al. 2010; Schrauwen et al. 2010; Chow et al. 2012;). The results of a study by Larysa et al. support this idea, as they demonstrated that de novo synthesis of ceramide is involved in palmitate-induced mtROS generation, mitochondrial dysfunction, and insulin signaling (Yuzefovych et al. 2010).

The effects of exercise and training on ceramide metabolism in human skeletal muscle have been previously studied (Helge et al. 2004). Thrush et al. observed an interesting phenomenon; although the inhibition of ceramide accumulation can prevent the detrimental effects of palmitate incubation, a single prior bout of exercise appears to protect the muscle against palmitate-induced IR, which may be independent of the variable ceramide concentration (Thrush et al. 2010). Further research has been performed by Skovbro et al. showing that human skeletal muscle ceramide content is not a major factor in muscle insulin sensitivity (Skovbro et al. 2008). In conclusion, the exact mechanism by which ceramide induces IR is still unclear, but it appears that this relationship might be influenced by skeletal muscle lipid composition.

EFFECTS OF LIPID COMPOSITION

As mentioned previously, DAG stereoisomers have different effects on skeletal muscle IR. The stereo structures, degree of fatty acid saturation, and the length of the fatty acid chains are all contributing factors to these effects. Oleate (18:1) and palmitate (16:0) are widely used in studies of DAGs and IR, since they are both prevalent plasma FFAs. Many studies suggest that saturated palmitate, but not monounsaturated oleate, induces inter-myocellular IR. In C2C12 myotubes, palmitate, but not oleate, inhibits insulin-stimulated glycogen synthesis, as well as the activation of Akt/PKB, an obligate intermediate in the regulation of anabolic metabolism. Palmitate also induces the accrual of ceramide and DAG, which have been confirmed to inhibit insulin signaling in cultured cells and to accumulate in IR tissues (Chavez and Summers 2003). Moreover, oleate protects rat skeletal muscle cell lines against palmitate-induced IR (Coll et al. 2008; Gao et al. 2009) and blocks palmitate-induced abnormal lipid distribution, endoplasmic reticulum expansion, and stress (Dimopoulos et al. 2006; Peng et al. 2012). These results have also been observed in other cell models (Listenberger et al. 2003). Furthermore, excluding monounsaturated fatty acid, the treatment of these cells with linoleate (C18:2) or n-6 polyunsaturated fatty acid does not alter DAG levels, ceramide levels, or glucose uptake, but increases myotube TAG levels compared with controls that lack the addition of fatty acids (Lee et al. 2006). These studies suggest that saturated fatty acid-induced IR occurs by a mechanism distinct from that of unsaturated fatty acids, and thus is related to the degree of saturation. Furthermore, this mechanism involves elevation of ceramide, which leads to PKB inhibition without affecting IRS-1 function (Schmitz-Peiffer et al. 1999). Unsaturated fatty acid serves as a protective functional factor through the promotion of TAG accumulation, and thereby decreases DAG and ceramide content (Listenberger et al. 2003). There are also other proposed mechanisms, including one involving mitochondria. It has been proposed that mtROS generation is the initial event in the induction of mitochondrial dysfunction and consequently apoptosis, and the inhibition of insulin signaling. The palmitate-induced mitochondrial dysfunction is ameliorated by oleate, which contributes to the prevention of palmitate-induced IR (Yuzefovych et al. 2010). Thus, it appears that the fatty acid concentration is not the only determinant in the induction of IR.

An in vivo study also suggested that dietary fat composition, rather than fatty acid over-supply, is the major determinant of fat-induced IR (Storlien et al. 1991, 2002). Human dietary fat consists of a range of fats including saturated, monounsaturated, polyunsaturated, and trans-unsaturated fatty acids. Previous studies found that high dietary intake of the monounsaturated fatty acid oleic acid, which is abundant in olive oil, is associated with improved insulin sensitivity in the general population, whereas saturated fatty acids (i.e., palmitate) show the opposite effect (Marshall et al. 1997; Soriguer et al. 2004). Furthermore, most animal and cell studies indicate that saturated fatty acids significantly increase IR, whereas n-3 polyunsaturated fatty acids prevent it (Storlien et al. 2002). Animals fed a diet high in n-6 polyunsaturated fat retained insulin sensitivity despite small increases in muscle DAG, which appears to be in a range somewhere between the effects of saturated and n-3 fatty acids (Storlien et al. 1991).

The most popular hypothesis of the complicated phenomenon of IR is that unsaturated FFA helps with the incorporation of saturated FFA into IMTG, accompanied by much less harmful lipid derivatives (e.g., DAG and ceramide) and therefore, muscle cells are protected from IR. Animal feeding or perfusion studies, as well as cell culture studies, have linked saturated fatty acid intake with elevated concentrations of specific lipid messengers in muscle (Lee et al. 2006; Coll et al. 2008; Peng et al. 2012; Sawada et al. 2012; Salvado et al. 2013). Another hypothesis is that the cell membrane fatty acid composition in turn affects cell membrane fluidity and rigidity. In humans and other species, the body is particularly efficient at regulating the components of cell membranes, such as the sarcolemma, which can be influenced by plasma FFA (Storlien et al. 1998). Since the efficiency of signal transduction is highly dependent on the orientation and position of various proteins within the membrane, the fatty acid composition of cellular membranes may play a pivotal role in adequate insulin sensitivity (Corcoran et al. 2007). Furthermore, there are also studies in humans suggesting that the fatty acid composition of phospholipids in the sarcolemma modulates insulin sensitivity (Storlien et al. 1991; Borkman et al. 1993). Animal studies seem to directly demonstrate that saturated FFA-containing membranes promote IR, whereas a high degree of unsaturation in the FFAs in the membranes protects against IR; this has also been noted in humans (Storlien et al. 1996).

CONCLUSION

Lipid and glucose metabolism are both important for human health, and skeletal muscle is the major organ responsible for balancing the use of lipid and glucose as fuel during complete rest and low-intensity activity. Research has considerably advanced our understanding of intramuscular lipid metabolism and its significance in human health. Although we have been aware of and gained considerable insight into the role of IMCL in metabolism and the development of IR, there are still complex questions to be answered. For now, it seems that maintaining the dynamic lipid balance may be one of the most important functions for human health. In the future, a better understanding of IMCL and IR will serve as a means to cure metabolic diseases and promote human health. Advanced lipidomics, combined with more specific and efficacious small molecule inhibitors, will accelerate research progress toward understanding how specific lipid metabolites influence metabolic homeostasis and finding applicable therapies for IR.

Acknowledgments This work was supported by the National Natural Science Foundation of China (31100854), the Importation and Development of High-Caliber Talents Project of Beijing Municipal Institutions (CIT&TCD201504086).

Conflicts of interest Yiran Li, Shimeng Xu, Xuelin Zhang, Zongchun Yi and Simon Cichello declare that they have no conflict of interest.

References

Adams JM, Pratipanawatr T, Berria R, Wang E, DeFronzo RA, Sullards MC, Mandarino LJ (2004) Ceramide content is increased in skeletal muscle from obese insulin-resistant humans. Diabetes 53(1):25–31

Amati F, Dube JJ, Alvarez-Carnero E, Edreira MM, Chomentowski P, Coen PM, Switzer GE, Bickel PE, Stefanovic-Racic M, Toledo FG, Goodpaster BH (2010) Skeletal muscle triglycerides, diacylglycerols, and ceramides in insulin resistance: another paradox in endurance-trained athletes. Diabetes 60(10):2588–2597

Anastasiou CA, Kavouras SA, Lentzas Y, Gova A, Sidossis LS, Melidonis A (2009a) Diabetes mellitus is associated with increased intramyocellular triglyceride, but not diglyceride, content in obese humans. Metabolism 58(11):1636–1642

Anastasiou CA, Kavouras SA, Lentzas Y, Gova A, Sidossis LS, Melidonis A (2009b) Diabetes mellitus is associated with increased intramyocellular triglyceride, but not diglyceride, content in obese humans. Metabolism 58(11):1636-1642

Bachmann OP, Dahl DB, Brechtel K, Machann J, Haap M, Maier T, Loviscach M, Stumvoll M, Claussen CD, Schick F, Haring HU, Jacob S (2001) Effects of intravenous and dietary lipid challenge on intramyocellular lipid content and the relation with insulin sensitivity in humans. Diabetes 50(11):2579-2584

Badin PM, Louche K, Mairal A, Liebisch G, Schmitz G, Rustan AC, Smith SR, Langin D, Moro C (2012) Altered skeletal muscle lipase expression and activity contribute to insulin resistance in humans. Diabetes 60(6):1734-1742

Bergman BC, Hunerdosse DM, Kerege A, Playdon MC, Perreault L (2012) Localisation and composition of skeletal muscle diacylglycerol predicts insulin resistance in humans. Diabetologia 55(4):1140-1150

Björnholm M, Zierath JR (2005) Insulin signal transduction in human skeletal muscle: identifying the defects in Type II diabetes. Biochem Soc Trans 33(2):354-357

Boni LT, Rando RR (1985) The nature of protein kinase C activation by physically defined phospholipid vesicles and diacylglycerols. J Biol Chem 260(19):10819-10825

Borkman M, Storlien LH, Pan DA, Jenkins AB, Chisholm DJ, Campbell LV (1993) The relation between insulin sensitivity and the fatty-acid composition of skeletal-muscle phospholipids. New Engl J Med 328(4):238-244

Chavez JA, Summers SA (2003) Characterizing the effects of saturated fatty acids on insulin signaling and ceramide and diacylglycerol accumulation in 3T3-L1 adipocytes and C2C12 myotubes. Arch Biochem Biophys 419(2):101-109

Chavez FP, Ryan J, Lluch-Cota SE, Ñiquen M (2003) From anchovies to sardines and back: multidecadal change in the Pacific Ocean. Science 299(5604):217-221

Chen HC, Smith SJ, Ladha Z, Jensen DR, Ferreira LD, Pulawa LK, McGuire JG, Pitas RE, Eckel RH, Farese RV (2002) Increased insulin and leptin sensitivity in mice lacking acyl CoA:diacylglycerol acyltransferase 1. J Clin Investig 109(8):1049-1055

Chen HC, Rao M, Sajan MP, Standaert M, Kanoh Y, Miura A, Farese RV, Farese RV (2004) Role of adipocyte-derived factors in enhancing insulin signaling in skeletal muscle and white adipose tissue of mice lacking Acyl CoA: diacylglycerol acyltransferase 1. Diabetes 53(6):1445-1451

Chibalin AV, Leng Y, Vieira E, Krook A, Bjornholm M, Long YC, Kotova O, Zhong Z, Sakane F, Steiler T, Nylen C, Wang J, Laakso M, Topham MK, Gilbert M, Wallberg-Henriksson H, Zierath JR (2008) Downregulation of diacylglycerol kinase delta contributes to hyperglycemia-induced insulin resistance. Cell 132(3):375-386

Chow L, From A, Seaquist E (2012) Skeletal muscle insulin resistance: the interplay of local lipid excess and mitochondrial dysfunction. Meta Clin Exp 59(1):70-85

Coen PM, Goodpaster BH (2012) Role of intramyocellular lipids in human health. Trends Endocrinol Metab 23(8):391-398

Coen PM, Dube JJ, Amati F, Stefanovic-Racic M, Ferrell RE, Toledo FG, Goodpaster BH (2010a) Insulin resistance is associated with higher intramyocellular triglycerides in type I but not type II myocytes concomitant with higher ceramide content. Diabetes 59(1):80-88

Coen PM, Dubé JJ, Amati F, Stefanovic-Racic M, Ferrell RE, Toledo FG, Goodpaster BH (2010b) Insulin resistance is associated with higher intramyocellular triglycerides in type I but not type II myocytes concomitant with higher ceramide content. Diabetes 59(1):80-88

Coll T, Eyre E, Rodriguez-Calvo R, Palomer X, Sanchez RM, Merlos M, Laguna JC, Vazquez-Carrera M (2008a) Oleate reverses palmitate-induced insulin resistance and inflammation in skeletal muscle cells. J Biol Chem 283(17):11107-11116

Coll T, Eyre E, Rodriguez-Calvo R, Palomer X, Sanchez RM, Merlos M, Laguna JC, Vazquez-Carrera M (2008b) Oleate reverses palmitate-induced insulin resistance and inflammation in skeletal muscle cells. J Biol Chem 283(17):11107-11116

Corcoran MP, Lamon-Fava S, Fielding RA (2007) Skeletal muscle lipid deposition and insulin resistance: effect of dietary fatty acids and exercise. Am J Clin Nutr 85(3):662-677

Denton RM, Randle PJ (1967) Concentrations of glycerides and phospholipids in rat heart and gastrocnemius muscles—effects of alloxan-diabetes and perfusion. Biochem J 104(2):416-419

Dimopoulos N, Watson M, Sakamoto K, Hundal H (2006) Differential effects of palmitate and palmitoleate on insulin action and glucose utilization in rat L6 skeletal muscle cells. Biochem J 399:473-481

Dube JJ, Amati F, Toledo FG, Stefanovic-Racic M, Rossi A, Coen P, Goodpaster BH (2011) Effects of weight loss and exercise on insulin resistance, and intramyocellular triacylglycerol, diacylglycerol and ceramide. Diabetologia 54(5):1147-1156

Eckel RH, Grundy SM, Zimmet PZ (2005) The metabolic syndrome. Lancet 365(9468):1415-1428

Erion DM, Shulman GI (2010) Diacylglycerol-mediated insulin resistance. Nat Med 16(4):400-402

Fleischman A, Kron M, Systrom DM, Hrovat M, Grinspoon SK (2009) Mitochondrial function and insulin resistance in overweight and normal-weight children. J Clin Endocrinol Metab 94(12):4923-4930

Fridlyand LE, Philipson LH (2006) Reactive species and early manifestation of insulin resistance in type 2 diabetes. Diabetes Obes Metab 8(2):136-145

Fritz IB, Davis DG, Holtrop RH, Dundee H (1958) Fatty acid oxidation by skeletal muscle during rest and activity. Am J Phys Leg Content 194(2):379-386

Gao Z, Wang Z, Zhang X, Butler AA, Zuberi A, Gawronska-Kozak B, Lefevre M, York D, Ravussin E, Berthoud HR, McGuinness O, Cefalu WT, Ye J (2007) Inactivation of PKC theta leads to increased susceptibility to obesity and dietary insulin resistance in mice. Am J Phys Endocrinol Metab 292(1):E84-E91

Gao D, Griffiths HR, Bailey CJ (2009) Oleate protects against palmitate-induced insulin resistance in L6 myotubes. Br J Nutr 102(11):1557-1563

Goodpaster BH, He J, Watkins S, Kelley DE (2001) Skeletal muscle lipid content and insulin resistance: evidence for a paradox in endurance-trained athletes. J Clin Endocrinol Metab 86(12):5755-5761

Hees AM, Jans A, Hul GB, Roche HM, Saris WH, Blaak EE (2001) Skeletal muscle fatty acid handling in insulin resistant men. Obesity 19(7):1350-1359

Helge JW, Dobrzyn A, Saltin B, Gorski J (2004) Exercise and training effects on ceramide metabolism in human skeletal muscle. Exp Phys 89(1):119-127

Hernandez-Alvarez MI, Thabit H, Burns N, Shah S, Brema I, Hatunic M, Finucane F, Liesa M, Chiellini C, Naon D, Zorzano A, Nolan JJ (2010) Subjects with early-onset type 2 diabetes show defective activation of the skeletal muscle PGC-1α/Mitofusin-2 regulatory pathway in response to physical activity. Diabetes Care 33(3):645-651

Himwich HE, Rose MI (1926) The respiratory quotient of exercising muscle. Exp Biol Med 24(2):169-170

Holland WL, Bikman BT, Wang LP, Yuguang G, Sargent KM, Bulchand S, Knotts TA, Shui G, Clegg DJ, Wenk MR, Pagliassotti MJ, Scherer PE, Summers SA (2012) Lipid-induced insulin resistance mediated by the proinflammatory receptor TLR4 requires saturated fatty acid-induced ceramide biosynthesis in mice. J Clin Investig 121(5):1858-1870

Itani SI, Ruderman NB, Schmieder F, Boden G (2002) Lipid-induced insulin resistance in human muscle is associated with changes in diacylglycerol, protein kinase C, and Ikappa B-alpha. Diabetes 51(7):2005–2011

Jacob S, Machann J, Rett K, Brechtel K, Volk A, Renn W, Maerker E, Matthaei S, Schick F, Claussen CD (1998) Association of increased intramyocellular lipid content with insulin resistance in lean nondiabetic offspring of type 2 diabetic subjects. Diabetes 48(5):1113–1119

James DE, Jenkins AB, Kraegen EW (1985) Heterogeneity of insulin action in individual muscles in vivo: euglycemic clamp studies in rats. Am J Phys Endocrinol Metab 248(5):E567–E574

Jimenez-Caballero PE, Mollejo-Villanueva M, Alvarez-Tejerina A (2008) Mitochondrial encephalopathy due to complex I deficiency. Brain tissue biopsy findings and clinical course following pharmacological. Revi de neur 47(1):27–30

Kelley DE, He J, Menshikova EV, Ritov VB (2002) Dysfunction of mitochondria in human skeletal muscle in type 2 diabetes. Diabetes 51(10):2944–2950

Kern M, Wells JA, Stephens JM, Elton CW, Friedman JE, Tapscott EB, Pekala PH, Dohm GL (1990) Insulin responsiveness in skeletal muscle is determined by glucose transporter [Glut4] protein level. Biochem J 270:397–400

Kim JK, Fillmore JJ, Sunshine MJ, Albrecht B, Higashimori T, Kim DW, Liu ZX, Soos TJ, Cline GW, O'Brien WR, Littman DR, Shulman GI (2004) PKC-theta knockout mice are protected from fat-induced insulin resistance. J Clin Investig 114(6):823–827

Kirwan JP (2013) Plasma ceramides target skeletal muscle in type 2 diabetes. Diabetes 62(2):352–354

Krogh A, Lindhard J (1920) The relative value of fat and carbohydrate as sources of muscular energy: with appendices on the correlation between standard metabolism and the respiratory quotient during rest and work. Biochem J 14(3–4):290

Lee JS, Pinnamaneni SK, Eo SJ, Cho IH, Pyo JH, Kim CK, Sinclair AJ, Febbraio MA, Watt MJ (2006) Saturated, but not n-6 polyunsaturated, fatty acids induce insulin resistance: role of intramuscular accumulation of lipid metabolites. J Appl Physiol 100(5):1467–1474

Levin MC, Monetti M, Watt MJ, Sajan MP, Stevens RD, Bain JR, Newgard CB, Farese RV, Farese RV (2007) Increased lipid accumulation and insulin resistance in transgenic mice expressing DGAT2 in glycolytic [type II] muscle. Am J Phys Endocrinol Metab 293(6):E1772–E1781

Listenberger LL, Han X, Lewis SE, Cases S, Farese RV, Ory DS, Schaffer JE (2003) Triglyceride accumulation protects against fatty acid-induced lipotoxicity. Proc Natl Acad Sci USA 100(6):3077–3082

Liu L, Zhang Y, Chen N, Shi X, Tsang B, Yu YH (2007) Upregulation of myocellular DGAT1 augments triglyceride synthesis in skeletal muscle and protects against fat-induced insulin resistance. J Clin Investig 117(6):1679–1689

Lowell BB, Shulman GI (2005) Oxidative capacity, lipotoxicity, and mitochondrial damage in type 2 diabetes. Diabetes 307(10):384–387

Marshall J, Bessesen D, Hamman R (1997) High saturated fat and low starch and fibre are associated with hyperinsulinaemia in a non-diabetic population: the San Luis Valley Diabetes Study. Diabetologia 40(4):430–438

Meex RC, Schrauwen-Hinderling VB, Moonen-Kornips E, Schaart G, Mensink M, Phielix E, van de Weijer T, Sels JP, Schrauwen P, Hesselink MK (2010) Restoration of muscle mitochondrial function and metabolic flexibility in type 2 diabetes by exercise training is paralleled by increased myocellular fat storage and improved insulin sensitivity. Diabetes 59(3):572–579

Moro C, Galgani JE, Luu L, Pasarica M, Mairal A, Bajpeyi S, Schmitz G, Langin D, Liebisch G, Smith SR (2009) Influence of gender, obesity, and muscle lipase activity on intramyocellular lipids in sedentary individuals. J Clin Endocrinol Metab 94(9):3440–3447

Pan DA, Lillioja S, Milner MR, Kriketos AD, Baur LA, Bogardus C, Storlien LH (1995) Skeletal muscle membrane lipid composition is related to adiposity and insulin action. J Clin Investig 96(6):2802–2808

Peng G, Li L, Liu Y, Pu J, Zhang S, Yu J, Zhao J, Liu P (2012) Oleate blocks palmitate-induced abnormal lipid distribution, endoplasmic reticulum expansion and stress, and insulin resistance in skeletal muscle. Endocrinology 152(6):2206–2218

Pette D, Peuker H, Staron R (1999) The impact of biochemical methods for single muscle fibre analysis. Acta Phys Scand 166(4):261–277

Pickersgill L, Litherland GJ, Greenberg AS, Walker M, Yeaman SJ (2007) Key role for ceramides in mediating insulin resistance in human muscle cells. J Biol Chem 282(17):12583–12589

Raben A, Mygind E, Astrup A (1998) Lower activity of oxidative key enzymes and smaller fiber areas in skeletal muscle of postobese women. Am J Phys Endocrinol Metab 275(3):E487–E494

Randle P, Garland P, Hales C, Newsholme E (1963) The glucose fatty-acid cycle its role in insulin sensitivity and the metabolic disturbances of diabetes mellitus. Lancet 281(7285):785–789

Russell AP, Gastaldi G, Bobbioni-Harsch E, Arboit P, Gobelet C, Deriaz O, Golay A, Witztum JL, Giacobino JP (2003) Lipid peroxidation in skeletal muscle of obese as compared to endurance-trained humans: a case of good vs. bad lipids? FEBS 551(1):104–106

Saltiel AR (2000) Series introduction: the molecular and physiological basis of insulin resistance: emerging implications for metabolic and cardiovascular diseases. J Clin Investig 106(2):163–164

Salvado L, Coll T, Gomez-Foix AM, Salmeron E, Barroso E, Palomer X, Vazquez-Carrera M (2013) Oleate prevents saturated-fatty-acid-induced ER stress, inflammation and insulin resistance in skeletal muscle cells through an AMPK-dependent mechanism. Diabetologia 56(6):1372–1382

Samuel VT, Shulman GI (2012) Mechanisms for insulin resistance: common threads and missing links. Cell 148(5):852–871

Sawada K, Kawabata K, Yamashita T, Kawasaki K, Yamamoto N, Ashida H (2012) Ameliorative effects of polyunsaturated fatty acids against palmitic acid-induced insulin resistance in L6 skeletal muscle cells. Lipids Health Dis 11:36

Schenk S, Horowitz JF (2007) Acute exercise increases triglyceride synthesis in skeletal muscle and prevents fatty acid-induced insulin resistance. J Clin Investig 117(6):1690–1698

Schmitz-Peiffer C, Craig DL, Biden TJ (1999) Ceramide generation is sufficient to account for the inhibition of the insulin-stimulated PKB pathway in C2C12 skeletal muscle cells pretreated with palmitate. J Biol Chem 274(34):24202–24210

Schrauwen P, Hesselink MK (2004) Oxidative capacity, lipotoxicity, and mitochondrial damage in type 2 diabetes. Diabetes 53(6):1412–1417

Schrauwen P, Schrauwen-Hinderling V, Hoeks J, Hesselink ML (2010) Mitochondrial dysfunction and lipotoxicity. Biochem Biophys Acta 1801(3):266–271

Serra C, Federici M, Buongiorno A, Senni MI, Morelli S, Segratella E, Pascuccio M, Tiveron C, Mattei E, Tatangelo L, Lauro R, Molinaro M, Giaccari A, Bouche M (2003) Transgenic mice with dominant negative PKC-theta in skeletal muscle: a new model of insulin resistance and obesity. J Cell Phys 196(1):89–97

Shaw CS, Shepherd SO, Wagenmakers AJ, Hansen D, Dendale P, van Loon LJ (2012) Prolonged exercise training increases intramuscular lipid content and perilipin 2 expression in type I muscle fibers of patients with type 2 diabetes. Am J Phys Endocrinol Metab 303(9):E1158–E1165

Skovbro M, Baranowski M, Skov-Jensen C, Flint A, Dela F, Gorski J, Helge JW (2008) Human skeletal muscle ceramide content is not a major factor in muscle insulin sensitivity. Diabetologia 51(7):1253–1260

Smith ME, Tippetts TS, Brassfield ES, Tucker BJ, Ockey A, Swensen AC, Anthonymuthu TS, Washburn TS, Kane DA, Prince JT, and Bikman BT (2013) Mitochondrial fission mediates ceramide-induced metabolic disruption in skeletal muscle. Biochem J:135–140

Soriguer F, Esteva I, Rojo-Martinez G, de Adana MR, Dobarganes M, Garcia-Almeida J, Tinahones F, Beltran M, Gonzalez-Romero S, Olveira G (2004) Oleic acid from cooking oils is associated with lower insulin resistance in the general population [Pizarra study]. Eur J Endocrinol 150(1):33–39

Storlien LH, Jenkins AB, Chisholm DJ, Pascoe WS, Khouri S, Kraegen EW (1991) Influence of dietary fat composition on development of insulin resistance in rats: relationship to muscle triglyceride and ω-3 fatty acids in muscle phospholipid. Diabetes 40(2):280–289

Storlien L, Pan D, Kriketos A, O'Connor J, Caterson I, Cooney G, Jenkins A, Baur L (1996) Skeletal muscle membrane lipids and insulin resistance. Lipids 31(1):S261–S265

Storlien LH, Hulbert AJ, Else PL (1998) Polyunsaturated fatty acids, membrane function and metabolic diseases such as diabetes and obesity. Curr Opin Clin Nutr Metab Care 1(6):559–563

Storlien LH, Higgins J, Thomas T, Brown MA, Wang H, Huang XF, Else P (2002) Diet composition and insulin action in animal models. Br J Nutr 83(s1):S85–S90

Stratford S, Hoehn KL, Liu F, Summers SA (2004) Regulation of insulin action by ceramide dual mechanisms linking ceramide accumulation to the inhibition of Akt/protein kinase B. J Biol Chem 279(35):36608–36615

Thrush AB, Harasim E, Chabowski A, Gulli R, Stefanyk L, Dyck DJ (2010) A single prior bout of exercise protects against palmitate-induced insulin resistance despite an increase in total ceramide content. Am J Phys Regu Integr Comp Phys 300(5):R1200–R1208

Timmers S, Schrauwen P, de Vogel J (2008) Muscular diacylglycerol metabolism and insulin resistance. Physiol Behav 94(2):242–251

Timmers S, de Vogel-van den Bosch J, Hesselink MK, van Beurden D, Schaart G, Ferraz MJ, Losen M, Martinez-Martinez P, De Baets MH, Aerts JM (2010) Paradoxical increase in TAG and DAG content parallel the insulin sensitizing effect of unilateral DGAT1 overexpression in rat skeletal muscle. Plos one 6(1):14500–14503

Turinsky J, O'Sullivan DM, Bayly BP (1990) 1, 2-Diacylglycerol and ceramide levels in insulin-resistant tissues of the rat in vivo. J Biol Chem 265(28):16880–16885

Ussher JR, Koves TR, Cadete VJ, Zhang L, Jaswal JS, Swyrd SJ, Lopaschuk DG, Proctor SD, Keung W, Muoio DM, Lopaschuk GD (2010) Inhibition of de novo ceramide synthesis reverses diet-induced insulin resistance and enhances whole-body oxygen consumption. Diabetes 59(10):2453–2464

Van Loon LJ (2004) Use of intramuscular triacylglycerol as a substrate source during exercise in humans. J Appl Phys 97(4):1170–1187

Watt MJ (2009) Storing up trouble: does accumulation of intramyocellular triglyceride protect skeletal muscle from insulin resistance. Clin Exp Pharmcol Phys 36(1):5–11

White LJ, Ferguson MA, McCoy SC, Kim H (2003) Intramyocellular lipid changes in men and women during aerobic exercise: a [1]H-magnetic resonance spectroscopy study. J Clin Endocrinol Metab 88(12):5638–5643

Yuzefovych L, Wilson G, Rachek L (2010) Different effects of oleate vs. palmitate on mitochondrial function, apoptosis, and insulin signaling in L6 skeletal muscle cells: role of oxidative stress. Am J Phys Endocinol Metab 299(E):1096–1105

Zechner R, Zimmermann R, Eichmann TO, Kohlwein SD, Haemmerle G, Lass A, Madeo G (2012) Fat signals-lipases and lipolysis in lipid metabolism and signaling. Cell Metab 15(3):279–291

Protocol for analyzing protein ensemble structures from chemical cross-links using DynaXL

Zhou Gong[1,2], Zhu Liu[3], Xu Dong[1,2], Yue-He Ding[4], Meng-Qiu Dong[5], Chun Tang[1,2]

[1] CAS Key Laboratory of Magnetic Resonance in Biological Systems, State Key Laboratory of Magnetic Resonance and Atomic Molecular Physics, and National Center for Magnetic Resonance at Wuhan, Wuhan Institute of Physics and Mathematics of the Chinese Academy of Sciences, Wuhan 430071, China

[2] National Center for Magnetic Resonance at Wuhan, Wuhan Institute of Physics and Mathematics of the Chinese Academy of Sciences, Wuhan 430071, China

[3] Department of Pharmacology, Institute of Neuroscience, Key Laboratory of Medical Neurobiology of the Ministry of Health of China, Zhejiang University School of Medicine, Hangzhou 310057, China

[4] RNA Therapeutics Institute, University of Massachusetts Medical School, 368 Plantation Street, Worcester, MA 01605, USA

[5] National Institute of Biological Sciences, Beijing 102206, China

Abstract Chemical cross-linking coupled with mass spectroscopy (CXMS) is a powerful technique for investigating protein structures. CXMS has been mostly used to characterize the predominant structure for a protein, whereas cross-links incompatible with a unique structure of a protein or a protein complex are often discarded. We have recently shown that the so-called over-length cross-links actually contain protein dynamics information. We have thus established a method called DynaXL, which allow us to extract the information from the over-length cross-links and to visualize protein ensemble structures. In this protocol, we present the detailed procedure for using DynaXL, which comprises five steps. They are identification of highly confident cross-links, delineation of protein domains/subunits, ensemble rigid-body refinement, and final validation/assessment. The DynaXL method is generally applicable for analyzing the ensemble structures of multi-domain proteins and protein–protein complexes, and is freely available at www.tanglab.org/resources.

Keywords Chemical cross-linking, DynaXL, Ensemble refinement, Solvent accessible surface distance, Multi-domain protein, Protein–protein complex

INTRODUCTION

Chemical cross-linking coupled with mass spectroscopy (CXMS) has been used to characterize protein structures (Lasker et al. 2012; Walzthoeni et al. 2013; Politis et al. 2014). Different cross-linkers with various lengths and chemical properties are widely used in CXMS experiments. The commonly used cross-linking reagents include bis-sulfosuccinimidyl suberate (BS^3), bis-sulfosuccinimidyl glutarate (BS^2G), pimelic acid dihydrazide (PDH), and Leiker. (Leitner et al. 2014; Ding et al. 2016; Tan et al. 2016). Cross-linking reagents can react with specific amino acids in a protein, and two amino acids separated by a distance shorter than the

✉ Correspondence: gongzhou@wipm.ac.cn (Z. Gong), tanglab@wipm.ac.cn (C. Tang)

length of the cross-linker can be theoretically cross-linked (Fig. 1). Mass spectrometry analysis is used to identify the cross-linked residues, which can be translated to inter-residue distance (Kahraman et al. 2013; Lossl et al. 2014). In addition, the protein does not have to be isotopically labeled, modified, or crystallized in CXMS experiments. The CXMS can also be used in conjunction with other methods, such as cryo-EM for characterizing the structures of protein machinery (Cheng et al. 2015a, b; Liu et al. 2015a, b).

Multi-domain proteins and protein–protein complexes often undergo conformational fluctuations (Liu et al. 2015a, b). As a result, the ensemble structures corresponding to multiple conformational states are required to fully depict protein dynamics. Protein structural characterization has been mostly focused on the predominant structure of a protein. Recently we and others have shown that the so-called over-length cross-links actually contain information about the alternative and often lowly populated conformational states of the protein (Shore et al. 2016). Based on the over-length cross-links, we have developed a computational approach called DynaXL to visualize protein dynamics. Using DynaXL, we were able to characterize the ensemble structures of protein–protein complexes, with the dissociation constant ranging from nanomolar to millimolar (Gong et al. 2015). We were also able to visualize open-to-closed movement of multi-domain proteins (Ding et al. 2017).

An important feature of DynaXL is the use of solvent accessible surface distance (SASD) to describe the spatial relationship between cross-linked residues. As illustrated in Fig. 2A, the Euclidean straight-line distance between $C\alpha$ atoms of Lys^{29} and Lys^6 of Ubiquitin is 15.1 Å, while at 31.4 Å the SASD is much longer. As the cross-linker cannot penetrate through the protein and can only be at the protein surface, the SASD is a more realistic representation of the cross-linker and affords more stringent distance restraint. Explicit modeling of the cross-linker is incorporated into the software DynaXL with graphical interface (Fig. 2B). Here we will explain how to use DynaXL step by step.

OVERVIEW OF DYNAXL ALGORITHM DESIGN

The DynaXL method contains four parts, as illustrated in Fig. 3.

(1) Obtaining highly reliable cross-links using pLink (Yang et al. 2012). The acceptance criteria are the following: each spectrum should have an E value $< 10^{-3}$, each cross-link must be identified by at least two spectra, and one of them should have an E value $< 10^{-8}$.

(2) Evaluating the conformational dynamics based on the CXMS data. The domains are defined based on the known structure of a protein and are treated as rigid bodies. For protein complexes, the monomers in the complex are treated as rigid bodies. The protein may solely exist in a single conformation if the known structure can already satisfy all experimental cross-links. Otherwise, there likely exist some alternative conformational states of the protein that give rise to the "over-length" cross-links.

(3) Performing ensemble rigid-body refinement. One of the rigid bodies is kept fixed, and the other rigid body is subjected to translation and rotation. The number of structures in the ensemble is gradually increased if an $N = 2$ ensemble still cannot satisfy all cross-links. The optimal ensemble size is reached when all CXMS restraints are satisfied. There can be additional conformational state present for the protein system, which however is not captured and manifested by over-length cross-links.

(4) Assessment of the ensemble structures, either by cross-validation or by corroboration from other experimental data.

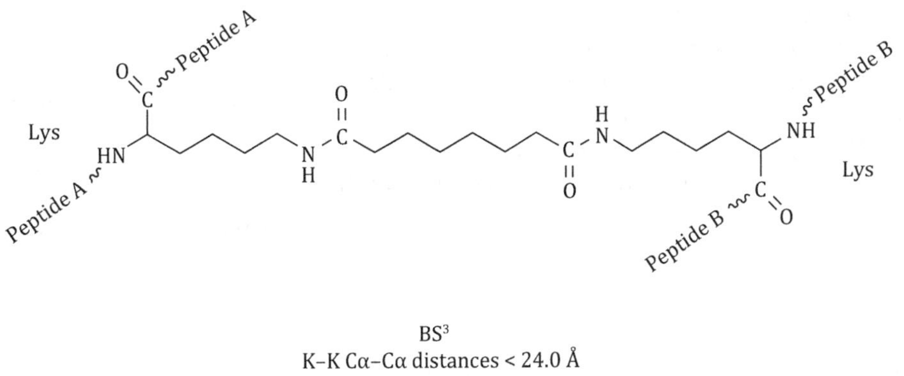

Fig. 1 Chemical structure for the BS^3 cross-link of two lysine residues in peptides A and B. The straight-line distance is less than 24 Å between the $C\alpha$ atoms, and less than 24 Å between the $N\zeta$ atoms. Note that the lysine side-chain amine group switches from sp3 hybridization to sp2 hybridization

BS^3
K–K $C\alpha$–$C\alpha$ distances < 24.0 Å

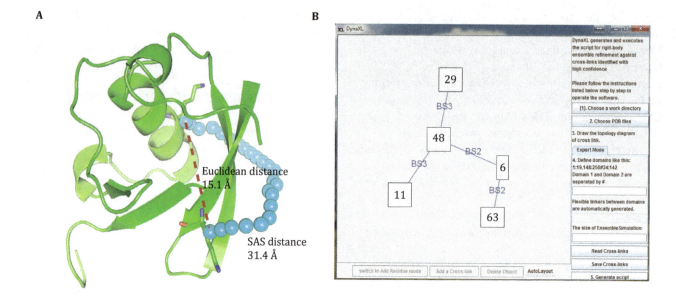

Fig. 2 Key features of DynaXL program. **A** Comparison of Euclidean distance (denoted with yellow dashed line) and solvent accessible surface distance (denoted with cyan sphere) for Cβ atoms of the two Lys residues in the protein. **B** The graphical user interface for DynaXL, in which one residue can be cross-linked to multiple residues with different cross-linking reagents

MATERIALS AND EQUIPMENT

Software for cross-link identification

pLink (Yang *et al.* 2012) is the program used for querying a database containing the protein sequence and for identifying the cross-linked peptides.

Software for protein structure modeling

Xplor-NIH (Schwieters *et al.* 2003), the software package for biomolecular structure refinement against experimental and knowledge-based restraints, is used here to identify an optimal ensemble structure that can account for all CXMS restraints.

AMBER 14 (Case *et al.* 2014), the molecular dynamic simulation package, is used here to refine the local conformation before ensemble refinement using Xplor-NIH.

Two programs, DynDom (Hayward *et al.* 1997; Hayward and Berendsen 1998) and ThreaDom (Xue *et al.* 2013; Wang *et al.* 2017), are used to define domain boundaries of multi-domain proteins.

PyMOL (the PyMOL Molecular Graphics System) is the software for illustrating and rendering protein structures.

SUMMARIZED PROCEDURE

(1) To identify high-confidence cross-links from CXMS experiment;
(2) To obtain the known structure from the PDB database;
(3) To define domain boundaries;
(4) To validate domain definition and evaluate local flexibility;
(5) To classify and identify the cross-links (intra-domain vs. inter-domain, intramolecular vs. inter-molecular cross-links), and prepare the CXMS restraints table;
(6) To prepare the starting structure files for Xplor-NIH;
(7) To patch the cross-linker to the protein structure;
(8) Ensemble rigid-body refinement against the CXMS restraints;
(9) Cross-validation with a subset of cross-links;
(10) To analyze and validate with other types of data.

PROCEDURE

Here, we use Ca^{2+}-loaded calmodulin as an example to illustrate how DynaXL is used to account for all CXMS data and to afford the ensemble structures.

Identification of cross-links

Intramolecular and inter-molecular cross-links can be differentiated by performing CXMS experiments on the sample containing equal molar amounts of unlabeled natural isotope abundance (light) and ^{15}N-labeled

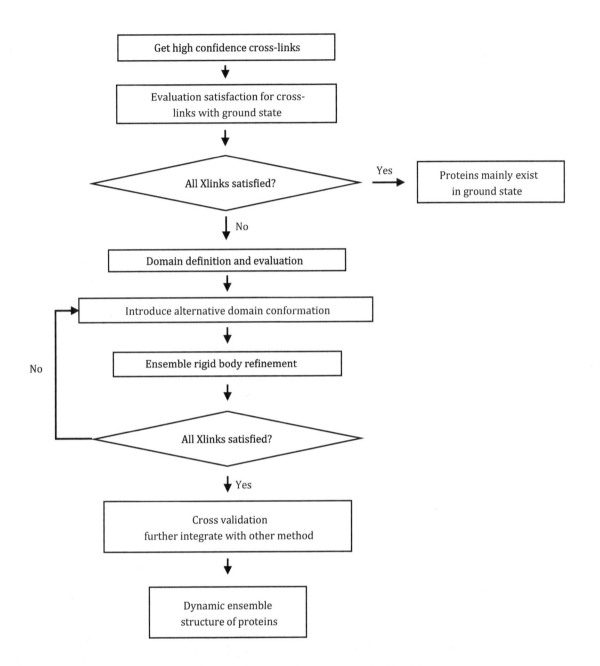

Fig. 3 The overall flowchart for the refinement of protein ensemble structures using DynaXL

proteins (heavy) (Ding et al. 2017). In this way, any cross-links arising from protein homodimer would contain both light and heavy peptides. Alternatively, the protein band corresponding to dimer (or monomer) can be excised from protein gel for mass spectrometry analysis. The intramolecular cross-links are further filtered with the following criteria using the software pLink.

(1) False discovery rate cutoff of 0.05 is applied and followed by an E value cutoff rate (Yang et al. 2012) of 10^{-3} at the spectrum level;
(2) Spectral count ≥ 2 and the best E value $< 10^{-8}$ for each pair of cross-link.

The cross-link spectra that pass the false discovery rate (FDR) cutoff are further filtered with these requirements: (A) each spectrum should have an E value of $< 10^{-3}$, and (B) each cross-link should be identified in at least two spectra.

Assessment of the predominant structure of the protein

The structure of the predominant conformational state of the protein under investigation can be downloaded from the PDB. For proteins without known structure, the structure can be modeled from homology modeling

(Marti-Renom et al. 2000), domain threading (Yang et al. 2015), or fragment splicing (Rohl et al. 2004).

The definition of protein domain boundary is performed with protein domain motion analysis (Hayward et al. 1997; Hayward and Berendsen 1998), multi-threading alignment (Xue et al. 2013; Belsom et al. 2016), or the assessment of evolutionary relationships (Cheng et al. 2014, 2015a, b). It should be noted that the definition of protein domain is not immutable, but will be amended based on further calculation and analysis (see below). For protein complexes, each subunit in the complex is treated as an individual rigid body.

Evaluation of protein local flexibility

(1) Structure completion. The atomic information is often missing in the PDB file, which is especially true for the X-ray structure. The missing parts include flexible loops and N- and C-terminal tails. In addition, hydrogen atoms are usually absent in relatively low-resolution structures. To complete the missing residues, e.g., the first three amino acids in calmodulin PDB structure 1CLL, the build-residue function in PyMOL software is used. To complete the missing atoms, e.g., side-chain atoms or hydrogen atoms, either PyMOL build-residue module or MD simulation software AMBER can be used.

(2) Flexibility evaluation. MD simulation using software AMBER can provide local flexibility for different parts of the protein upon assessing the fluctuation over time. The flexibility can also be assessed from crystal B-factors (Shore et al. 2016) and from NMR heteronuclear NOE values (Shore et al. 2016).

Identification and classification of the cross-links

Based on the known structure and domain definitions, the intramolecular cross-links can be classified into two categories including intra-domain and inter-domain ones. The intra-domain cross-links can also be used to confirm the definition of domains. For a protein complex, the cross-links are categorized as intramolecular and inter-molecular cross-links. If the known structure cannot satisfy all intramolecular inter-domain cross-links, there can be two possibilities:

(1) The discrepancy between the theoretical solvent-accessible inter-residue surface distance and the maximum length of the cross-linker is small, and the over-length cross-links can be attributed to local dynamic of the protein.

(2) The discrepancy between the theoretical solvent-accessible inter-residue surface distance and the maximum length of the cross-linker is large. Local dynamics alone cannot account for all the over-length cross-links. Therefore, the protein has to undergo collective domain movement, and further computational analysis is warranted.

Ensemble rigid-body refinement against CXMS restraints

The ensemble rigid-body refinement against CXMS restraints is performed when the predominant conformation or the known structure of the protein cannot satisfy all intramolecular inter-domain cross-links identified with high confidence. The protein domains are treated as rigid bodies, and their relative orientations are optimized on the basis of explicitly represented CXMS distance restraints. Similar approach is used to optimize the ensemble structures of protein–protein complexes. The details of the process are as follows.

Prepare the initial structure for Xplor-NIH

The structure refinement process against the CXMS restraints is conducted with the use of Xplor-NIH software. The Xplor-NIH requires a PDB file (providing atomic coordinate information) and a PSF file (providing structural connection information and other parameters) as the initial input. The user should pay attention to the following when preparing the input files:

(A) Different programs may have different atomic naming rules (especially for hydrogen atoms). Therefore, one should first remove all the hydrogen atoms, and use the Xplor-NIH script to re-protonate the protein.

(B) For assessment of protein local dynamics and domain boundaries, it may be necessary to re-number the residues of the protein. The first residue handled by AMBER software is always 1. Please refer to the respective manual for the software used.

(C) The Xplor-NIH will add an extra oxygen atom for the last residue and rename the last two oxygen atoms as OT1 and OT2 in the PSF file. As a result, the PSF file may be inconsistent with the PDB file provided. A quick solution is to duplicate the last line in the PDB file and to name the last two atoms as OT1 and OT2 as follows:

```
ATOM   2261  C    LYS  148   12.612   14.659   -4.066  1.00  0.00
ATOM   2262  OT1  LYS  148   12.190   15.449   -4.923  1.00  0.00
ATOM   2263  OT2  LYS  148   12.190   15.449   -4.923  1.00  0.00
```

Patch the cross-linker to protein

In DynaXL, the cross-linker is explicitly modeled onto the protein structure. For a pair of cross-linked lysine residues, the cross-linker is patched to one of the residues with the formation of an isopeptide bond. Similar approach can also be used to patch other cross-linking reagent with different reactivity to different types of protein residues. The patching process is done in these steps:

(A) Presented in this Protocol, we use two common cross-linkers BS^2G and BS^3, whose PDB and PSF files are provided in "Supplementary material." For other types of cross-linking reagents, the PDB files can be generated using the build function in PyMOL, and the corresponding parameter file can be obtained from online servers like HIC-Up (Kleywegt 2007) or PRODRG (Schuttelkopf and van Aalten 2004).

(B) For subsequent structure optimization, the cross-linker is only patched to one of the two cross-linked residues or to one of the protein domains. We have found that patching the cross-linker to either domain affords essentially the same results. The peptide bonds are formed between the side-chain of the protein and the cross-linkers, thus requiring the modification of the corresponding atoms. For example, as the nitrogen atom (N_ζ) of Lys side-chain is connected with three hydrogen atoms, it is necessary to remove the two extra hydrogen atoms, and the atom types for the remaining nitrogen ant hydrogen atoms are modified accordingly.

(C) The segment ID is another important distinguisher in Xplor-NIH in addition to the residue ID, *i.e.*, residues with the same residue ID values (residue number) and with different segment ID values correspond to different residues. It happens when a residue can be cross-linked to different residues in the opposite domain. Thus, we assign the cross-links at the same residue with different segment ID values. Physically, the multiple cross-links involving same residues should take place one at a time, and accordingly the *iso*-residue cross-links can overlap with each other without incurring van der Waals clashes during the refinement.

Preparation of the CXMS restraints table

With the cross-linker patched to one domain (or one subunit), the cross-linking process is simulated by enforcing a distance restraint between the end of the cross-linker and the reactive group of the other cross-linked residue. Specifically, it is achieved by constraining the distance between the carbonyl atom in the cross-linker and Lys N_ζ atom ranging from 1.3 Å (covalent bond length) to 5 Å (the sum of the VDW radius of both the carbon atom and the nitrogen).

Simulated annealing refinement

When all the input files and constraint files are prepared, a user can start the ensemble rigid-body refinement against the CXMS restraints. As mentioned above, when the given structure cannot satisfy all experimental high-confidence cross-links, the ensemble refinement based on CXMS restraints can be performed. The over-length cross-links capture one or more alternative conformational states. The ensemble refinement process starts with an $N = 2$ ensemble that comprises the predominant conformation and the alternative one. An additional conformer is included if the $N = 2$ ensemble cannot satisfy all the cross-links. The process is repeated until the experimental data are fully accounted for.

Here, we treat the different domains of the protein as rigid bodies, and the local conformational changes within each domain are not considered. The connecting loop residues between the domains are given full torsion angle freedom. As the domain movement is relative, one domain is kept fixed, and the other domain(s) are grouped together and are allowed to freely rotate and to translate with respect to the fixed domain. The fixing/grouping is implemented using the following script in Xplor-NIH.

```
dyn.fix ("""  segid " " and resi 1:77 """)
dyn.group ("""  segid ALT0 and resi 82:148 """)
dyn.group ("""  segid ALT1 and resi 82:148 """)
```

Here, the N-terminal domain including residues 1–77 of the calmodulin is fixed, and the C-terminal domain including residues 82–148 moves as a rigid body. In addition, the flexible loop region between the including residues 78–81 has full torsional freedom. In the Xplor-NIH script shown above, note that there are two different conformers for the C-terminal domain, marked with segment ID ALT0 and ALT1. The two conformers correspond to the two conformational states of calmodulin. To speed up the computation, the non-bonded van der Waals interactions within each rigid

body are not considered and calculated. This is implemented using the following statement.
Constraints:

inter = (segid " ") (segid ALT0 or resi 82:148)
inter = (segid " ") (segid ALT1 or resi 82:148)
inter = (segid ALT0 and resi 78:82) (segid " " and segid ALT0)
inter = (segid ALT1 and resi 78:82) (segid " " and segid ALT1)
weights * 1 end end

In the ensemble refinement, ambiguous distance restraints are employed, and the back-calculated distance R is defined as

$$R = \left(\sum_{k=1}^{N} r_k^{-6}\right)^{-1/6}.$$

In which r_k is the distance between the two atoms (nitrogen atom from the lysine residue and carbonyl atom from the cross-linker as discussed before) in conformer k, and N is the number of conformers in the ensemble. An exponential factor of -6 is used here, and as a result the $<r^{-6}>$ averaged distance is heavily biased towards the shortest distance.

The ensemble refinement is carried out by simulated annealing. The system is heated to a relatively high temperature and then slowly cooled down. The structure is refined against the CXMS restraints during the cooling process. The computational process is repeated many times for effective sampling. Finally, the structures with no CXMS violation (satisfying all cross-links) and low energy (no atomic overlap) are selected for further analysis.

We have found that the explicit representation of cross-linker not only provides more realistic and stringent restraints, but also allows better convergence for the ensemble structures, as compared to straight-line Euclidean distance restraints.

Cross-validation with a subset of cross-links

The cross-validation process is performed to verify the accuracy of the ensemble structures. In detail, a subset of CXMS restraints is removed, and the remaining cross-links are used for the ensemble refinement as described above. The ensemble structures generated with a subset of the restraints are evaluated and the CXMS restraints excluded in the refinement should be cross-validated.

Analysis and validation with other types of experimental data

The over-length cross-links capture protein alternative conformations in solution. The ensemble structures obtained by refining against the CXMS restraints may be compared to those obtained from other biochemical and biophysical methods, such as paramagnetic relaxation enhancement (Tang et al. 2006) and small-angle X-ray scattering (Schneidman-Duhovny et al. 2012; Kikhney and Svergun 2015).

LIMITATIONS OF THE DYNAXL METHOD

The ensemble structures obtained based on the CXMS restraints may suffer from certain limitations as described below.

False identifications

Due to the quality of the mass spectra, false identification of the cross-links may occur. In other words, the experiment may identify incorrect cross-links, even though stringent criteria are applied when selecting high-confidence cross-links. Multiple technical and biological repeats are necessary to minimize false identifications.

Insufficient number of restraints

There may not be a large number cross-links identified with high confidence that can be used as the restraints. Certainly more restraints would enable a researcher to better refine the structure and to discover discrepancy within the restraints. However, it has been shown that the structural model of a protein complex can be obtained from just a single inter-molecular cross-linking restraint (Gong et al. 2015). Thus, the DynaXL approach may only identify the minimum number of ensemble structures that can account for all available CXMS restraints. Should there are more conformational states that elude cross-linking reactions, DynaXL cannot uncover.

Over-fitting problem

The ensemble size may have to be increased to account for all the cross-links. Additional conformers introduce additional parameters, which may lead to over-fitting. It is also possible that some over-length cross-links can be satisfied by intra-domain dynamics without the invocation of domain movement. Thus, cross-validation is important.

FUTURE PERSPECTIVE

CXMS has been increasingly used for protein structure modeling. Here, we present the detailed protocol using DynaXL for explicitly modeling the cross-links and characterization of protein ensemble structures. The chemical cross-linking as well photo-cross-linking are rapidly evolving (Chiang et al. 2016), and new types of cross-linking reagents (Brodie et al. 2016) with various linker lengths and reactivity are becoming increasingly available, which can afford more spatial information between protein residues. In the age of integrative structural biology, protein ensemble structures can be better visualized with the joint refinement against multiple types of experimental inputs including but not limited to NMR, cryo-EM, and FRET.

Acknowledgements This work was supported by a Grant from the National Key R&D Program of China (2016YFA0501200), Chinese Ministry of Science and Technology (2013CB910200), and National Natural Science Foundation of China (31225007, 31400735 and 31500595).

Compliance with Ethical Standards

Conflict of interest Zhou Gong, Zhu Liu, Yue-He Ding, Meng-Qiu Dong, and Chun Tang declare that they have no conflict of interest.

Human and animal rights and informed consent This article does not contain any studies with human or animal subjects performed by any of the authors.

References

Belsom A, Schneider M, Fischer L, Brock O, Rappsilber J (2016) Serum albumin domain structures in human blood serum by mass spectrometry and computational biology. Mol Cell Proteom 15:1105–1116
Brodie NI, Petrotchenko EV, Borchers CH (2016) The novel isotopically coded short-range photo-reactive crosslinker 2,4,6-triazido-1,3,5-triazine (TATA) for studying protein structures. J Proteom 149:69–76
Case DA, Babin V, Berryman JT, Betz RM, Cai Q, Cerutti DS, Cheatham TE III, Darden TA, Duke RE, Gohlke H, Goetz AW, Gusarov S, Homeyer N, Janowski P, Kaus J, Kolossváry I, Kovalenko A, Lee TS, LeGrand S, Luchko T, Luo R, Madej B, Merz KM, Paesani F, Roe DR, Roitberg A, Sagui C, Salomon-Ferrer R, Seabra G, Simmerling CL, Smith W, Swails J, Walker RC, Wang J, Wolf RM, Wu X, Kollman PA (2014) AMBER 14. University of California, San Francisco
Cheng H, Schaeffer RD, Liao Y, Kinch LN, Pei J, Shi S, Kim BH, Grishin NV (2014) ECOD: an evolutionary classification of protein domains. PLoS computational biol 10:e1003926
Cheng H, Liao Y, Schaeffer RD, Grishin NV (2015a) Manual classification strategies in the ECOD database. Proteins 83:1238–1251
Cheng Y, Grigorieff N, Penczek PA, Walz T (2015b) A primer to single-particle cryo-electron microscopy. Cell 161:438–449
Chiang MY, Hsu YW, Hsieh HY, Chen SY, Fan SK (2016) Constructing 3D heterogeneous hydrogels from electrically manipulated prepolymer droplets and crosslinked microgels. Sci Adv 2:e1600964
Ding YH, Fan SB, Li S, Feng BY, Gao N, Ye K, He SM, Dong MQ (2016) Increasing the depth of mass-spectrometry-based structural analysis of protein complexes through the use of multiple cross-linkers. Anal Chem 88:4461–4469
Ding YH, Gong Z, Dong X, Liu K, Liu Z, Liu C, He SM, Dong MQ, Tang C (2017) Modeling protein excited-state structures from "over-length" chemical cross-links. J Biol Chem 292:1187–1196
Gong Z, Ding Y-H, Dong X, Liu N, Zhang EE, Dong M-Q, Tang C (2015) Visualizing the ensemble structures of protein complexes using chemical cross-linking coupled with mass spectrometry. Biophys Rep 1:127–138
Hayward S, Berendsen HJC (1998) Systematic analysis of domain motions in proteins from conformational change: new results on citrate synthase and T4 lysozyme. Prot-Struct Funct Genet 30:144–154
Hayward S, Kitao A, Berendsen HJC (1997) Model-free methods of analyzing domain motions in proteins from simulation: a comparison of normal mode analysis and molecular dynamics simulation of lysozyme. Prot-Struct Funct Genet 27:425–437
Kahraman A, Herzog F, Leitner A, Rosenberger G, Aebersold R, Malmstrom L (2013) Cross-link guided molecular modeling with ROSETTA. PLoS ONE 8:e73411
Kikhney AG, Svergun DI (2015) A practical guide to small angle X-ray scattering (SAXS) of flexible and intrinsically disordered proteins. FEBS Lett 589:2570–2577
Kleywegt GJ (2007) Crystallographic refinement of ligand complexes. Acta Crystallogr Sect D: Biol Crystallogr 63:94–100
Lasker K, Forster F, Bohn S, Walzthoeni T, Villa E, Unverdorben P, Beck F, Aebersold R, Sali A, Baumeister W (2012) Molecular architecture of the 26S proteasome holocomplex determined by an integrative approach. In: Proceedings of the National Academy of Sciences of the United States of America 109: 1380–1387
Leitner A, Joachimiak LA, Unverdorben P, Walzthoeni T, Frydman J, Forster F, Aebersold R (2014) Chemical cross-linking/mass spectrometry targeting acidic residues in proteins and protein complexes. In: Proceedings of the National Academy of Sciences of the United States of America 111: 9455–9460
Liu F, Rijkers DT, Post H, Heck AJ (2015a) Proteome-wide profiling of protein assemblies by cross-linking mass spectrometry. Nat Method 12:1179–1184
Liu Z, Gong Z, Jiang WX, Yang J, Zhu WK, Guo DC, Zhang WP, Liu ML, Tang C (2015b) Lys63-linked ubiquitin chain adopts multiple conformational states for specific target recognition. Elife 4:e05767
Lossl P, Kolbel K, Tanzler D, Nannemann D, Ihling CH, Keller MV, Schneider M, Zaucke F, Meiler J, Sinz A (2014) Analysis of

nidogen-1/laminin gamma1 interaction by cross-linking, mass spectrometry, and computational modeling reveals multiple binding modes. PLoS ONE 9:e112886

Marti-Renom MA, Stuart AC, Fiser A, Sanchez R, Melo F, Sali A (2000) Comparative protein structure modeling of genes and genomes. Annu Rev Biophys Biomol Struct 29:291–325

Politis A, Stengel F, Hall Z, Hernandez H, Leitner A, Walzthoeni T, Robinson CV, Aebersold R (2014) A mass spectrometry-based hybrid method for structural modeling of protein complexes. Nat Method 11:403–406

Rohl CA, Strauss CEM, Misura KMS, Baker D (2004) Protein structure prediction using rosetta. Numer Comput Method, Pt D 383:66–93

Schneidman-Duhovny D, Kim SJ, Sali A (2012) Integrative structural modeling with small angle X-ray scattering profiles. BMC Struct Biol 12:17

Schuttelkopf AW, van Aalten DM (2004) PRODRG: a tool for high-throughput crystallography of protein-ligand complexes. Acta Crystallogr Sect D: Biol Crystallogr 60:1355–1363

Schwieters CD, Kuszewski JJ, Tjandra N, Clore GM (2003) The Xplor-NIH NMR molecular structure determination package. J Magn Reson 160:65–73

Shore S, Henderson JM, Lebedev A, Salcedo MP, Zon G, McCaffrey AP, Paul N, Hogrefe RI (2016) Small RNA library preparation method for next-generation sequencing using chemical modifications to prevent adapter dimer formation. PLoS ONE 11:e0167009

Tan D, Li Q, Zhang MJ, Liu C, Ma C, Zhang P, Ding YH, Fan SB, Tao L, Yang B, Li X, Ma S, Liu J, Feng B, Liu X, Wang HW, He SM, Gao N, Ye K, Dong MQ, Lei X (2016) Trifunctional cross-linker for mapping protein-protein interaction networks and comparing protein conformational states. Elife 5:e12509

Tang C, Iwahara J, Clore GM (2006) Visualization of transient encounter complexes in protein-protein association. Nature 444:383–386

The PyMOL Molecular Graphics System, Version 1.8 Schrödinger, LLC

Walzthoeni T, Leitner A, Stengel F, Aebersold R (2013) Mass spectrometry supported determination of protein complex structure. Curr Opin Struct Biol 23:252–260

Wang Y, Wang J, Li R, Shi Q, Xue Z, Zhang Y (2017) ThreaDomEx: a unified platform for predicting continuous and discontinuous protein domains by multiple-threading and segment assembly. Nucl acids res 45:400–407

Xue ZD, Xu D, Wang Y, Zhang Y (2013) ThreaDom: extracting protein domain boundary information from multiple threading alignments. Bioinformatics 29:247–256

Yang B, Wu YJ, Zhu M, Fan SB, Lin J, Zhang K, Li S, Chi H, Li YX, Chen HF, Luo SK, Ding YH, Wang LH, Hao Z, Xiu LY, Chen S, Ye K, He SM, Dong MQ (2012) Identification of cross-linked peptides from complex samples. Nat Method 9:904–906

Yang J, Yan R, Roy A, Xu D, Poisson J, Zhang Y (2015) The I-TASSER Suite: protein structure and function prediction. Nat Method 12:7–8

Symmetry-mismatch reconstruction of genomes and associated proteins within icosahedral viruses using cryo-EM

Xiaowu Li[1], Hongrong Liu[1], Lingpeng Cheng[2]

[1] College of Physics and Information Science, Synergetic Innovation Center for Quantum Effects and Applications, Hunan Normal University, Changsha 410081, China
[2] School of Life Sciences, Tsinghua University, Beijing 100084, China

Abstract Although near-atomic resolutions have been routinely achieved for structural determination of many icosahedral viral capsids, structures of genomes and associated proteins within the capsids are still less characterized because the genome information is overlapped by the highly symmetric capsid information in the virus particle images. We recently developed a software package for symmetry-mismatch structural reconstruction and determined the structures of the genome and RNA polymerases within an icosahedral virus for the first time. Here, we describe the protocol used for this structural determination, which may facilitate structural biologists in investigating the structures of viral genome and associated proteins.

Keywords Viral genome, Cryo-EM reconstruction, Symmetry-mismatch

INTRODUCTION

Three-dimensional (3D) structural determination of viruses aids in elucidating viral molecular mechanism and pathogenesis. Cryo-electron microscopy (cryo-EM) and X-ray crystallography are two major methods used for the 3D structural determination of viruses. The major advantages of cryo-EM are that the specimens require no crystallization and are embedded in a thin vitrified ice of buffer thus preserving their near-native structure.

Recent instrumental and computational developments of cryo-EM have enabled the structural determination of viruses and other biological assemblies at near-atomic to atomic resolutions (Grigorieff and Harrison 2011; Bai et al. 2015; Bartesaghi et al. 2015), comparable to the structures determined using X-ray crystallography. A new generation of direct electron detection (DED) camera has significantly improved cryo-EM image quality. Compared with the traditional charge-coupled device (CCD) camera, DED camera allows reconstruction of higher resolution structures of biological complexes including viruses, while using fewer particle images (Veesler et al. 2013; McMullan et al. 2014; Bartesaghi et al. 2015). Icosahedral viral capsids were the first biological assemblies whose structures have been determined at near-atomic resolution using cryo-EM (Jiang et al. 2008; Liu et al. 2010; Cheng et al. 2011) due to their high (60-fold) symmetry and relatively large size, which promise a more accurate orientation determination and higher contrast of the particle images than other smaller low-symmetry biological assemblies.

Although the 3D structures of icosahedral viral capsids have been studied extensively by using both EM and X-ray crystallography for decades (Crowther 1971;

✉ Correspondence: hrliu@hunnu.edu.cn (H. Liu), lingpengcheng@mail.tsinghua.edu.cn (L. Cheng)

Harrison et al. 1978; Baker et al. 1999; Thuman-Commike and Chiu 2000), precise structures of viral genomes and associated proteins within the capsids are still less characterized. The difficulty can be attributed to the fact that the viral genome and associated proteins are encapsidated in a high symmetric layered (or multi-layered) capsid. The high symmetry of the capsid is an advantage for determination of orientations of virus particle images using common-line based reconstruction algorithm (Crowther 1971; Fuller et al. 1996). However, when it comes to the structural determination of the genome and proteins within the capsid, the orientation determination of the non-symmetric genome information will be biased by the overlapping high-symmetry capsid information in the virus particle images. Here, we present our recently developed protocol and software package for the structural determination of icosahedral virus genome and associated proteins.

Development of the layer-based cryo-EM image processing and symmetry-mismatch reconstruction method

Symmetry mismatches are present between viral capsids and genomes as well as within capsids. First attempts of symmetry-mismatch reconstruction made at virus structure determination were to reconstruct a unique tail at an icosahedral vertex of bacteriophages (Tao et al. 1998; Jiang et al. 2006). In the two previous studies, Jiang et al. and Tao et al. reconstructed the bacteriophage head structures taking advantage of the icosahedral symmetry, and then the icosahedral symmetry was further relaxed to lower symmetries to generate the reconstruction without symmetry imposition. Briggs et al. and Morais et al. reconstructed non-icosahedral structures at a Kelp fly virus vertex (Briggs et al. 2005) and a bacteriophage tail (Morais et al. 2001). They first boxed the vertex images from raw cryo-EM particle images with known icosahedral orientations, and then classified and reconstructed these boxed vertex images. However, their structural resolutions of non-symmetric regions of the viruses are relatively low, probably because the orientation determination of the target structural information was biased as mentioned above. In order to reduce the effect of the capsid information, Huiskonen et al. tried to remove the icosahedrally ordered capsid parts from the raw cryo-EM images of Cystovirus Φ8 by subtracting equivalent projections of the icosahedral model from the raw images (Huiskonen et al., 2007). The images of the RNA packaging motor boxed from the resulting image after subtraction were then classified and reconstructed to obtain a 15 Å resolution structure. The reconstruction resolution is low because the contrast transfer function (CTF) modulation of the images was not considered. In addition, the structures of encapsidated non-symmetric viral genomes and associated proteins remained unresolved.

In order to solve this problem, we developed a new symmetry-mismatch reconstruction method and for the first time we determined the structures of genome and polymerase within an icosahedral dsRNA cypovirus (Liu and Cheng 2015). In the subsequent sections, we have described a detailed protocol of this reconstruction method by using the cypovirus as an example.

Applications and advantages of the protocol

The method has been released as a software package running under Microsoft Windows. The package is designed to reconstruct the 3D structure of lower or non-symmetry viral genome and associated proteins enclosed in higher symmetry viral capsid from 2D cryo-EM images. Nevertheless, it can be used for the reconstruction of other biological assemblies exhibiting similar symmetry mismatches. We believe that viruses have a functional state with a relatively organized genome structure that can be determined, and the key for the structure determination is how to catch this state biochemically.

Limitations of the protocol

This protocol requires that the viral capsid is icosahedral and has been reconstructed at a high resolution from cryo-EM images of virus particles, of which the orientations and centers used for the capsid reconstruction have been determined by virus reconstruction software packages, for example, RELION (Scheres 2012), EMAN (Ludtke et al. 1999), and FREALIGN (Grigorieff 2007).

EXPERIMENTAL DESIGN

Before reading the following protocol, we suggest the readers to refer to the basic principle of the protocol in the supplementary materials of our published paper (Liu and Cheng, 2015). The protocol includes two modules: the first module describes the steps to extract the non-symmetric genome structure information from the cryo-EM images by eliminating the symmetric capsid information. According to the approximation of weak phase objects, a cryo-EM image of a virus particle can be considered the sum of a genome image and a capsid image. Therefore, the genome image can be obtained by subtracting the capsid image, which can be obtained by projecting the 3D density map of the capsid on the capsid orientation and applying CTF modulation on the projection, from the cryo-EM image of a virus particle. The second module describes the steps to iteratively determine orientations of all genome images and reconstruction. Since the 60 equivalent orientations of the icosahedral capsid are known and the asymmetric genome structure has a fixed orientation related to the symmetric capsid, the correct orientation of the genome can be determined by searching the 60 equivalent orientations of the capsid (Liu and Cheng 2015). By using this protocol and software package, we could determine structure of genome and RNA polymerases within the cypoviruses.

MATERIALS

Totally, 420 cryo-EM micrographs of a cypovirus were collected using Tecnai Arctica 200 kV electron microscope equipped with Falcon II camera at a magnification of 78,000 corresponding to a pixel size of 0.932 Å/pixel. Approximate 5000 virus particle images with a size of 1024×1024 pixels were boxed out from the 420 micrographs. The defocus value for each micrograph was determined using the CTFIT program in the EMAN package (Ludtke et al. 1999). In order to speed up the computational process, the images were subsampled by 2. All 2D image stacks and 3D density maps used in this protocol were stored in the MRC format.

EQUIPMENT AND SOFTWARE SETUP

The following procedure requires a computer running Microsoft Windows and our symmetry-mismatch reconstruction software package. Unpacking the software package to a folder and setting a path to this folder in the Windows system are required before using the software package. This procedure assumes that users have experiences in cryo-EM image processing and single particle reconstruction.

PROCEDURE

This section describes the step-by-step image processing procedure (Fig. 1). The lines in italic font are command lines, which must be input in the command prompt window in a directory where a stack of raw images are stored.

Genome image extraction

(1) Moving the virus particle centers to the images centers. In order to avoid different definitions of the particle center in other reconstruction software packages, the centers of the virus particle images are moved to the center of the images.

img2d rawImgBin2.stck outputimgstck=virusImgBin2.stck inputort=ort0.dat outputort=ort.dat norm=0,1 trans=y

"rawImgBin2.stck" is a stack of the subsampled particle images with an image size of 512×512. "virusImgBin2.stck" is the output stack of the virus particle images whose particles are centered in the images. "ort0.dat" is a text file containing particle orientations, centers, and image defocus values (astigmatism) of these particle images (hereafter orientation file). "ort.dat" is a generated orientation file containing the same orientations and defocus values as those of "ort0.dat" but with the center parameters updated to 0,0. "norm=0,1" indicates that the images in "rawImgBin2.stck" are normalized to a mean value of 0 and a standard deviation of 1; "trans=y" is an option used to center the particles in the images.

The format of orientation file in this protocol is uniform. The first column contains the serial numbers of the virus particle images. The second, third, and fourth columns contain the Euler angles defining the orientations of these virus particles. The fifth and sixth columns contain centers defining the central positions of the virus particles in the images. The seventh column contains the correlation coefficients of the cross-correlation between the virus images and the model images. The eighth column contains serial numbers of the micrographs. The ninth, 10th, and 11th columns contain defocus values X, Y, and astigmatism angles of the particle images, respectively.

(2) Reconstructing a high-resolution density map of the capsid.

recCartesian_fast virusImgBin2.stck ort.dat virusMap.mrc maxFR=220 imgmask=212 apix=1.864 applyCTF=y sym=I mincc=0.2 Cs=2.7 vol=200, ampw=0.1, norm=0,1

"virusImgBin2.stck" and "ort.dat" are stack of virus particle images and its orientation file generated in Step (1), respectively. "virusMap.mrc" is the output density map of virus; "maxFR=220", "imagemask=212", "apix=1.864", "applyCTF=y" and "sym=I" specify the maximum Fourier radius (here, 220 corresponds to a resolution of 4.3 Å) used for reconstruction, mask radius for the virus particle images (in pixel), pixel size of the virus particle images, application of CTF

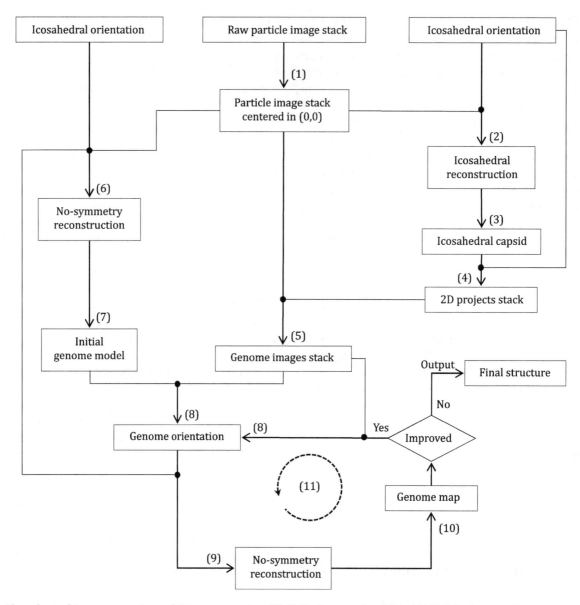

Fig. 1 Flow chart of image processing and 3D reconstruction. (1)–(11) correspond to Steps (1)–(11) in the procedure

correction, and applied symmetry during reconstruction (icosahedral), respectively. "Cs" and "vol" are the objective lens spherical aberration coefficient in mm and accelerating voltage of the electron microscope. "ampw=0.1" specifies the ratio of amplitude contrast. For cross-correlation (CC)-based particle orientation and center determination, "mincc=0.2" indicates that only those particle images with CC values higher than or equal to 0.2 are included in the reconstruction. For common-line-based particle orientation and center determination (Crowther 1971; Thuman-Commike and Chiu 2000), "mincc=0.2" indicates that only those particle images with cosine values of phase residuals higher than or equal to 0.2 are included in the reconstruction.

The reconstructed density map of the cypovirus capsid is shown in Fig. 2.

(3) Masking the inner genome densities within the density map.

img3d   virusMap.mrc   outputmap=capsidMap.mrc imask=136 mask=212

"virusMap.mrc" is the density map of virus generated in Step (2). "imask" and "mask" specify radii of the genome and the capsid. This command eliminates the density within and outside the capsid and generates a density map of the capsid ("capsidMap.mrc").

(4) Projecting the masked capsid density map according to the icosahedral orientation of each particle image.

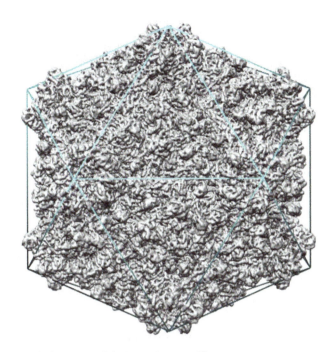

Fig. 2 Structure of the cypovirus capsid

cent_slice_icos2f capsidMap.mrc ort.dat proj.stck mask=212 norm=0,1

"capsidMap.mrc" is the capsid density map generated in Step (3). "proj.stck" is a stack of the projection images of the capsid map generated in this step. "ort.dat", which is generated in Step (1), indicates the projection orientations; therefore, these projection images have orientations identical to their corresponding virus particle images.

(5) Extracting genome images from the virus images.

cc proj.stck stck2=virusImgBin2.stck ort2=ort.dat lp=220 norm=0,1 mask=212 applyCTF=y imgsubproj=genomeImg.stck trans=1,1

"proj.stck" is the stack of capsid projection images generated in Step (4) and "virusImgBin2.stck" is the stack of virus images. "genomeImg.stck" is a stack of genome images generated in this step by subtracting the projection images from the virus particle images. Before subtracting, the program applies CTF to the projection images ("proj.stck") and scales the pixel values on the projection images to the same grayscale with capsid information on the virus particle image. Pixel size, accelerating voltage, spherical aberration, ratio of amplitude contrast, B factor, and inner and outer radii values need to be input when the program is running. Amplitudes of 0.1 and B factor of 0 are recommended here. The two input radii define a circular region used for scaling the capsid information in virus particle images ("virusImgBin2.stck") to the same grayscale with the corresponding capsid projections ("proj.stck") (see Supplementary Fig. 13 of our previous paper (Liu and Cheng, 2015)). The inner and outer radii of 136 and 160 (in pixel) are recommended here. "trans=1,1" indicates that all genome will be moved to the center of the images. The extracted genome images ("genomeImg.stck") will be used to determine the orientations of the genome images in Step (8).

Initial model generation

(6) Reconstructing a complete virus structure (including the capsid and genome) using the virus particle images. The initial orientation for each virus particle image is randomly selected from the 60 icosahedral equivalent orientations of its capsid because the orientations computed from the virus capsid are all located within an asymmetric unit.

icos2f_randort ort.dat ortRand.dat

"ort.dat" contains orientations of virus particle images within an asymmetric unit. "ortRand.dat" contains orientations, each of which is randomly selected from the 60 icosahedral equivalent orientations of each orientation in the "ort.dat".

recCartesian_fast virusImgBin2.stck ortRand.dat capsidGenome.mrc maxFR=40 imgmask=212 norm=0,1 applyCTF=y Cs=2.7 vol=200 apix=1.864 ampw=0.1

"virusImgBin2.stck" is the stack of virus images. "ortRand.dat" contains orientations of the virus images. "capsidGenome.mrc" is a generated density map of the complete virus structure.

(7) Generating a density map of the genome (initial model).

img3d capsidGenome.mrc outputmap=genomeMap.mrc mask=136

"mask=136" indicates that the densities out of radius 136 pixels will be masked. "genomeMap.mrc" is the generated density map of the genome.

Orientation determination and 3D reconstruction of genome

(8) Projecting the 3D genome model to generate 60 projection images for each genome image according to the 60 equivalent icosahedral orientations of the capsid. The cross coefficients between each genome image and the 60 projection images are calculated here. The orientation of the projection image best-matched with each genome image is assigned the corresponding orientation to the genome image for further analysis.

align_core genomeMap.mrc genomeImg.stck ort.dat genomeOrt.dat lp=15 mask=136 applyCTF=y norm=0,1 mode=cpc centshift=y apix=1.864 Cs=2.7 vol=200 ampw=0.1

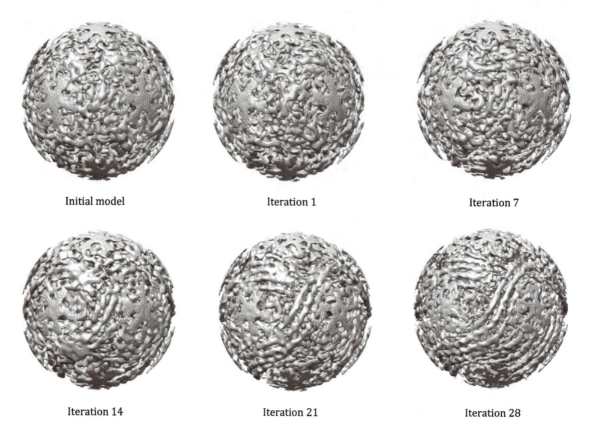

Fig. 3 Density maps of the genome and RdRps from the initial model to reconstruction of 28 cycles. The dsRNA fragments are visible in the reconstruction of the 28th cycle

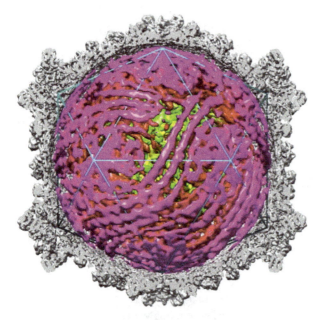

Fig. 4 Structure of the genome within the capsid. Half of the icosahedral capsid (*gray*) is removed to show the structures of genomic dsRNA and RdRps (*purple*)

"genomeMap.mrc" is the structural model for the orientation determination of the genome images ("genomeImg.stck"). The orientations of the best-matched projection images are stored in "genomeOrt.dat". "mode=cpc" indicates that the cross coefficient is calculated using cross phase correlation. "centshift=y" indicates that the center of the genome is allowed to shift when calculating cross coefficient. "lp=15" specifies the low-pass filter (in Fourier radius). "mask=136" indicates that only image region within a radius of 136 pixels in the genome images ("genomeImg.stck") is used to calculate cross phase correlation.

This step is time intensive. Users can write a script to run it in parallel. For example, if the "genomeImg.stck" contains 5000 images, the script can be written as follows:

start align_core genomeMap.mrc genomeImg.stck ort.-dat genomeOrt_1.dat lp=15 mask=136 applyCTF=y norm=0,1 mode=cpc centshift=y apix=1.864 Cs=2.7 vol=200 ampw=0.1 first1 last=1000

start align_core genomeMap.mrc genomeImg.stck ort.-dat genomeOrt_2.dat lp=15 mask=136 applyCTF=y norm=0,1 mode=cpc centshift=y apix=1.864 Cs=2.7 vol=200 ampw0.1 first=1001 last=2000

start align_core genomeMap.mrc genomeImg.stck ort.-dat genomeOrt_3.dat lp=15 mask=136 applyCTF=y norm=0,1 mode=cpc centshift=y apix=1.864 Cs=2.7 vol=200 ampw=0.1 first=2001 last=3000

start align_core genomeMap.mrc genomeImg.stck ort.-dat genomeOrt_4.dat lp15 mask=136 applyCTF=y norm=0,1 modecpc centshift=y apix=1.864 Cs=2.7 vol=200 ampw=0.1 first=3001 last=4000

start align_core genomeMap.mrc genomeImg.stck ort.-dat genomeOrt_5.dat lp=15 mask=136 applyCTF=y norm=0,1 mode=cpc centshift=y apix=1.864 Cs=2.7 vol=200 ampw=0.1 first=4001 last=5000

In each command, "first" and "last" indicate the first and last images to be processed in the "ort.dat". Users can generate the combined "genomeOrt.dat" of all genome images by running *type genomeOrt_1.dat genomeOrt_2.dat genomeOrt_3.dat genomeOrt_4.dat genomeOrt_5.dat > genomeOrt.dat*.

(9) Reconstructing the genome using the newly assigned orientations and centers ("genomeOrt.dat").

recCartesian_fast virusImgBin2.stck genomeOrt.dat virusMap-1.mrc maxFR=40 imgmask=212 norm=0,1 applyCTF=y mincc=0.09 bound=3 Cs=2.7 vol=200 apix=1.864

"bound=3" indicates that the genome images, whose centers shift more than 3 pixels, are not included in the reconstruction.

(10) Masking the capsid structure surrounding the genome structure.

img3d virusMap-1.mrc outputmap=genomeMap-1.mrc mask=136

The generated genome structure ("genomeMap.mrc") serves as structural model for next round of orientation determination of the genome images.

(11) Iterating Steps (8) to (10) until the orientations of each genome image stabilizes and no further improvement of the genome structure can be obtained. For each of the iteration, the output map generated in Step (10) is used as the structural model in Step (8). The "lp" of "align_core" and "maxFR" of "recCartesian_fast" can be improved steadily when the genome orientations in the "genomeOrt.dat" does not change during the iterations. The intermediate results during iterations are shown in Fig. 3.

ANTICIPATED RESULTS

On performing this protocol, users can reconstruct a cryo-EM density map of the cypovirus genome structure (Fig. 4). The double helices of both dsRNA fragments located close to the inner capsid surface and interacting with the RNA polymerase can be observed. This protocol is also applicable for the genome structure determination of other icosahedral viruses with structurally homogenous genomes.

Acknowledgments All EM data were collected at Tsinghua University Branch of the National Center for Protein Sciences (Beijing). We thank Bin Xu and Tao Yang for data collection and computational support. This research was supported by the National Natural Science Foundation of China (Grant Nos. 91530321, 31570727, 31570742, 31370736, and 31170697).

Compliance with Ethical Standards

Conflict of Interest Xiaowu Li, Hongrong Liu and Lingpeng Cheng declare that they have no conflict of interest.

Human and Animal Rights and Informed Consent This article does not contain any studies with human or animal subjects performed by any of the authors.

References

Bai XC, McMullan G, Scheres SH (2015) How cryo-EM is revolutionizing structural biology. Trends Biochem Sci 40:49–57

Baker TS, Olson NH, Fuller SD (1999) Adding the third dimension to virus life cycles: three-dimensional reconstruction of icosahedral viruses from cryo-electron micrographs. Microbiol Mol Biol Rev 63:862–922

Bartesaghi A, Merk A, Banerjee S, Matthies D, Wu X, Milne JL, Subramaniam S (2015) 2.2 Å resolution cryo-EM structure of beta-galactosidase in complex with a cell-permeant inhibitor. Science 348:1147–1151

Briggs JA, Huiskonen JT, Fernando KV, Gilbert RJ, Scotti P, Butcher SJ, Fuller SD (2005) Classification and three-dimensional reconstruction of unevenly distributed or symmetry mismatched features of icosahedral particles. J Struct Biol 150:332–339

Cheng L, Sun J, Zhang K, Mou Z, Huang X, Ji G, Sun F, Zhang J, Zhu P (2011) Atomic model of a cypovirus built from cryo-EM structure provides insight into the mechanism of mRNA capping. Proc Natl Acad Sci USA 108:1373–1378

Crowther RA (1971) Procedures for three-dimensional reconstruction of spherical viruses by Fourier synthesis from electron micrographs. Philos Trans R Soc Lond B Biol Sci 261:221–230

Fuller SD, Butcher SJ, Cheng RH, Baker TS (1996) Three-dimensional reconstruction of icosahedral particles—the uncommon line. J Struct Biol 116:48–55

Grigorieff N (2007) FREALIGN: high-resolution refinement of single particle structures. J Struct Biol 157:117–125

Grigorieff N, Harrison SC (2011) Near-atomic resolution reconstructions of icosahedral viruses from electron cryo-microscopy. Curr Opin Struct Biol 21:265–273

Harrison SC, Olson AJ, Schutt CE, Winkler FK, Bricogne G (1978) Tomato bushy stunt virus at 2.9 Å resolution. Nature 276:368–373

Huiskonen JT, Jaalinoja HT, Briggs JA, Fuller SD, Butcher SJ (2007) Structure of a hexameric RNA packaging motor in a viral polymerase complex. J Struct Biol 158:156–164

Jiang W, Chang J, Jakana J, Weigele P, King J, Chiu W (2006) Structure of epsilon15 bacteriophage reveals genome organization and DNA packaging/injection apparatus. Nature 439:612–616

Jiang W, Baker ML, Jakana J, Weigele PR, King J, Chiu W (2008)

Backbone structure of the infectious epsilon 15 virus capsid revealed by electron cryomicroscopy. Nature 451:1130–1134

Liu H, Cheng L (2015) Cryo-EM shows the polymerase structures and a nonspooled genome within a dsRNA virus. Science 349:1347–1350

Liu H, Jin L, Koh SBS, Atanasov I, Schein S, Wu L, Zhou ZH (2010) Atomic structure of human adenovirus by cryo-EM reveals interactions among protein networks. Science 329:1038–1043

Ludtke SJ, Baldwin PR, Chiu W (1999) EMAN: semiautomated software for high-resolution single-particle reconstructions. J Struct Biol 128:82–97

McMullan G, Faruqi AR, Clare D, Henderson R (2014) Comparison of optimal performance at 300 keV of three direct electron detectors for use in low dose electron microscopy. Ultramicroscopy 147:156–163

Morais MC, Tao Y, Olson NH, Grimes S, Jardine PJ, Anderson DL, Baker TS, Rossmann MG (2001) Cryoelectron-microscopy image reconstruction of symmetry mismatches in bacteriophage phi29. J Struct Biol 135:38–46

Scheres SH (2012) RELION: implementation of a Bayesian approach to cryo-EM structure determination. J Struct Biol 180:519–530

Tao Y, Olson NH, Xu W, Anderson DL, Rossmann MG, Baker TS (1998) Assembly of a tailed bacterial virus and its genome release studied in three dimensions. Cell 95:431–437

Thuman-Commike PA, Chiu W (2000) Reconstruction principles of icosahedral virus structure determination using electron cryomicroscopy. Micron 31:687–711

Veesler D, Campbell MG, Cheng A, Fu CY, Murez Z, Johnson JE, Potter CS, Carragher B (2013) Maximizing the potential of electron cryomicroscopy data collected using direct detectors. J Struct Biol 184:193–202

Choosing proper fluorescent dyes, proteins, and imaging techniques to study mitochondrial dynamics in mammalian cells

Xingguo Liu[1], Liang Yang[1], Qi Long[1], David Weaver[2], György Hajnóczky[2]

[1] Key Laboratory of Regenerative Biology, Guangdong Provincial Key Laboratory of Stem Cell and Regenerative Medicine, South China Institute for Stem Cell Biology and Regenerative Medicine, Guangzhou Institutes of Biomedicine and Health, Chinese Academy of Sciences, Guangzhou 510530, China
[2] Department of Pathology, MitoCare Center, Anatomy and Cell Biology, Thomas Jefferson University, Philadelphia, PA 19107, USA

Abstract Mitochondrial dynamics refers to the processes maintaining mitochondrial homeostasis, including mitochondrial fission, fusion, transport, biogenesis, and mitophagy. Mitochondrial dynamics is essential for maintaining the metabolic function of mitochondria as well as their regulatory roles in cell signaling. In this review, we summarize the recently developed imaging techniques for studying mitochondrial dynamics including: mitochondrial-targeted fluorescent proteins and dyes, live-cell imaging using photoactivation, photoswitching and cell fusion, mitochondrial transcription and replication imaging by in situ hybridization, and imaging mitochondrial dynamics by super-resolution microscopy. Moreover, we discuss examples of how to choose and combine proper fluorescent dyes and/or proteins.

Keywords Imaging, Photoactivation, Photoswitching, Mitochondrial dynamics, Super-resolution microscope

INTRODUCTION

The dynamic properties of mitochondria (mitochondrial fission, fusion, transport, biogenesis, and degradation) are critical for maintaining electron transfer chain function and electrical connectivity of mitochondria, preventing build-up of damaged proteins, protecting mitochondrial DNA (mtDNA) integrity, controlling mitochondrial turnover, and regulating many signaling pathways (McBride et al. 2006; Detmer and Chan 2007; Tatsuta and Langer 2008; Liu et al. 2009; Youle and van der Bliek 2012; Liesa and Shirihai 2013; Friedman and Nunnari 2014). Mitochondria continually exchange contents including proteins, mitochondrial RNA (mtRNA), and mtDNA through mitochondrial fusion. Mitochondrial fusion–fission dynamics, transport, and biogenesis/mitophagy also determine mitochondrial morphology, position, and amount. Mitochondria have distinctive features including their own genome, two layers of membranes, and the existence of inner membrane potential, which determine the unique features of mitochondrial dynamics including sequential fusion of two layers of mitochondrial membranes and selective fusion controlled by mitochondrial membrane potential. These features make mitochondrial imaging different from many other organelles.

Fluorescent sensors have attracted considerable attention in mitochondrial dynamics as they enable real-time monitoring and imaging of sub-cellular structures. Improvements in microscopy and the development of mitochondria-specific imaging techniques have greatly advanced the study of mitochondrial dynamics. For example, fluorescence recovery after photobleaching (FRAP) studies have been developed to understand

✉ Correspondence: liu_xingguo@gibh.ac.cn (X. Liu), gyorgy.hajnoczky@jefferson.edu (G. Hajnóczky)

mitochondrial dynamics (Mitra and Lippincott-Schwartz 2010). Several studies have investigated mitochondrial dynamics in plants (Logan and Leaver 2000; Matsushima et al. 2008; Ekanayake et al. 2015); here, we will highlight the ongoing research into mitochondrial dynamics in mammalian cells. We summarize the recently developed imaging techniques and discuss examples of how to properly choose and combine fluorescent dyes and/or proteins.

MITOCHONDRIA-SPECIFIC FLUORESCENT PROTEINS AND DYES

Mitochondrial-targeted fluorescent proteins (FPs) have been widely used. Specific localization is achieved by fusing the FP gene to the mitochondrial targeting sequence of an endogenous protein. The targeting sequence of cytochrome *c* oxidase subunit VIII, which localizes in the mitochondrial matrix, is the most widely used one. According the demands of experiments, one can choose different mitochondrial targeting sequences using software prediction algorithms (Fukasawa et al. 2015; Martelli et al. 2015). Several effective targeting sequences have also been found for the outer surface of the outer membrane, based on AKAP1, OMP25, and TOM70 (Huang et al. 1999; Suzuki et al. 2002; Malka et al. 2005), and the outer surface of the inner membrane using the Cox8a subunit of complex IV (Wilkens et al. 2013).

Fluorescent dyes have also been used to mark mitochondria. Several fluorescent probes including Rhodamine 123 (R123), tetramethylrhodamine methyl ester (TMRM), and JC-1 can be used to monitor mitochondrial membrane potential ($\Delta\Psi_m$) in various cell types (Joshi and Bakowska 2011), making them suitable for marking functional mitochondria. TMRM did not suppress respiration when it was used at low concentrations, making it a good and popular choice for marking functional mitochondria (Scaduto and Grotyohann 1999). The membrane-permeant JC-1 dye exhibits potential-dependent accumulation in mitochondria, indicated by a fluorescence emission shift from green (~ 529 nm) to red (~ 590 nm). Due to the spectral overlap with red- and green-fluorescent dyes and proteins, it is difficult to use JC-1 in a multi-parameter measurement. TMRM and mtGFP can be combined to study high and low $\Delta\Psi_m$ of mitochondria in cells under normal and stress conditions (Twig et al. 2008). However, these $\Delta\Psi_m$ sensors are not optimal for marking mitochondrial morphology since they are easily washed out of cells once the mitochondria experience a loss in $\Delta\Psi_m$. They are also unsuitable for experiments that require cells to be treated with aldehyde fixatives or with other agents that affect the metabolic state of the mitochondria. MitoTracker® probes including MitoTracker Green and MitoTracker Red are fixable probes that are not dependent on $\Delta\Psi_m$ (Chazotte 2011). Although they cannot be used to study $\Delta\Psi_m$, they are good markers for mitochondrial morphology.

To overcome some of the deficiencies in the above-mentioned stains, including poor photostability and cytotoxicity, new alternatives have recently been developed, including NPA-TPP (Huang et al. 2015b) and Splendor (Carvalho et al. 2014), which not only have high specificity to mitochondria, but also exhibit high photostability, negligible cytotoxicity, good water solubility, better fluorescence intensity (FI), and chemical stability in living cells. The extraordinarily high photostability of NPA-TPP is attributed to its extremely stable structure owing to the high π-conjugated macrocycle and strong fluorescent response (Huang et al. 2015a), and Splendor is selected from a series of new rationally designed 2,1,3-benzothiadiazole (BTD) fluorescent derivatives with excellent fluorescence intensity and almost no background signal (Carvalho et al. 2014). Another probe, named MitoBADY, can be also used to visualize mitochondria in living cells by Raman microscopy (Yamakoshi et al. 2015). These stains showed great potential for marking mitochondria and enable real-time and long-term tracking mitochondrial dynamics. The properties and applications of these dyes are summarized in Table 1.

VISUALIZING MITOCHONDRIAL DYNAMICS BY LIVE-CELL IMAGING

Photoactivation

Photoactivatable fluorescent proteins including PAGFP (Patterson and Lippincott-Schwartz 2002), Kindling fluorescent protein (KFP) (Chudakov et al. 2003), and photoactivatable mCherry (PAmCherry) (Subach et al. 2009) are widely used to visualize mitochondrial dynamics (Fig. 1). Using a microscope system with region-of-interest illumination capability, it is possible to mark a subset of mitochondria within a cell. Real-time imaging of photoactivatable fluorescent proteins has allowed the visualization of mitochondrial fusion by time-lapse imaging (Karbowski et al. 2014). Studies of the fusion dynamics of different mitochondrial subcompartments, inner membrane (IMM), outer membrane (OMM), and inter-membrane space (IMS) have been achieved by fusing the photoactivatable FPs to the appropriate targeting signals or proteins. Single-molecule tracking in mitochondria was used to study

Table 1 The properties and applications of mitochondria-specific dyes

Fluorescent dyes	Properties	Applications
Rhodamine 123	Negligible cytotoxicity and potential-sensitive	Measure $\Delta\Psi_m$ and mark functional mitochondria
JC-1	Dual-emission and potential-sensitive	Measure $\Delta\Psi_m$
TMRM/TMRE	Potential-sensitive	Measure $\Delta\Psi_m$, mark functional mitochondria and image time-dependent $\Delta\Psi_m$
MitoTracker® probes	Fixable	Mark mitochondria
NPA-TPP	High photostability, negligible cytotoxicity, and good water solubility	Long-term tracking
Splendor	Excellent fluorescence intensity and negligible background signal	Long-term tracking
MitoBADY	High sensitivity and specificity	Long-term tracking

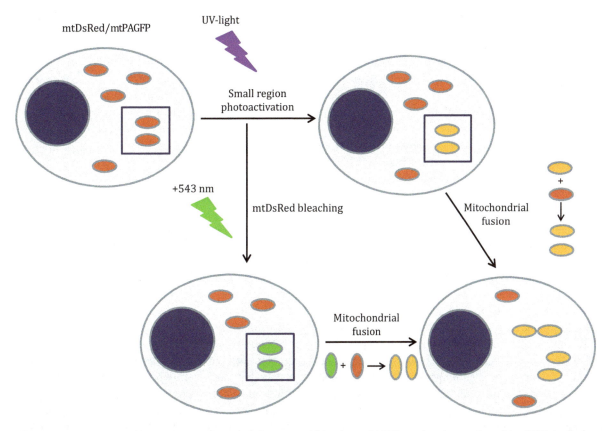

Fig. 1 The combination of mtPAGFP photoactivation with mtDsRed bleaching. PAGPF can be photoactivated by UV-light from a non-fluorescent to a green-fluorescent state. Mitochondria expressing mtPAGFP and mtDsRed are marked in *red* and the nucleus is marked in *blue* in this figure. After small region photoactivation, mitochondria in the photoactivated region are labeled as *yellow* due to the overlapping of *red* and *green*. The combination of mtPAGFP photoactivation with mtDsRed bleaching by 543 nm laser can mark the photoactivated mitochondria with single green fluorescence. When the non-photoactivated mitochondria labeled as *red* fuse with *yellow* mitochondria or *green* mitochondria, the color of both mitochondria become *yellow* for the acquirement of GFP fluorescence

mitochondrial dynamics including the diffusion or movement of mitochondrial membrane proteins. However, its application is limited (Kuzmenko *et al.* 2011; Appelhans *et al.* 2012). In most cases, a combination of one photoactivatable FP and an ordinary FP of another color is used to divide the mitochondria to two classes.

Photoactivatable FPs exhibit weak fluorescence before photoactivation and strong fluorescence after photoactivation. Mitochondrial fusion can be determined by the transfer of photoactivated FP from a subset of mitochondria throughout the mitochondrial network. For instance, the cyan-photoactivated, green-fluorescent PAGFP (mtPAGFP) can be combined with mtRFP or mtDsRed to visualize the exchange of matrix contents between individual mitochondria in real time (Fig. 1). Using high-resolution confocal imaging with region-of-interest scanning, an irreversible photoactivation within the mitochondria is achieved by illumination with a pulse of UV-light (Patterson and Lippincott-Schwartz 2002). As the result of fusion, activated mtPAGFP molecules transfer to unlabeled mitochondria, also leading to a decrease in PAGFP FI of the mitochondria containing activated mtPAGFP. The combination of mitochondrial matrix-targeted, green-photoactivated, red-fluorescent KFP (mtKFP), and green or yellow fluorescent protein (mtGFP, mtYFP) may also be used to study mitochondrial dynamics (Liu et al. 2009), though the newer PAmCherry has faster maturation, better pH stability, faster photoactivation, higher photoactivation contrast, and better photostability (Subach et al. 2009).

The bleaching of one FP can be combined with the photoactivation of another to improve the visibility of fusion, for instance, using the combination described above of PAGFP and DsRed. Using two-photon excitation for photoactivation at ∼750 nm effectively simultaneously bleaches DsRed and photoactivates PAGFP. Alternatively, high intensity from the green, DsRed excitation laser (typically 543 or 561 nm) can be used simultaneously with UV illumination to achieve the photoactivation/bleaching effect, which is illustrated in Fig. 1. This technique has helped show that transfer of matrix contents during mitochondrial fusion is un-directed, i.e. occurs by passive diffusion (Liu et al. 2009).

Combined photoactivation and bleaching is also used to study the transfer of mitochondrial proteins of the IMM, OMM, IMS, and matrix and of nucleoids between mitochondria, the targeted compartment can be marked with red fluorescence in combination with mtPAGFP, or alternatively PAGFP targeted to the sub-compartment of interest can be combined with mtDsRed. In the former case, mitochondrial fusion events can be recognized by the change of PAGFP FI, and the transfer of targeted proteins can be studied by observing the red fluorescence signal among the two mitochondria undergoing mitochondrial fusion. For instance, Tfam-DsRed marking mtDNA and mtPAGFP were combined to show the transfer of mtDNA nucleoids among mitochondria (Yang et al. 2015). Also, TMRM measurements of $\Delta\Psi_m$ and mtPAGFP were combined to show that mitochondria with lower $\Delta\Psi_m$ are less likely to undergo fusion (Twig et al. 2008).

The rate at which the activated mtPAGFP molecules equilibrate across the entire mitochondrial population is used as a measure of fusion activity. In most cases, measurements were performed using a single-cell time-lapse approach, quantifying the equilibration in one cell over a certain time period. Another approach is to use a programmable, motorized stage to alternate among fields and collect multiple cells simultaneously at lower temporal resolution for each cell. This provides a significant increase in the number of cells that can be acquired, but sacrifices the ability to resolve individual fusion events, while maintaining the highly resolved sub-cellular quantification (Lovy et al. 2012).

Photoactivatable fluorescent proteins have also been used to study the fate of mitochondria as they age in cells. First, mtPAGFP was expressed in stem cells. After exposure to a pulse of UV-light, mtPAGFP continued to be synthesized. Subsequently, cells accumulated unlabeled 'young' mitochondria which could be distinguished from the 'old' ones by different PAGFP FI. By following the fates of old and young organelles during the division of human mammary stem-like cells, it was shown that cells apportion the mitochondria asymmetrically between daughter cells according to age (Katajisto et al. 2015).

Photoswitching

Photoswitching fluorescent proteins, which change color upon selective illumination, including Kaede (Ando et al. 2002; Owens and Edelman 2015), Dendra2 (Pham et al. 2012) and Eos (Wiedenmann et al. 2004) have also been used to visualize mitochondrial dynamics. Similar to the experiments described for monitoring mitochondrial fusion by photoactivation, a subset of mitochondria may be irreversibly photoswitched from green to red emitting by targeted illumination of labeled organelles; however, expression of only one protein is required. These color-changing properties provide a simple and powerful technique for regional optical marking (Ando et al. 2002). Mitochondrial-targeted Kaede (MitoKaede) has been used to study mitochondrial dynamics (Owens and Edelman 2015), in the manner illustrated in Fig. 2. MitoKaede was also used to observe mitochondrial dynamics within neuronal processes to show that there is a widespread exchange of mitochondrial components throughout a neuron as a result of organelle fusion (Owens and Walcott 2012). A mouse line expressing mitochondrial Dendra2 (mito-Dendra2) has been generated to study mitochondrial dynamics in primary tissue (Pham et al. 2012).

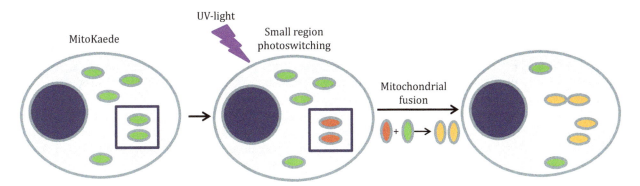

Fig. 2 Photoswitching of MitoKaede. Mitochondria expressing MitoKaede are marked in *green*, and nucleus is marked in *blue* in this figure. After small region of photoswitching with UV-light, mitochondria change color from *green* to *red*. When photoswitched mitochondria labeled as *red* fuse with mitochondria labeled as *green*, the color of mitochondria become *yellow* for the acquirement of GFP fluorescence

Mitochondrial fusion assayed by cell fusion

Cell fusion by treatment with polyethylene glycol (PEG) provides another way to study mitochondrial fusion dynamics. As illustrated in Fig. 3, a cell fusion model is generated using cells with mitochondria labeled in two different colors. Practically, cells expressing mtGFP and others with mtDsRed are mixed and plated commingled.

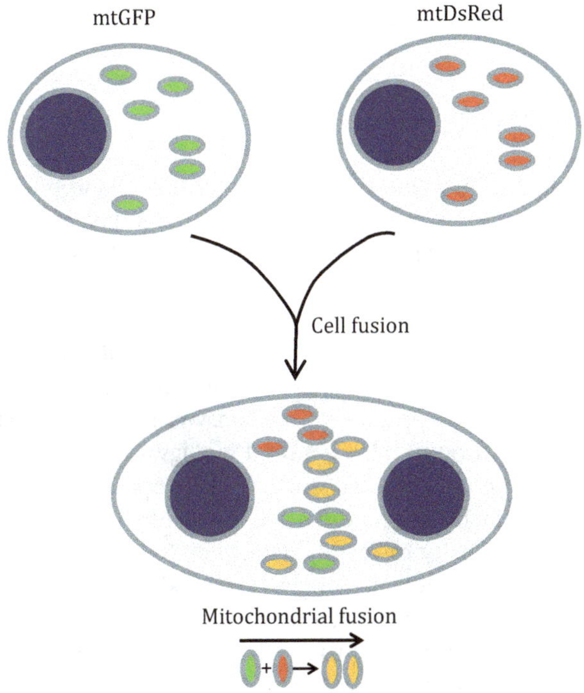

The cells are kept in the presence of cycloheximide from before PEG treatment to block any de novo FP synthesis. Fused cells were imaged after 7 h to visualize the dramatic loss of fusion activity from deletion of mitofusin proteins (Chen *et al.* 2003; Liu *et al.* 2009), but to visualize fusion events in normal cells, imaging should be done within 2 h (Liu *et al.* 2009).

MITOCHONDRIAL TRANSCRIPTION AND REPLICATION IMAGING BY IN SITU HYBRIDIZATION

Understanding the replication and transcription of mtDNA is important to a complete picture of mitochondrial dynamics. The mitochondrial transcription and replication imaging protocol (mTRIP) is a modified fluorescence in situ hybridization approach that specifically targets replicating mtDNA nucleoids and can simultaneously be used to exhibit transcription activity by targeting mtRNA. When combined with labeling of the mitochondria by immunofluorescence or Mito-Tracker loading, mTRIP allows classification of the organelles by replication and/or transcription activity, which has shown to change depending on physiological conditions (Chatre and Ricchetti 2015).

IMAGING MITOCHONDRIAL DYNAMICS BY SUPER-RESOLUTION MICROSCOPY

Conventional confocal microscopy techniques have a limited resolving power of ∼250 nm, due to the properties of light. This diffraction limit is close to the size of mitochondria, which are typically 0.5–1 μm in diameter. However, newly developed super-resolution imaging technologies supply new ways to analyze mitochondrial

Fig. 3 Cell fusion model using cells expressing mtDsRed and cells expressing mtGFP. Mitochondria expressing mtGFP or mtDsRed are marked in *green* or *red*, and nucleus is marked in *blue* in this figure. In cell hybrids between cells expressing mtGFP and cells mtDsRed, mitochondria become *yellow* after *green*-labeled mitochondria fuse with *red*-labeled mitochondria

dynamics and the interactions among sub-cellular organelles. The following four methods of super-resolution microscopy have been commercialized: SIM (Gustafsson 2005), STED (Hell 2003), PALM(Betzig et al. 2006)/STORM (Rust et al. 2006), and Zeiss Airyscan. Any ordinary fluorescent probes and dyes are in principle usable in SIM and Airyscan imaging. Both of these techniques provide resolution of ~130 nm in commercial devices, making it possible to visualize that mitochondria contain different compartments, i.e., distinguish outer membrane from matrix (Gustafsson 2000, 2005). However, with these techniques, it is still not possible to visualize the dynamics of the two membranes of mitochondria, including the events of protein interactions and membrane detaching (Kukat et al. 2011; Jans et al. 2013). STED and PALM technologies can attain better resolution, 50 and 20 nm, respectively (Rust et al. 2006; Shim et al. 2012; Jans et al. 2013). However, not all fluorescent dyes targeted to mitochondria could be used in these two methods and combinations of multiple colors remain especially difficult. However, PALM imaging of mitochondrial dynamics in live cells using a single color has been achieved (Fernandez-Suarez and Ting 2008; Shim et al. 2012; Shcherbakova et al. 2014).

Fluorophores applied for PALM imaging of mitochondrial dynamics must have appropriate ability to switch between dark and fluorescing states, and should have a high quantum efficiency to minimize photodamage of the sample (Patterson et al. 2010; Shcherbakova et al. 2014). Three fluorescent dyes have been used in PALM imaging of mitochondrial dynamics: MitoTracker Orange, MitoTracker Red, and MitoTracker Deep Red (Shim et al. 2012). The cells are labeled according to the ordinary protocol, but should be imaged with PALM system with an extremely high-speed camera (such as sCMOS) to avoid artifacts from the high speed of diffusion of the dyes. The SNAP-tag and CLIP-tag (NEB, MA) or HaloTag (Promega) systems could also be used in super-resolution imaging of mitochondrial dynamics (Klein et al. 2011; Lukinavicius et al. 2013). Mitochondria-targeted proteins carrying these tags can be imaged with PALM after conjugation with proper fluorescence dyes, such as TMR-Star (Klein et al. 2011) and SIR (Lukinavicius et al. 2013). Multi-color super-resolution images of mitochondrial dynamic events can also be achieved in this way.

There are three kinds of FPs that could be applicable in super-resolution imaging, including photoactivatable FPs (PAFPs), photoswitchable FPs (PSFPs), and reversely photoswitchable FPs (rsFPs) (Shcherbakova et al. 2014). PAFPs are not ideal for single-color imaging with PALM as these FPs are invisible in pre-imaging, making it difficult to find the interesting areas and events. They are, however, in principle useful for multi-color super-resolution imaging since each occupies only one color. Both PAGFP and PAmCherry can be used in PALM imaging, though their optical properties are far from optimal.

Many PSFPs have already been used in time-lapse imaging of mitochondrial dynamics as described above, including mEOS family proteins, mKiKGR, and Dendra2. Each of these could also in principle be used in PALM imaging. PSmOrange and PSmOrange2 have potential for use in multi-color PALM imaging using the far-red spectrum (Subach et al. 2011).

rsFPs can be switched on or off by an illumination at different wavelengths. They are potentially useful for studying mitochondrial dynamics with PALM. However, there are few rsFPs that could be applicable in PALM imaging due to the requirement for high quantum efficiency and long half-time period. Dronpa family members such as Dronpa2 and Dronpa3 have been successfully applied in PALM imaging of mitochondria (Rocha et al. 2014). Newly developed mGeos-M is derived from mEos2 and was reported to also be effective in PALM imaging (Chang et al. 2012). Thus, it is another candidate for mitochondrial dynamics with PALM imaging.

DISCUSSION

While the promise of spectrally diverse fluorophores and reporter strains is exciting for the study of mitochondrial dynamics, the choice of proper FPs or probes is crucial for success in most applications. These probes and live imaging approaches can be used to understand basic biological mechanisms of mitochondrial dynamics, and have already given ground-breaking insight into mitochondrial dynamics and answered some basic questions around mitochondrial biology.

Impediments to the further development of live imaging include: (1) the stability of FPs and probes, (2) the intensity of readily detectable fluorescence that is required for live imaging, (3) the toxicity of many FP fusion proteins when expressed at readily detectable levels and probes that may affect mitochondrial metabolism, and (4) the availability of reagents for directing lineage-specific FP expression.

Advances in live imaging are facilitating the visualization of development at single-cell resolution in living zebrafish embryos (Andersen et al. 2010) or C. elegans (Cornaglia et al. 2015). As reflected in these studies, FPs have immense potential for live imaging applications. We can predict that high-resolution microscopy will give a deeper understanding of mitochondrial dynamics in living mammalian.

Super-resolution microscopy has already made headways in mitochondrial biology (Jakobs and Wurm 2014) but significant limitations remain in live and multi-color applications of the higher resolution approaches of STED, PALM, and STORM. Progress with new fluorophores/fluorescent proteins, especially ones that can utilize longer excitation wavelengths, are expected to move this field forward. Furthermore, development of protocols for quantitative data analysis will greatly enhance the information that can be harvested from STORM/PALM datasets.

Acknowledgements This work was financially supported by the "973" program (2013CB967403, 2012CB721105, and 2016YFA0100302), the "Frontier Science Key Research Program" of the Chinese Academy of Sciences (QYZDB-SSW-SMC001), the National Natural Science Foundation projects of China (31622037, 31271527, 81570520, 31601176, and 31601088), Guangzhou Science and Technology Program (2014Y2-00161), Guangzhou Health Care and Cooperative Innovation Major Project (201604020009), Guangdong Natural Science Foundation for Distinguished Young Scientists (S20120011368), Guangdong Province Science and Technology Innovation The Leading Talents Program (2015TX01R047), Guangdong Province Science and Technology Innovation Young Talents Program (2014TQ01R559), Guangdong Province Science and Technology Program (2015A020212031), the PhD Start-up Fund of Natural Science Foundation of Guangdong Province (2014A030310071), and by NIH Grants (AA017773 and DK051526) to G.H.

Compliance with Ethical Standards

Conflict of interest Xingguo Liu, Liang Yang, Qi Long, David Weaver and György Hajnóczky declare that they have no conflict of interest.

Human and Animal Rights and Informed Consent This article does not contain any studies with human or animal subjects performed by any of the authors.

References

Andersen E, Asuri N, Clay M, Halloran M (2010) Live imaging of cell motility and actin cytoskeleton of individual neurons and neural crest cells in zebrafish embryos. J Vis Exp 36:1726

Ando R, Hama H, Yamamoto-Hino M, Mizuno H, Miyawaki A (2002) An optical marker based on the UV-induced green-to-red photoconversion of a fluorescent protein. Proc Natl Acad Sci USA 99:12651–12656

Appelhans T, Richter CP, Wilkens V, Hess ST, Piehler J, Busch KB (2012) Nanoscale organization of mitochondrial microcompartments revealed by combining tracking and localization microscopy. Nano Lett 12:610–616

Betzig E, Patterson GH, Sougrat R, Lindwasser OW, Olenych S, Bonifacino JS, Davidson MW, Lippincott-Schwartz J, Hess HF (2006) Imaging intracellular fluorescent proteins at nanometer resolution. Science 313:1642–1645

Carvalho PH, Correa JR, Guido BC, Gatto CC, De Oliveira HC, Soares TA, Neto BA (2014) Designed benzothiadiazole fluorophores for selective mitochondrial imaging and dynamics. Chemistry 20:15360–15374

Chang H, Zhang M, Ji W, Chen J, Zhang Y, Liu B, Lu J, Zhang J, Xu P, Xu T (2012) A unique series of reversibly switchable fluorescent proteins with beneficial properties for various applications. Proc Natl Acad Sci USA 109:4455–4460

Chatre L, Ricchetti M (2015) mTRIP: an imaging tool to investigate mitochondrial DNA dynamics in physiology and disease at the single-cell resolution. Methods Mol Biol 1264:133–147

Chazotte B (2011) Labeling mitochondria with MitoTracker dyes. Cold Spring Harb Protoc 2011:990–992

Chen H, Detmer SA, Ewald AJ, Griffin EE, Fraser SE, Chan DC (2003) Mitofusins Mfn1 and Mfn2 coordinately regulate mitochondrial fusion and are essential for embryonic development. J Cell Biol 160:189–200

Chudakov DM, Belousov VV, Zaraisky AG, Novoselov VV, Staroverov DB, Zorov DB, Lukyanov S, Lukyanov KA (2003) Kindling fluorescent proteins for precise in vivo photolabeling. Nat Biotechnol 21:191–194

Cornaglia M, Mouchiroud L, Marette A, Narasimhan S, Lehnert T, Jovaisaite V, Auwerx J, Gijs MA (2015) An automated microfluidic platform for C. elegans embryo arraying, phenotyping, and long-term live imaging. Sci Rep 5:10192

Detmer SA, Chan DC (2007) Functions and dysfunctions of mitochondrial dynamics. Nat Rev Mol Cell Biol 8:870–879

Ekanayake SB, El Zawily AM, Paszkiewicz G, Rolland A, Logan DC (2015) Imaging and analysis of mitochondrial dynamics in living cells. Methods Mol Biol 1305:223–240

Fernandez-Suarez M, Ting AY (2008) Fluorescent probes for super-resolution imaging in living cells. Nat Rev Mol Cell Biol 9:929–943

Friedman JR, Nunnari J (2014) Mitochondrial form and function. Nature 505:335–343

Fukasawa Y, Tsuji J, Fu SC, Tomii K, Horton P, Imai K (2015) MitoFates: improved prediction of mitochondrial targeting sequences and their cleavage sites. Mol Cell Proteom 14:1113–1126

Gustafsson MG (2000) Surpassing the lateral resolution limit by a factor of two using structured illumination microscopy. J Microsc 198:82–87

Gustafsson MG (2005) Nonlinear structured-illumination microscopy: wide-field fluorescence imaging with theoretically unlimited resolution. Proc Natl Acad Sci USA 102:13081–13086

Hell SW (2003) Toward fluorescence nanoscopy. Nat Biotechnol 21:1347–1355

Huang LJ, Wang L, Ma Y, Durick K, Perkins G, Deerinck TJ, Ellisman MH, Taylor SS (1999) NH2-Terminal targeting motifs direct dual specificity A-kinase-anchoring protein 1 (D-AKAP1) to either mitochondria or endoplasmic reticulum. J Cell Biol 145:951–959

Huang S, Han R, Zhuang Q, Du L, Jia H, Liu Y, Liu Y (2015a) New photostable naphthalimide-based fluorescent probe for mitochondrial imaging and tracking. Biosens Bioelectron 71:313–321

Huang S, Han R, Zhuang Q, Du L, Jia H, Liu Y, Liu Y (2015b) New photostable naphthalimide-based fluorescent probe for mitochondrial imaging and tracking. Biosens Bioelectron 71:313–321

Jakobs S, Wurm CA (2014) Super-resolution microscopy of mitochondria. Curr Opin Chem Biol 20:9–15

Jans DC, Wurm CA, Riedel D, Wenzel D, Stagge F, Deckers M, Rehling P, Jakobs S (2013) STED super-resolution microscopy reveals an array of MINOS clusters along human mitochondria. Proc Natl Acad Sci USA 110:8936–8941

Joshi DC, Bakowska JC (2011) Determination of mitochondrial membrane potential and reactive oxygen species in live rat cortical neurons. J Vis Exp 51:e2704

Karbowski M, Cleland MM, Roelofs BA (2014) Photoactivatable green fluorescent protein-based visualization and quantification of mitochondrial fusion and mitochondrial network complexity in living cells. Methods Enzymol 547:57–73

Katajisto P, Dohla J, Chaffer CL, Pentinmikko N, Marjanovic N, Iqbal S, Zoncu R, Chen W, Weinberg RA, Sabatini DM (2015) Stem cells. Asymmetric apportioning of aged mitochondria between daughter cells is required for stemness. Science 348:340–343

Klein T, Loschberger A, Proppert S, Wolter S, van de Linde S, Sauer M (2011) Live-cell dSTORM with SNAP-tag fusion proteins. Nat Methods 8:7–9

Kukat C, Wurm CA, Spahr H, Falkenberg M, Larsson NG, Jakobs S (2011) Super-resolution microscopy reveals that mammalian mitochondrial nucleoids have a uniform size and frequently contain a single copy of mtDNA. Proc Natl Acad Sci USA 108:13534–13539

Kuzmenko A, Tankov S, English BP, Tarassov I, Tenson T, Kamenski P, Elf J, Hauryliuk V (2011) Single molecule tracking fluorescence microscopy in mitochondria reveals highly dynamic but confined movement of Tom40. Sci Rep 1:195

Liesa M, Shirihai OS (2013) Mitochondrial dynamics in the regulation of nutrient utilization and energy expenditure. Cell Metab 17:491–506

Liu X, Weaver D, Shirihai O, Hajnoczky G (2009) Mitochondrial 'kiss-and-run': interplay between mitochondrial motility and fusion-fission dynamics. EMBO J 28:3074–3089

Logan DC, Leaver CJ (2000) Mitochondria-targeted GFP highlights the heterogeneity of mitochondrial shape, size and movement within living plant cells. J Exp Bot 51:865–871

Lovy A, Molina AJ, Cerqueira FM, Trudeau K, Shirihai OS (2012) A faster, high resolution, mtPA-GFP-based mitochondrial fusion assay acquiring kinetic data of multiple cells in parallel using confocal microscopy. J Vis Exp 65:e3991

Lukinavicius G, Umezawa K, Olivier N, Honigmann A, Yang G, Plass T, Mueller V, Reymond L, Correa IR Jr, Luo ZG, Schultz C, Lemke EA, Heppenstall P, Eggeling C, Manley S, Johnsson K (2013) A near-infrared fluorophore for live-cell super-resolution microscopy of cellular proteins. Nat Chem 5:132–139

Malka F, Guillery O, Cifuentes-Diaz C, Guillou E, Belenguer P, Lombes A, Rojo M (2005) Separate fusion of outer and inner mitochondrial membranes. EMBO Rep 6:853–859

Martelli PL, Savojardo C, Fariselli P, Tasco G, Casadio R (2015) Computer-based prediction of mitochondria-targeting peptides. Methods Mol Biol 1264:305–320

Matsushima R, Hamamura Y, Higashiyama T, Arimura S, Sodmergen Tsutsumi N, Sakamoto W (2008) Mitochondrial dynamics in plant male gametophyte visualized by fluorescent live imaging. Plant Cell Physiol 49:1074–1083

McBride HM, Neuspiel M, Wasiak S (2006) Mitochondria: more than just a powerhouse. Curr Biol: CB 16:R551–R560

Mitra K, Lippincott-Schwartz J (2010) Analysis of mitochondrial dynamics and functions using imaging approaches. Curr Protoc Cell Biol Chapter 4: Unit 4 25 21–21

Owens GC, Edelman DB (2015) Photoconvertible fluorescent protein-based live imaging of mitochondrial fusion. Methods Mol Biol 1313:237–246

Owens GC, Walcott EC (2012) Extensive fusion of mitochondria in spinal cord motor neurons. PLoS ONE 7:e38435

Patterson GH, Lippincott-Schwartz J (2002) A photoactivatable GFP for selective photolabeling of proteins and cells. Science 297:1873–1877

Patterson G, Davidson M, Manley S, Lippincott-Schwartz J (2010) Superresolution imaging using single-molecule localization. Ann Rev Phys Chem 61:345–367

Pham AH, McCaffery JM, Chan DC (2012) Mouse lines with photoactivatable mitochondria to study mitochondrial dynamics. Genesis 50:833–843

Rocha S, De Keersmaecker H, Uji-i H, Hofkens J, Mizuno H (2014) Photoswitchable fluorescent proteins for superresolution fluorescence microscopy circumventing the diffraction limit of light. Methods Mol Biol 1076:793–812

Rust MJ, Bates M, Zhuang X (2006) Sub-diffraction-limit imaging by stochastic optical reconstruction microscopy (STORM). Nat Methods 3:793–795

Scaduto RC Jr, Grotyohann LW (1999) Measurement of mitochondrial membrane potential using fluorescent rhodamine derivatives. Biophys J 76:469–477

Shcherbakova DM, Sengupta P, Lippincott-Schwartz J, Verkhusha VV (2014) Photocontrollable fluorescent proteins for super-resolution imaging. Ann Rev Biophys 43:303–329

Shim SH, Xia C, Zhong G, Babcock HP, Vaughan JC, Huang B, Wang X, Xu C, Bi GQ, Zhuang X (2012) Super-resolution fluorescence imaging of organelles in live cells with photoswitchable membrane probes. Proc Natl Acad Sci USA 109:13978–13983

Subach FV, Patterson GH, Manley S, Gillette JM, Lippincott-Schwartz J, Verkhusha VV (2009) Photoactivatable mCherry for high-resolution two-color fluorescence microscopy. Nat Methods 6:153–159

Subach OM, Patterson GH, Ting LM, Wang Y, Condeelis JS, Verkhusha VV (2011) A photoswitchable orange-to-far-red fluorescent protein, PSmOrange. Nat Methods 8:771–777

Suzuki H, Maeda M, Mihara K (2002) Characterization of rat TOM70 as a receptor of the preprotein translocase of the mitochondrial outer membrane. J Cell Sci 115:1895–1905

Tatsuta T, Langer T (2008) Quality control of mitochondria: protection against neurodegeneration and ageing. EMBO J 27:306–314

Twig G, Elorza A, Molina AJ, Mohamed H, Wikstrom JD, Walzer G, Stiles L, Haigh SE, Katz S, Las G, Alroy J, Wu M, Py BF, Yuan J, Deeney JT, Corkey BE, Shirihai OS (2008) Fission and selective fusion govern mitochondrial segregation and elimination by autophagy. EMBO J 27:433–446

Wiedenmann J, Ivanchenko S, Oswald F, Schmitt F, Rocker C, Salih A, Spindler KD, Nienhaus GU (2004) EosFP, a fluorescent marker protein with UV-inducible green-to-red fluorescence conversion. Proc Natl Acad Sci USA 101:15905–15910

Wilkens V, Kohl W, Busch K (2013) Restricted diffusion of OXPHOS complexes in dynamic mitochondria delays their exchange between cristae and engenders a transitory mosaic distribution. J Cell Sci 126:103–116

Yamakoshi H, Palonpon A, Dodo K, Ando J, Kawata S, Fujita K, Sodeoka M (2015) A sensitive and specific Raman probe based on bisarylbutadiyne for live cell imaging of mitochondria. Bioorg Med Chem Lett 25:664–667

Yang L, Long Q, Liu J, Tang H, Li Y, Bao F, Qin D, Pei D, Liu X (2015) Mitochondrial fusion provides an 'initial metabolic complementation' controlled by mtDNA. Cell Mol Life Sci 72:2585

Youle RJ, van der Bliek AM (2012) Mitochondrial fission, fusion, and stress. Science 337:1062–1065

Crystal structures of MdfA complexed with acetylcholine and inhibitor reserpine

Ming Liu[1], Jie Heng[2], Yuan Gao[2], Xianping Wang[2]✉

[1] College of Biotechnology, Tianjin University of Science and Technology, Tianjin 300457, China
[2] National Laboratory of Macromolecules, National Center of Protein Science - Beijing, Institute of Biophysics, Chinese Academy of Sciences, Beijing 100101, China

Abstract The DHA12 family of transporters contains a number of prokaryotic and eukaryote membrane proteins. Some of these proteins share conserved sites intrinsic to substrate recognition, structural stabilization and conformational changes. For this study, we chose the MdfA transporter as a model DHA12 protein to study some general characteristics of the vesicular neurotransmitter transporters (VNTs), which all belong to the DHA12 family. Two crystal structures were produced for *E. coli* MdfA, one in complex with acetylcholine and the other with potential reserpine, which are substrate and inhibitor of VNTs, respectively. These structures show that the binding sites of these two molecules are different. The Ach-binding MfdA is mainly dependent on D34, while reserpine-binding site is more hydrophobic. Based on sequence alignment and homology modelling, we were able to provide mechanistic insights into the association between the inhibition and the conformational changes of these transporters.

Keywords MdfA, Reserpine, DHA12, Antiporter, Acetylcholine

INTRODUCTION

MdfA, as a typical antiporter of the major facilitator superfamily (MFS), has been the subject of extensive study, especially in the research of multidrug transport mechanisms (Edgar and Bibi 1997). Many small molecules are substrates of MdfA, including neutral compounds such as chloramphenicol (Cm) and thiamphenicol, lipophilic cations such as tetraphenylphosphonium (TPP$^+$) and ethidium bromide (EtBr), and the zwitterionic drug—ciprofloxacin (Adler and Bibi 2004). Crystal structures of *E. coli* MdfA (ecMdfA) have recently been reported by our laboratory (Heng et al. 2015). Based on the parsed structures of ecMdfA, the key location for the binding of a variety of substrates was determined to be the large central cavity within MdfA, which is lined with mostly hydrophobic residues. An acidic residue, D34, is also present deep within this cavity, and is proposed to be critical for the binding of certain substrates (Heng et al. 2015). Also, the substrate-bound crystal structure of MdfA provided a base for further structural and functional study of other homologous proteins, such as MdtM, another MFS transporter, which share high sequence identity with MdfA (Paul et al 2014).

In mammalian cells, vesicular monoamine transporter (VMAT) and vesicular acetylcholine transporter (VAChT) are homologous proteins with MdfA and transport mono-positively charged amine neurotransmitters and hormones from the cytoplasm and concentrate them within secretory vesicles by alternating their access mechanism (Parsons 2000). They are subclassified into the SLC18 group of MFS in eukaryotes (Lawal and Krantz 2013) and also belong to the DHA12 family (Paulsen et al. 1996; Putman et al. 2000; Heng et al. 2015). Two isoforms of VMAT, 1 and 2, have been

✉ Correspondence: wangxp@moon.ibp.ac.cn (X. Wang)

cloned and identified (Erickson et al. 1992). VMATs and VAChT are predicted to possess 12 transmembrane helices (TM) consisting of two pseudo-symmetrical domains, and use the proton motive force (PMF) of the transmembrane electrochemical proton gradient as the driving force for the uptake of cytoplasmic substrates into the vesicles (Zhang et al. 2015). Extensive pharmacological studies have shown that VMATs and VAChT have broad substrate specificity, and a large number of substrates and inhibitors of VMATs and VAChT have been identified (Yelin and Schuldiner 1995; Erickson et al. 1996; Bravo et al. 2005). Several residues in these proteins have been identified to contribute to substrate recognition (Merickel et al. 1995; Finn and Edwards 1997), protonation and energy coupling (Yaffe et al. 2013).

To study the transport mechanisms of VMATs and VAChT, a homology model based on LacY was initially used to investigate the alternating access mechanism of VMAT2 (Vardy et al. 2004; Yaffe et al. 2013). However, owing to their low sequence similarity and the fact that LacY is a symporter rather than antiporter, LacY is not suitable for studying the proton/substrate antiport mechanism of VMATs and VAChT. Unlike previous passable symporter LacY, it is reasonable to assume that the proton/substrate antiporter MdfA would be a better reference for homology model and functional study of VMATs and VAChT. Although MdfA and its eucaryotic homologs share low sequence identity (Fig. S1), those conserved motifs sharing in DHA12 family reflect the fact that they possess similar proton-substrate antiporting mechanism as discussed at length in previous paper(Heng et al. 2015). More importantly, several well-known substrates of MdfA are also transported by VAChT, including tetraphenylphosphonium, ethidium and rhodamine (Bravo et al. 2005). An inhibitor of VMATs, reserpine (Erickson et al. 1996), is also reported to inhibit the translocation cycle of MdfA (Edgar and Bibi 1999).

Reserpine is an antipsychotic and antihypertensive drug used for the relief of psychotic symptoms, and irreversibly blocks VMATs in the presynaptic neurons (Preskorn 2007). Reserpine was proposed to directly bind and competitively inhibit the efflux pump during proton/substrate antiport (Holler et al. 2012).

To elucidate the transportation mechanism of VMATs and VAChT and the potential inhibition mechanism of reserpine towards multiple transporters, we therefore used the ecMdfA structure as a model to investigate the binding of acetylcholine (ACh) and reserpine molecules. Here, we report the crystal structures of ecMdfA complexed with ACh at 2.8-Å resolution using the soaking method, and with possible reserpine at 3.5-Å resolution using co-crystallization. Both of these structures were captured in the inward-facing conformation, albeit under different crystallization conditions. The reserpine-MdfA complex structure shows more extensive interactions between the transporter and substrate than that observed within the ACh-MdfA complex. Based on these ecMdfA structures, we could simulate the docking of reserpine into a model of VAChT (Fig. 1). We used this docked model to postulate a mechanism for substrate transport and inhibitor binding for VAChT.

RESULTS

Acetylcholine (ACh)-bound structure of ecMdfA

The complexed structures of ecMdfA with ACh and reserpine were obtained under two distinct crystallization conditions. Their space groups were $C2$ and $P3_121$, respectively, and the structures were resolved with the molecular replacement method (Table S1). In the two crystal structures, the final refined structural models contained the intact peptide chains of residues 10–400 and 14–400, respectively, and were both in the inward-facing conformation consistent with the previously reported Dxc-ecMdfA structure (Heng et al. 2015). The

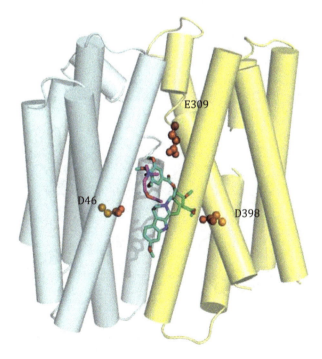

Fig. 1 Possible inhibition mechanism for VAChT. Three conserved, negatively charged residues of human VAChT (D46, E309 and D398) line the pocket of the predicted VAChT structure. ACh and reserpine are, respectively, represented with *magenta* and *cyan* sticks. The N-terminal domain is shown in *blue*, and the C-terminal domain is shown in *yellow*

ACh-complex structure was obtained with the soaking method, during which the pH value was increased from 5.4 to 8.0, using the same conditions under which crystals of Dxc-ecMdfA were obtained and supplemented with 5 mmol/L ACh. We postulated that Dxc could be replaced by ACh when the pH of crystallized condition increases since key amino acids around the binding cavity determine ecMdfA prefers to bind positively charged Ach rather than negatively charged Dxc. Results of the structural studies showed no differences between these two complexed structures, except for the electron density of Dxc being replaced with that of ACh (Fig. 2).

As expected, the resolved ACh-complex structure shows that ACh indeed binds in the vicinity of the D34 residue. The distance between the positively charged head group of ACh and the negatively charged side chain of D34 is ~4.5 Å, while residues Y30, L263, F358 and M358 form hydrophobic interactions with ACh (Fig. 2). We overlaid the potential protonation sites of the VMATs or VAChT on the structure of ecMdfA based on sequence alignment (Fig. S1). We found that three acidic residues of VAChT, D33, E309 and D398, correspond with residues E26, I239 and N331 of ecMdfA, respectively, which are located near the ACh-binding pocket. E309 is located at the bottom of the substrate-binding cavity, and may play the same critical protonation-deprotonation role as D34 in ecMdfA. Residue D33, which corresponds with E26 on ecMdfA, has been shown to take part in the recognition of positively charged substrates (Merickel et al. 1995).

Reserpine-bound structure of ecMdfA

The second crystal structure we generated was of reserpine-complexed ecMdfA. Reserpine is an effective inhibitor of MdfA (Edgar and Bibi 1999). There were two ecMdfA molecules in one asymmetric unit of the $P3_121$ structure. The interfaces are consisted of the twelfth helices of ecMdfA, and comprised mostly hydrophobic amino acids, such as L382, V386 and I389. Both of the two transporters in the unit adopted an inward-facing conformation (Fig. S2).

In this structure, we identified a suspected density attributable to the presence of reserpine in the central cavity. We built a reserpine molecule into the model at this position to analyse its possible inhibition mechanism. Reserpine occupies a more hydrophobic position than ACh because it is a larger molecule. The trimethoxybenzoyl in the tail of reserpine is near residue D34. Hydrophobic interactions were localized on the trimethoxybenzoyl group and other groups of the reserpine molecule (Fig. 3). For comparison, in the

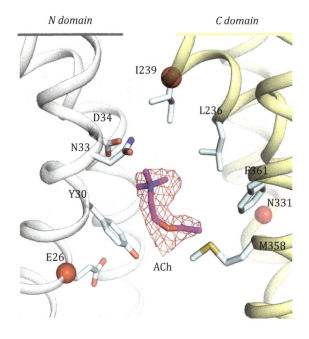

Fig. 2 ACh-binding sites of MdfA. ACh (*magenta sticks*) binds in the pocket between the N- and C-terminal domains, which are illustrated in *white* and *yellow*, respectively. The omit Fo-Fc density (in *red*) for ACh is contoured at 3σ. Three equivalent conserved acid residues which may be important in substrate binding and protonation in VAChT are marked with *red spheres*

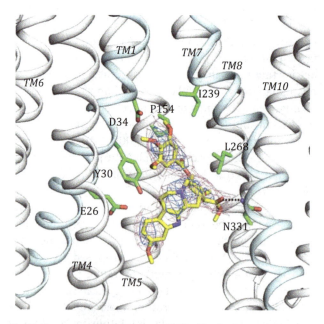

Fig. 3 Reserpine-binding site of MdfA. The backbone of MdfA is shown in cartoon representation. Two important cavity helices, TM1 and TM7, are coloured in *cyan*, while the other helices are in *white*. The reserpine molecule is illustrated with *yellow sticks*. Amino acid residues near reserpine are shown with *green sticks*. The omit 2Fo-Fc density for reserpine is contoured in *marine* (1σ) and *pink* (0.4σ). Hydrogen bonds between reserpine and N331 are shown as *dotted lines*

previous 2.0-Å resolution Dxc-complex structure of ecMdfA, residue D34 was observed to form hydrogen bond with Dxc, which exerts an inhibitory effect on the Cm resistance of ecMdfA (Fig. 4). In contrast, three methoxyls of reserpine were in proximity to the D34 residue, potentially forming hydrogen bonds for structure stabilization. In addition, both Dxc and reserpine directly interact with residue P154 in the conserved motif C of MdfA. The mechanism of reserpine inhibition is therefore likely to be associated with motif C.

DISCUSSION

According to the multiple sequence alignment of the 12 transmembrane proton-dependent bacterial multidrug transporters of the MFS family (also known as DHA12), these transporters contain a number of highly conserved motifs, which indicates that they may share a similar transportation mechanism (Putman et al. 2000). Here, we show the sequence alignments of VMATs and VAChT from *Homo sapiens*, *Mus musculus* and *Rattus norvegicus* (Fig. S1), including their conserved motifs. Motif A is in the cytoplasmic loop between TM 2 and 3 of the MFS, with the most conserved residue being G73, which has previously been clearly demonstrated in the structure of YajR (Jiang et al. 2013). Motif C, motif D and motif G (or motif C′) all contain a conserved proline residue in their consensus sequences of TMs 5, 1 and 11, respectively. Proline, which is incapable of acting as a hydrogen bond donor, often plays a role of helix "breaker" in the transmembrane helices of α-helical membrane proteins (Chandrasekaran et al. 2006).

Based on the structures of the ecMdfA-ACh and ecMdfA-reserpine complexes, we modelled a homology structure of VAChT overlaid with ACh and reserpine (Fig. 1). The homology VAChT model was cytoplasm-facing and consisted of 12 transmembrane helices, similar to the structure of ecMdfA. In addition, there were three carboxyl residues, D46, E309 and D398, located in TM1, 7 and 8, respectively, all within the central cavity. In the ACh-VAChT complexed model, ACh binds to E309 through negative–positive charge interaction. Residue E309 is located at the bottom of the large hydrophobic substrate-binding pocket, corresponding with the positioning of residue D34 of ecMdfA. This confirmed that D34 of MdfA is likely deprotonated during the process of substrate binding as previously proposed (Heng et al. 2015), as upon ligand binding, E309 is likely to become similarly deprotonated. This buried acidic residue is necessary for the recognition of monovalent, positively charged substrates. In the VAChT-reserpine-complexed model, it appeared that residues D398 and E309 may form a hydrogen bond network with reserpine. This observation may explain the inhibition mechanism of reserpine, because this hydrogen bond network would stabilize VAChT in the cytoplasm-facing conformation, whereby inactivating it (Darchen et al. 1989). Furthermore, Khare et al. (2010) showed that both D398 and E309 directly bind with the allosteric inhibitor—vesamicol, but only the E309 residue can bind ACh, while D398 does not. These observations are consistent with our model.

Most members of the DHA12 family recognize multiple substrates and trigger conformational change upon substrate binding. Why the binding of a variety of substrates can drive the transportation cycle, while inhibitor binding may stabilize them in only one of their two possible conformations. The conserved proline in VMATs and VAChTs offers a clue as to the mechanism of substrate binding and the subsequent conformational change exerted. Highly conserved prolines are located in motif C, which is also known as the "antiporter motif" (De Jesus et al. 2005). Several conformationally sensitive residues, including some prolines, have been identified in motif C in VMAT2 (Ugolev et al. 2013) and VAChT (Luo and Parsons 2010). Here, we presented structural information of motif C from ecMdfA (Fig. 5A, B). Several prolines from TM1, 5 and 7 were observed to cluster together to form a "bottleneck" in the 3D structure. Similar features of other DHA12 members have been shown to inhibit solute leak inside the cell (De Jesus et al. 2005). We named this structural motif "3D-motif C", which consists of four helices each from the

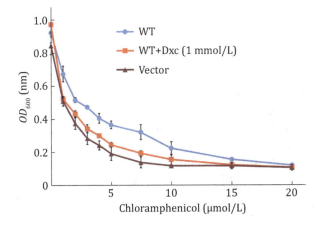

Fig. 4 Inhibition of MdfA chloramphenicol resistance by Dxc. *E. coli* C43 (DE3) cells harbouring wild-type MdfA or vector only were grown in the presence of increasing concentrations of chloramphenicol supplemented with Dxc (1 mmol/L). Relative growth was calculated from the cell density and measured by culture absorption at 600 nm. Assays were done in quadruplicate, plotting the average with error bars of ±1 SD

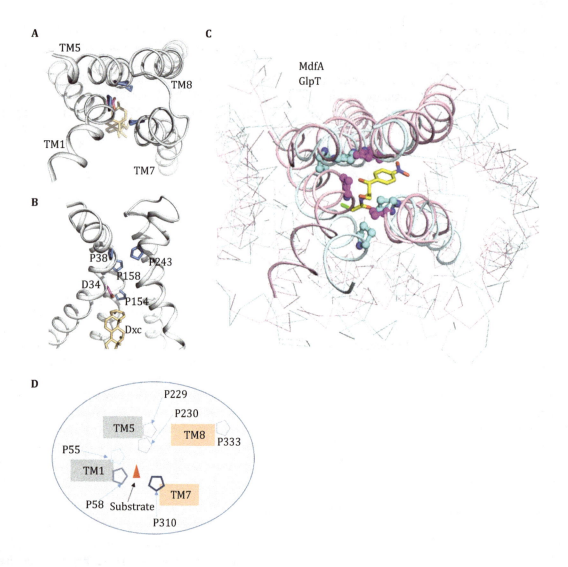

Fig. 5 Structural features of the 3D-motif C of MdfA and a schematic of the transportation model. **A** 3D-motif C of MdfA (PDB ID: 4ZP0) viewed from the periplasmic side. Prolines are shown as *blue sticks* and Dxc is in *yellow*. **B** 3D-motif C of MdfA viewed parallel to the membrane. **C** Superposition of the 3D-motif C of MdfA and GlpT. **D** Schematic of the 3D-motif C of mVAChT

N-terminal and C-terminal domains. We also found another inward-facing MFS antiporter GlpT, an organic phosphate/inorganic phosphate antiporter, which contained a similar prolines cluster in the same location of antiporter motif when GlpT was superposed on the ecMdfA structure (Huang et al. 2003) (Fig. 5C). Molecular dynamics simulations of GlpT are consistent with the proposed mechanism that proline-induced flexibility in the TM helices is critical to the conformational change of MFS pseudo-rigid body motions (D'Rozario and Sansom 2008). Based on this, a sketch map of the 3D-motif C in mVAChT (mouse VAChT) could be postulated on the basis of multiple sequence alignment and homology modelling (Fig. 5D). Residue P333 of mVAChT differs from 3D-motif C in MdfA and may facilitate the bending of TM 8 towards the proline cluster, similarly to the role of P263 in MdfA. Mutations near 3D-motif C of VMATs have been shown to influence the transport rate by affecting the rate-limiting step of the transport cycle (Ugolev et al. 2013). Evidence of interactions between P154 and inhibitors in the ecMdfA-Dxc and ecMdfA-reserpine complexes points to the close association of 3D-motif C with inhibition. In short, the conserved sequence of 3D-motif C may participate in the gating-like movements of these transporters.

In conclusion, the structures reported here of ecMdfA in complex with ACh and potential reserpine improve our understanding of the mechanisms of multiple substrate recognition of MdfA. Using sequence alignment of the VMATs, VAChT and ecMdfA, several conserved residues and motifs were identified. Therefore, we were able to model a homology structure of VAChT based on the ecMdfA structure, and the model had credibility based on previously published biochemical results. The 3D-motif C was determined to likely play a critical role in substrate transport and conformational change. Study of ecMdfA-complexed structures and predicted models may facilitate further research on the structure and function of vesicular neurotransmitter transporters. Certainly, the authentic structures of VNTs are necessary to explain the mechanisms of substrates recognition and transport.

MATERIALS AND METHODS

Protein expression and purification

The full-length *MdfA* gene was subcloned into the pET-28a vector (Novagen) with a C-terminal His_6-tag(Heng et al. 2015). For protein expression, *E. coli* C43(DE3) strain transformed with the recombinant plasmid was cultured in Terrific Broth supplemented with 25 μg/mL kanamycin at 37 °C and induced at an OD_{600nm} of 0.8 with 0.5 mmol/L isopropyl-D-thiogalactoside (IPTG) at 16 °C for 18 h. The cells were harvested by centrifugation at 4000 g for 30 min, resuspended in buffer A (20 mmol/L HEPES pH 7.2, 300 mmol/L NaCl, 10% (v/v) glycerol and 5 mmol/L β-mercaptoethanol) and then disrupted at 10,000–15,000 psi using a JN-R2C homogenizer (JNBio, China). Cell debris was removed by centrifugation at 17,000 g for 15 min. The supernatant was ultracentrifuged at 100,000 g for 1 h. Membrane fraction was harvested and solubilized with 0.5% (w/v) n-decyl-β-D-maltopyranoside (DM; Anatrace) for 15 min at 4 °C. We eluted the target proteins from 2 mL Ni^{2+}-nitrilotriacetate affinity resin (Ni–NTA; Qiagen) using 15 mL buffer A containing 300 mmol/L imidazole and 0.2% (w/v) DM, and concentrated to about 10–15 mg/mL. The concentrated sample was incubated with 0.8 mmol/L Cm and subsequently loaded onto a Superdex-200 10/30 column (GE Healthcare) pre-equilibrated with buffer B (20 mmol/L HEPES pH 7.2, 100 mmol/L NaCl and 5 mmol/L β-mercaptoethanol), supplemented with mixed detergents of 0.2% (w/v) n-nonyl-β-D-glucopyranoside (NG; Anatrace) and 0.025% (w/v) n-dodecyl-N,N-dimethylamine-N-oxide (LDAO; Anatrace). Cm (0.8 mmol/L) was added to the collected protein, which was then concentrated to 20 mg/mL and mixed with 0.5 mmol/L reserpine for crystallization.

Crystallization

Crystal screening was performed using the hanging drop vapour-diffusion method (1 μL plus 1 μL over 200 μL) at 16 °C and obtained under the conditions of 0.1 mol/L Tris (pH 8.0), 0.22 mol/L sodium citrate and 35% (v/v) PEG400 from the MemGold screening kit (molecular dimensions). The crystals grew in ~2 months and were flash-cooled in liquid nitrogen for storage and data collection.

Data collection and structure determination

X-ray diffraction datasets were collected at Shanghai Synchrotron Radiation Facility (SSRF) and processed with the HKL-2000 software package (Otwinowski and Minor 1997). The space group of the reserpine structure was $P3_121$, while the space group of ACh was $C2$. All ligand-complex structures were resolved by molecular replacement. There were two MdfA molecules per crystallographic asymmetric unit for $P3_121$. The model was further refined using the program Coot (Emsley and Cowtan 2004). Model validation was carried out using the web-based program MolProbity (Davis et al. 2004).

Inhibition of drug resistance assays

The *E. coli* C43 (DE3) strain was transformed with the *MdfA* gene-containing pET28a plasmid. A single clone was picked from LB-agar plates for inoculation into 5 mL LB supplemented with kanamycin (30 μg/mL) and grown at 37 °C. The cultures were induced with 0.5 mmol/L IPTG once an OD_{600nm} of 0.6 was obtained, and the cultures were incubated overnight for protein expression. The cells were then diluted into 48-well plates containing 1 mL LB with increasing concentrations of the test drug (Cm) and kanamycin (30 μg/mL). At the beginning of each typical experiment, the cell density in the wells was 0.05 OD_{600nm} units. Plates were incubated at 37 °C with shaking, and the cell density was measured with a Varioskan Flash reader (Thermo Fisher Scientific) by following the absorption at 600 nm over 12 h.

Abbreviations

Ach Acetylcholine
Cm   Chloramphenicol
Dxc Deoxycholate
TM   Transmembrane (helix)

Acknowledgments The authors thank the staff of the Protein Research Core Facility at the Institute of Biophysics, Chinese Academy of Sciences for their excellent technical support. We are grateful to staff members of the SSRF (China) synchrotron facilities for their assistance with crystal screening and data collection.

This work was supported by the "973" program from the Ministry of Science and Technology, China (2011CB910301).

Compliance with ethical standards

Conflict of Interest Ming Liu, Jie Heng, Yuan Gao, and Xianping Wang declare that they have no conflict of interest.

Human and Animal Rights and Informed Consent This article does not contain any studies with human or animal subjects performed by any of the authors.

References

Adler J, Bibi E (2004) Determinants of substrate recognition by the *Escherichia coli* multidrug transporter MdfA identified on both sides of the membrane. J Biol Chem 279:8957–8965

Bravo DT, Kolmakova NG, Parsons SM (2005) New transport assay demonstrates vesicular acetylcholine transporter has many alternative substrates. Neurochem Int 47:243–247

Chandrasekaran A, Ojeda AM, Kolmakova NG, Parsons SM (2006) Mutational and bioinformatics analysis of proline- and glycine-rich motifs in vesicular acetylcholine transporter. J Neurochem 98:1551–1559

Darchen F, Scherman D, Henry JP (1989) Reserpine binding to chromaffin granules suggests the existence of two conformations of the monoamine transporter. Biochemistry 28:1692–1697

Davis IW, Murray LW, Richardson JS, Richardson DC (2004) MOLPROBITY: structure validation and all-atom contact analysis for nucleic acids and their complexes. Nucleic Acids Res 32:W615–619

De Jesus M, Jin J, Guffanti AA, Krulwich TA (2005) Importance of the GP dipeptide of the antiporter motif and other membrane-embedded proline and glycine residues in tetracycline efflux protein Tet(L). Biochemistry 44:12896–12904

D'Rozario RS, Sansom MS (2008) Helix dynamics in a membrane transport protein: comparative simulations of the glycerol-3-phosphate transporter and its constituent helices. Mol Membr Biol 25:571–583

Edgar R, Bibi E (1997) MdfA, an Escherichia coli multidrug resistance protein with an extraordinarily broad spectrum of drug recognition. J Bacteriol 179:2274–2280

Edgar R, Bibi E (1999) A single membrane-embedded negative charge is critical for recognizing positively charged drugs by the *Escherichia coli* multidrug resistance protein MdfA. EMBO J 18:822–832

Emsley P, Cowtan K (2004) Coot: model-building tools for molecular graphics. Acta Crystallogr D Biol Crystallogr 60:2126–2132

Erickson JD, Eiden LE, Hoffman BJ (1992) Expression cloning of a reserpine-sensitive vesicular monoamine transporter. Proc Natl Acad Sci USA 89:10993–10997

Erickson JD, Schafer MK, Bonner TI, Eiden LE, Weihe E (1996) Distinct pharmacological properties and distribution in neurons and endocrine cells of two isoforms of the human vesicular monoamine transporter. Proc Natl Acad Sci USA 93:5166–5171

Finn JP, Edwards RH (1997) Individual residues contribute to multiple differences in ligand recognition between vesicular monoamine transporters 1 and 2. J Biol Chem 272:16301–16307

Heng J, Zhao Y, Liu M, Liu Y, Fan J, Wang X, Zhao Y, Zhang XC (2015) Substrate-bound structure of the *E. coli* multidrug resistance transporter MdfA. Cell Res 25:1060–1073

Holler JG, Christensen SB, Slotved HC, Rasmussen HB, Guzman A, Olsen CE, Petersen B, Molgaard P (2012) Novel inhibitory activity of the *Staphylococcus aureus* NorA efflux pump by a kaempferol rhamnoside isolated from *Persea lingue* Nees. J Antimicrob Chemother 67:1138–1144

Huang Y, Lemieux MJ, Song J, Auer M, Wang DN (2003) Structure and mechanism of the glycerol-3-phosphate transporter from *Escherichia coli*. Science 301:616–620

Jiang D, Zhao Y, Wang X, Fan J, Heng J, Liu X, Feng W, Kang X, Huang B, Liu J, Zhang XC (2013) Structure of the YajR transporter suggests a transport mechanism based on the conserved motif A. Proc Natl Acad Sci USA 110:14664–14669

Khare P, Ojeda AM, Chandrasekaran A, Parsons SM (2010) Possible important pair of acidic residues in vesicular acetylcholine transporter. Biochemistry 49:3049–3059

Lawal HO, Krantz DE (2013) SLC18: vesicular neurotransmitter transporters for monoamines and acetylcholine. Mol Aspects Med 34:360–372

Luo J, Parsons SM (2010) Conformational propensities of peptides mimicking transmembrane Helix 5 and Motif C in wild-type and mutant vesicular acetylcholine transporters. ACS Chem Neurosci 1:381–390

Merickel A, Rosandich P, Peter D, Edwards RH (1995) Identification of residues involved in substrate recognition by a vesicular monoamine transporter. J Biol Chem 270:25798–25804

Otwinowski Z, Minor W (1997) Processing of X-ray diffraction data collected in oscillation mode. Method Enzymol 276:307–326

Parsons SM (2000) Transport mechanisms in acetylcholine and monoamine storage. FASEB J 14:2423–2434

Paul S, Alegre KO, Holdsworth SR, Rice M, Brown JA, McVeigh P, Kelly SM, Law CJ (2014) A single-component multidrug transporter of the major facilitator superfamily is part of a network that protects *Escherichia coli* from bile salt stress. Mol Microbiol 92:872–884

Paulsen IT, Brown MH, Skurray RA (1996) Proton-dependent multidrug efflux systems. Microbiol Rev 60:575–608

Preskorn SH (2007) The evolution of antipsychotic drug therapy: reserpine, chlorpromazine, and haloperidol. J Psychiatr Pract 13:253–257

Putman M, van Veen HW, Konings WN (2000) Molecular properties of bacterial multidrug transporters. Microbiol Mol Biol Rev 64:672–693

Ugolev Y, Segal T, Yaffe D, Gros Y, Schuldiner S (2013) Identification of conformationally sensitive residues essential for inhibition of vesicular monoamine transport by the noncompetitive inhibitor tetrabenazine. J Biol Chem 288:32160–32171

Vardy E, Arkin IT, Gottschalk KE, Kaback HR, Schuldiner S (2004) Structural conservation in the major facilitator superfamily as revealed by comparative modeling. Protein Sci 13:1832–1840

Yaffe D, Radestock S, Shuster Y, Forrest LR, Schuldiner S (2013) Identification of molecular hinge points mediating alternating access in the vesicular monoamine transporter VMAT2. Proc Natl Acad Sci USA 110:E1332–E1341

Yelin R, Schuldiner S (1995) The pharmacological profile of the vesicular monoamine transporter resembles that of multidrug transporters. FEBS Lett 377:201–207

Zhang XC, Zhao Y, Heng J, Jiang D (2015) Energy coupling mechanisms of MFS transporters. Protein Sci 24:1560–1579

Structural roles of lipid molecules in the assembly of plant PSII–LHCII supercomplex

Xin Sheng[1,2], Xiuying Liu[1,2], Peng Cao[1], Mei Li[1], Zhenfeng Liu[1,2]✉

[1] National Laboratory of Biomacromolecules, CAS Center for Excellence in Biomacromolecules, Institute of Biophysics, Chinese Academy of Sciences, Beijing 100101, China
[2] University of Chinese Academy of Sciences, Beijing 100049, China

Abstract In plants, photosystem II (PSII) associates with light-harvesting complexes II (LHCII) to form PSII–LHCII supercomplexes. They are multi-subunit supramolecular systems embedded in the thylakoid membrane of chloroplast, functioning as energy-converting and water-splitting machinery powered by light energy. The high-resolution structure of a PSII–LHCII supercomplex, previously solved through cryo-electron microscopy, revealed 34 well-defined lipid molecules per monomer of the homodimeric system. Here we characterize the distribution of lipid-binding sites in plant PSII–LHCII supercomplex and summarize their arrangement pattern within and across the membrane. These lipid molecules have crucial roles in stabilizing the oligomerization interfaces of plant PSII dimer and LHCII trimer. Moreover, they also mediate the interactions among PSII core subunits and contribute to the assembly between peripheral antenna complexes and PSII core. The detailed information of lipid-binding sites within PSII–LHCII supercomplex may serve as a framework for future researches on the functional roles of lipids in plant photosynthesis.

Keywords Lipid, Photosystem II, Light-harvesting complex II, Membrane protein, Photosynthesis

INTRODUCTION

In plants, algae and cyanobacteria, photosystem II (PSII) cooperates with photosystem I (PSI) and cytochrome b_6f (Cyt b_6f) to carry out the light-driven electron transport process during the energy conversion process in oxygenic photosynthesis (Nelson and Ben-Shem 2004). The two photosystems and Cyt b_6f are multi-subunit membrane protein complexes embedded within the thylakoid membrane, and their proper functions rely on the amphipathic lipid bilayer environments of the membrane. Moreover, specific lipid molecules may participate in the assembly of photosynthetic complexes, serving as their intrinsic components essential for the optimal activity and stability of these membrane protein complexes (Kern and Guskov 2011). There are four major types of lipids in the thylakoid membranes of cyanobacteria and plant chloroplast, namely monogalactosyldiacylglycerol (MGDG), digalactosyldiacylglycerol (DGDG), phosphatidylglycerol (PG), and sulfoquinovosyldiacylglycerol (SQDG) (Mizusawa and Wada 2012). Each of these lipids has distinct polar head group and fulfills specific role in the assembly of photosynthetic complexes (Kern and Guskov 2011; Leng et al. 2008). Among them, MGDG and DGDG are the bulk lipids of the thylakoid membranes of cyanobacteria and chloroplasts, and provide amphipathic membrane environments to host photosynthetic complexes (Holzl and Dormann 2007; Mizusawa and Wada 2012). As the predominant neutral glycoglycerolipids of thylakoid membranes, they account for approximately 50% and 30% (mol%) of plant thylakoid membrane lipids, respectively (Siegenthaler 1998). While SQDG and PG are less abundant than MGDG and DGDG, they are

✉ Correspondence: liuzf@sun5.ibp.ac.cn (Z. Liu)

anionic lipids contributing negative charges on the surface of the thylakoid membrane and important for photoautotrophic growth of cyanobacteria and plants (Frentzen 2004; Sato 2004).

The functions of these four types of lipid molecules have been well studied through genetic and biochemical approaches. MGDG serves to stimulate functional interaction between plant PSII core complexes and the major light-harvesting complexes II (LHCII), presumably by enhancing physical interactions between the two complexes (Fujii et al. 2014; Zhou et al. 2009). Moreover, it has an essential role in photoprotection during photosynthesis by supporting the activity of violaxanthin de-epoxidase in the xanthophyll cycle (Jahns et al. 2009). MGDG is synthesized by MGDG synthase, and there are three isoforms of MGDG synthases in *Arabidopsis thaliana*, namely MGD1, MGD2, and MGD3 (Kobayashi et al. 2013). Among them, MGD1 is the dominant isoform for galactolipid synthesis, and the content of MGDG in *Arabidopsis* MGD1 mutant was reduced to 42% compared to the wild type (Jarvis et al. 2000). The reduction of MGDG content, through T-DNA insertion in the MGD1 promoter region or an artificial microRNA targeting MGD1, showed a severe defect in thylakoid membrane development and impaired photosynthetic electron transport (Fujii et al. 2014; Zhou et al. 2009). Lipase treatment of the PSII sample from *Thermosynechococcus vulcanus* led to degradation of half of the total MGDG in the sample and 16% reduction of the oxygen evolution activity of PSII (Leng et al. 2008).

DGDG is a bilayer-forming glycolipid which may be responsible for the formation and stabilization of thylakoid membrane (Lee 2000). There are two isoforms of DGDG synthases (DGD1 and DGD2) involved in DGDG synthesis in *Arabidopsis thaliana*. The content of DGDG in the DGD1 mutant of *Arabidopsis thaliana* is reduced by more than 90% and the mutant plant showed severe growth retardation and altered chloroplast structure when compared to the wild type (Kelly et al. 2003). DGDG is required for the functional and structural integrity of the oxygen-evolving complex and thermal stability of plant PSII (Reifarth et al. 1997). Deficiency of DGDG lowers the thermal stability of the LHCII–PSII-containing macrodomains and PSI complexes (Krumova et al. 2010). Moreover, DGDG is also involved in stabilization of plant LHCII trimers and mediate the interactions between adjacent LHCII trimers at the luminal side (Holzl and Dormann 2007; Liu et al. 2004). In cyanobacteria, DGDG may be involved in binding of extrinsic proteins to PSII and stabilizing the oxygen-evolving complex (Sakurai et al. 2007). The *dgda* mutant of cyanobacteria contains no detectable DGDG, and also shows growth retardation under high-light stress and high temperature (Mizusawa et al. 2009a, b). The mutant exhibits increased sensitivity to photoinhibition and it was suggested that DGDG may have an important role in the repair cycle of photosynthetic complexes (Mizusawa et al. 2009b).

As the major phospholipid in thylakoid membranes, PG has important roles in various photosynthetic complexes (Jones 2007; Sato 2004; Wada and Murata 2007). Degradation of PG by enzymatic treatment with phospholipase A_2 or phospholipase C significantly inhibits the photosynthetic electron transport activities (Jordan et al. 1983). The PSII sample purified from *Thermosynechococcus vulcanus* has the content of PG decreased by 59% and the oxygen evolution activity reduced by 40% after being treated with phospholipase A_2 (Leng et al. 2008). The *pgsA* (gene encoding a PG phosphate synthase) mutant of *Synechocystis sp.* PCC 6803 is deficient in biosynthesis of PG, could only grow in the presence of exogenously supplied PG, and the photosynthetic oxygen-evolving activity of the mutant cells was reduced by 40% after a 3-day depletion of PG (Hagio et al. 2000). Further study indicated that the content of PSII dimer decreased significantly in the *pgsA* mutant grown under high-light condition, indicating that PG is indispensable for maintaining the dimeric state of PSII (Sakurai et al. 2003). A previous biochemical study revealed that spinach PSII core dimer dissociated into monomers when it was treated with phospholipase A_2, and the PG molecules with trans-hexadecanoic fatty acyl chains can induce dimerization of isolated PSII monomers (Kruse et al. 2000). Furthermore, PG is also involved in mediating trimerization of LHCII in plants (Liu et al. 2004).

The functional role of SQDG in photosynthesis varies among different species. In *Chlamydomonas*, a SQGD-deficient mutant shows slightly reduced growth rate and 32%–46% decrease in PSII activity compared to the wild type (Sato et al. 1995). SQDG is involved in maintaining the structural integrity and thermal stability of PSII from *Chlamydomonas* (Sato et al. 2003). In *Arabidopsis*, SQGD-deficient *sqd2* mutants seem to have little impact on photosynthesis when compared to the wild type (Essigmann et al. 1998). SQDG may be required for photosynthetic electron transport with limited availability of PG in plants (Kobayashi et al. 2016). The SQDG requirement for PSII is species-dependent in cyanobacteria. SQDG-null mutant of *Synechocystis* sp. PCC6803 has decreased photosynthetic and PSII activities, whereas the deficiency of SQDG in PCC7942 strain does not affect PSII activity (Aoki et al. 2004).

An in-depth understanding on the pivotal functional roles of lipid molecules in photosynthesis can be achieved by solving the structures of photosynthetic complexes at high resolution (better than 3 Å). Accurate assignment of the binding sites and identities of various lipid molecules in the photosynthetic complexes is indispensible for revealing the specific interactions between lipid molecules and proteins/cofactors within the complexes, as exemplified in the high-resolution crystal structures of cyanobacterial PSI (Jordan et al. 2001) and PSII (Umena et al. 2011), cytochrome b_6f complex (Hasan and Cramer 2014), plant LHCII (Liu et al. 2004), and plant PSI (Mazor et al. 2017). Previously, crystal structure of a cyanobacterial PSII from *Thermosynechococcus elongatus* (TePSII) was solved at 2.9 Å (Guskov et al. 2009) and revealed the presence of 25 lipid molecules per monomer of the PSII homodimer, including 11 MGDG, seven DGDG, five SQDG, and two PG molecules (Kern and Guskov 2011). These lipid molecules serve as multifunctional cofactors involved in the assembly and functional regulation of PSII (Mizusawa and Wada 2012). The high-resolution structure of PSII from *T. vulcanus* (TvPSII) at 1.9 Å contains 20 lipid molecules in each monomer, including six MGDG, five DGDG, four SQDG, and five PG molecules (Umena et al. 2011). While all DGDG and SQDG binding sites found in TvPSII are conserved in TePSII, TvPSII contains more PG but less MGDG binding sites than TePSII. As for plant PSII, its core complex exhibits high similarity with cyanobacterial PSII, and it is assembled with peripheral light-harvesting complexes (LHCII, CP29, CP26, and CP24) to form PSII–LHCII supercomplexes (Su et al. 2017; Wei et al. 2016). Recently, the structures of C_2S_2 and $C_2S_2M_2$-type (C: core; S: strongly associated LHCII; M: moderately associated LHCII) PSII–LHCII supercomplexes have been solved through single-particle cryo-electron microscopy at overall resolutions of 3.2 and 2.7 Å, respectively (Su et al. 2017; Wei et al. 2016). The assembly between peripheral antennae LHCII/CP29/CP26/CP24 and PSII core relies on the specific interactions between adjacent complexes. Besides protein and pigment molecules, numerous lipid molecules have been located in each monomer of the $C_2S_2M_2$-type PSII–LHCII supercomplexes. In this review, we focus on discussing the structural roles of lipid molecules in the $C_2S_2M_2$-type PSII–LHCII supercomplex basing on the 2.7-Å resolution structure and compare them with those found in cyanobacterial and red algal PSII. We have also performed a detailed analysis on the lipid-binding sites in the supercomplex and explain their roles in the assembly of individual complexes and the formation of supercomplex.

TOWARD A HIGH-RESOLUTION STRUCTURE OF PLANT PSII–LHCII SUPERCOMPLEXES

Back in 1990s, Boekema et al. applied single-particle electron microscopy method to observe the negatively stained sample of spinach PSII and obtained two-dimensional (2D) projection images of the PSII complexes at 15–26 Å resolution (Boekema et al. 1995). It was proposed that two LHCII trimers are linked to PSII complex through CP29, CP26, and CP24 to form dimeric PSII–LHCII supercomplex. Subsequently, they were able to identify two different types of LHCII trimers (strongly and moderately associated LHCII, S and M) being associated with PSII core (C), and classify the PSII–LHCII supercomplexes as C_2S, C_2S_2, C_2SM, C_2S_2M and $C_2S_2M_2$ types using the negatively stained sample (Boekema et al. 1998). In 2009, Caffarri et al. optimized the purification method for isolating the different classes of *Arabidopsis* PSII–LHCII supercomplexes through sucrose density gradient ultracentrifugation method and improved the 2D projection map of $C_2S_2M_2$ supercomplex to 12-Å resolution (Caffarri et al. 2009). The approximate locations of S-LHCII, M-LHCII, CP29, CP26, and CP24 around the PSII core were assigned in the 2D map.

As the peripheral antenna complexes are weakly associated with plant PSII core complex, the PSII–LHCII supercomplexes are highly unstable and heterogeneous when they are extracted from the membrane and purified in detergent solution. Such a property is unfavorable for growing high-quality three-dimensional (3D) crystal samples. Thus, the attempts to solve the 3D structure of PSII–LHCII supercomplex through X-ray crystallography were unsuccessful, despite that crystals of the reaction center complex of spinach PSII were obtained in the presence of detergent mixtures (Adir 1999). On the other hand, progresses have been made through electron crystallography and single-particle cryo-electron microscopy (cryo-EM) in solving the 3D structures of plant PSII and PSII–LHCII supercomplexes. Rhee et al. reported an 8-Å structure of a spinach PSII core complex solved through electron crystallography and assigned the locations of D1, D2 and CP47 subunits (Rhee et al. 1997, 1998). Subsequently, Hankamer et al. applied the method to solve the structure of spinach PSII core dimer at ~10-Å resolution, found that CP43 and CP47 are located on opposite sides of the central D1-D2 heterodimer and further located the transmembrane helices of major subunits and low-molecular-weight subunits (Hankamer et al. 1999, 2001). In 2000, Nield et al. reported a 3D model of spinach C_2S_2 PSII–LHCII supercomplex basing on a 24-Å map obtained through single-particle cryo-EM method (Nield et al.

2000), and further improved the resolution to 17 Å after refinement (Nield and Barber 2006). The approximate locations of PSII core proteins and peripheral antenna complexes were assigned in the low-resolution 3D map, but the detailed features at the interfaces between adjacent complexes remained unknown.

The first near-atomic resolution (3.2 Å) structure of plant PSII–LHCII supercomplex was solved by Wei *et al.* through single-particle cryo-EM method in 2016 (Wei *et al.* 2016). The structure provides detailed information for the locations of most amino acid residues from 25 protein subunits, 105 chlorophylls, and 28 carotenoid molecules within the spinach C_2S_2-type PSII–LHCII supercomplex. The specific interactions between adjacent peripheral antenna complexes and between peripheral antenna and core antenna complexes were described in detail, and the potential energy transfer pathways from the antenna complex to the PSII core have been identified (Wei *et al.* 2016). In 2017, Bezouwen *et al.* reported the cryo-EM structure of *Arabidopsis* $C_2S_2M_2$-type PSII–LHCII supercomplex at 5.3-Å resolution and discussed the subunit and chlorophyll organization within the supercomplex (Bezouwen *et al.* 2017). Furthermore, Su *et al.* solved the structures of stacked and unstacked forms of $C_2S_2M_2$-type PSII–LHCII supercomplex from *Pisum sativum* (PsPSII–LHCII) at 2.7 and 3.2 Å respectively, and revealed near-atomic details of the protein subunits and various cofactors within the supercomplex (Su *et al.* 2017).

OVERALL ARCHITECTURE OF PSII–LHCII SUPERCOMPLEX

In the 2.7-Å resolution structure of plant $C_2S_2M_2$ supercomplex, there are 28 protein subunits, 157 chlorophyll, two pheophytin, 44 carotenoid, one Mn_4CaO_5 cluster, 34 lipids, and numerous other cofactors in each monomer. The core complex includes four large intrinsic membrane proteins (D1, D2, CP43, and CP47), 12 small intrinsic proteins (PsbE, PsbF, PsbH, PsbI, PsbJ, PsbK, PsbL, PsbM, PsbTc, PsbW, PsbX, and PsbZ) and three extrinsic proteins (PsbO, PsbP, and PsbQ) on the luminal side (Fig. 1A, B). In the C_2S_2 region of the supercomplex, there are strongly associated LHCII trimer and a CP26 monomer attached to two different sides of the CP43 complex, and a CP29 monomer bound to CP47 in the core complex from the other side (Wei *et al.* 2016). To form the $C_2S_2M_2$ supercomplex, the C_2S_2 complex is further assembled with two moderately associated LHCII trimers (M-LHCII) and two CP24 monomers on the sides nearby S-LHCII and CP29 (Fig. 1A). The M-LHCII binds to the concaved groove between S-LHCII and CP29, and also forms extensive interactions with CP24. Furthermore, CP24 interacts closely with CP29 to form a heterodimer, and it is separated from CP47 by a large ($\sim 30 \times 55$ Å) void region potentially filled by lipid molecules or unobserved subunits of PSII. The M-LHCII-CP24 subcomplex exhibits high mobility and appears to adopt different tilt positions in the stacked and unstacked $C_2S_2M_2$ structures (Su *et al.* 2017). The observed off-plane tilting of the subcomplex may facilitate its detachment from the C_2S_2 supercomplex and provide a means for the regulation of light harvesting in response to the increase of light intensity (Su *et al.* 2017).

Chlorophyll molecules are the major light-harvesting pigments and there are 108 chlorophyll *a* (Chl *a*) and 49 chlorophyll *b* (Chl *b*) molecules in the structure of $C_2S_2M_2$ supercomplex. Among them, Chl *a* molecules are widely distributed in both PSII core complexes (CP47, CP43, D1, and D2) and the peripheral antenna complexes (LHCII, CP29, CP26, and CP24). On the other hand, Chl *b* molecules exist only in the peripheral antenna complexes but not in the core complexes (Fig. 1C). At the interfaces between adjacent antenna complexes, there are numerous pairs of chlorophyll molecules (Mg-to-Mg distance at 13–25 Å) forming the potential energy transfer pathways from the peripheral antenna complexes to the core complexes of PSII (Fig. 1C). Besides chlorophylls, carotenoid molecules have important functions in photoprotection, maintain the structural stability of photosynthetic complexes, and may also contribute to light harvesting (Hashimoto *et al.* 2016). There are 44 carotenoid molecules in the $C_2S_2M_2$ supercomplex, and among them, lutein, violaxanthin and neoxanthin only exist in the peripheral antenna domains. In contrast, β-carotene mainly exists in the PSII core region and there might be one located in CP24 complex (Fig. 1D) (Su *et al.* 2017). These carotenoid molecules are mostly distributed in regions enriched with chlorophylls and form close interactions with them so as to fulfill the photoprotective function during photosynthesis.

DISTRIBUTION OF LIPID-BINDING SITES IN THE PSII–LHCII SUPERCOMPLEX

Within the $C_2S_2M_2$-type PSII–LHCII supercomplex, the lipid molecules located in each monomer include 18 PG, seven MGDG, five DGDG, and four SQDG (Fig. 2A). Among them, PG molecules are distributed more or less evenly throughout the supercomplex. They are found in the peripheral antenna complexes (LHCII, CP29, CP26 and CP24), at the interfaces between LHCII/CP26/CP29 and CP43/CP47, and within the core complex (at the interfaces between core subunits and around the

Structural roles of lipid molecules in the assembly of plant PSII–LHCII supercomplex

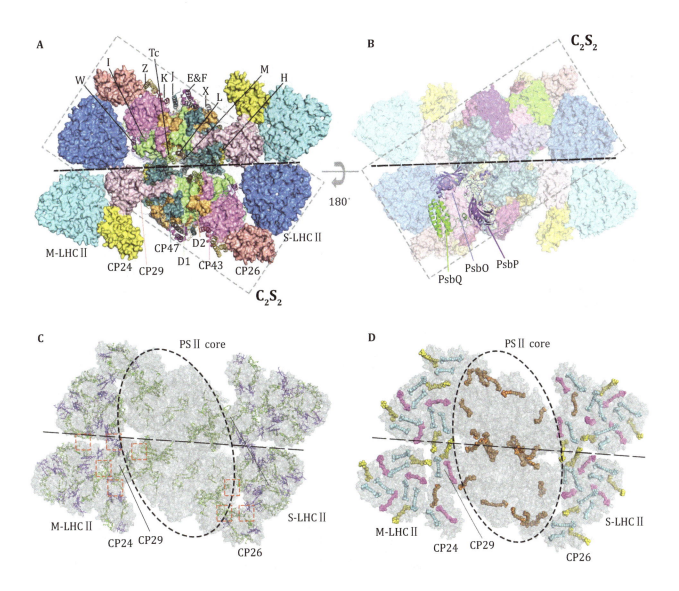

Fig. 1 Overall structure of the $C_2S_2M_2$-type PSII–LHCII supercomplex from *Pisum sativum*. **A, B** The $C_2S_2M_2$ PSII–LHCII supercomplex viewed from stromal side (**A**) and luminal side (**B**) along the membrane normal. The small intrinsic subunits (labeled by the single-letter or two-letter codes) and extrinsic subunits are presented as cartoon models, while the other subunits are shown as surface models. PDB code: 5XNL. **C, D** Arrangement of chlorophyll (**C**) and carotenoid (**D**) molecules in the $C_2S_2M_2$ PSII–LHCII supercomplex. The red dashed boxes indicate the interfaces between adjacent peripheral complexes and between peripheral and core complexes. The chlorophyll and carotenoid molecules are presented as stick and sphere models, respectively. Color codes: green, Chl a; blue, Chl b; orange, β-carotene; cyan, lutein; magenta, violaxanthin; yellow, neoxanthin

dimerization interface). Meanwhile, MGDG, DGDG, and SQDG molecules are located mostly in the PSII core region. MGDG molecules are widely distributed around CP43, CP47, D1 and D2 subunits, whereas DGDG molecules are concentrated at the inner core region around D1 and D2 subunits. Three of the SQDG molecules line at the dimerization interface of the PSII core, and one is located at a peripheral cavity surrounded by CP43, PsbK, PsbJ, Cyt $b559$ (PsbE and PsbF), and D1. Curiously, the lipid molecules exhibit an evidently asymmetric distribution pattern across the membrane (Fig. 2C–F). For instance, 16 of the PG molecules have their polar head groups located on the stromal side and only two PG molecules are located on the luminal side (at the interface between CP29 and CP47) (Fig. 2C). All DGDG and MGDG molecules have their head groups positioned at the luminal surface (Fig. 2E, F), while the head groups of all four SQDG molecules are on the stromal side (Fig. 2D). Thereby, the enrichment of anionic lipid molecules (PG and SQDG) on the stromal side contributes significant amount of negative charges on the stromal surface of PSII–LHCII supercomplex.

Fig. 2 Overall distribution of lipid-binding sites within the $C_2S_2M_2$ PSII–LHCII supercomplex from *Pisum sativum*. **A** Top view of the supercomplex from the stromal side. The lipid molecules located within the supercomplex are highlighted as colored sphere models, while the proteins are shown as silver cartoon models. The pigments and other cofactors are omitted for clarity. **B** Side view of the supercomplex along the membrane plane. **C–F** Asymmetric distributions of PG (**C**), SQDG (**D**), DGDG (**E**) and MGDG (**F**) across the membrane-embedded region within the supercomplex

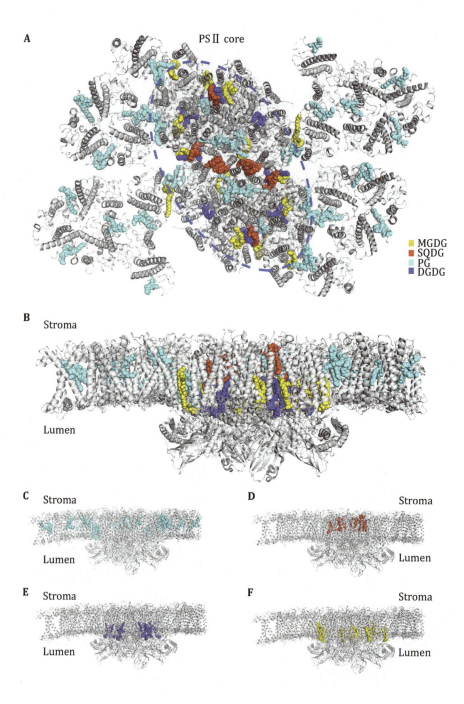

Similar asymmetric distribution of lipid molecules was also observed in the crystal structures of cyanobacterial PSII (TvPSII and TePSII) (Umena *et al.* 2011).

ROLE OF LIPID MOLECULES IN STABILIZING THE PSII CORE ASSEMBLY

At the PSII core region of the $C_2S_2M_2$ supercomplex, the lipid-binding sites are grouped into two clusters (Clusters 1 and 2) and two local binding sites (Sites 1 and 2) buried within CP43 and CP47 (Figs. 3, 4). Cluster 1 is located at the dimerization interface of PSII core, and nine lipid molecules per monomer including three SQDG, three PG, one DGDG and two MGDG are located at this region (Fig. 3A). The PsbTc–PsbL–PsbM trimer is connected with the symmetry-related PsbTc'–PsbL'–PsbM' trimer through extensive hydrophobic interactions between PsbM and PsbM'. These single-transmembrane-helix small subunits are located at the central dimerization interface of PSII core dimer, and are stabilized by three SQDG (SQD621/SQD623/SQD

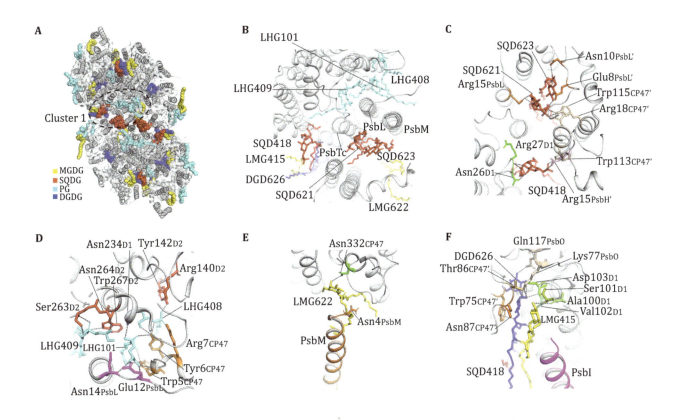

Fig. 3 Distribution of lipid-binding sites around the dimerization interface of PsPSII. **A** Top view of the PsPSII core dimer from the stromal side. The lipid molecules are highlighted as colored sphere models. The peripheral antenna complexes (LHCII, CP29, CP26 and CP24) are omitted for clarity. The region around Cluster-1 lipids is labeled by the dashed elliptical ring. **B** Zoom-in view of the Cluster-1 lipid molecules from the stromal side. **C** The binding sites of three SQDG molecules at Cluster 1 region viewed from the stromal side. **D** The binding sites of three PG molecules at Cluster 1 region viewed from the stromal side. **E** The binding sites of a MGDG molecule at Cluster 1 region viewed from luminal side. **F** The binding sites of one MGDG and one DGDG molecule around CP47, D1, PsbO and PsbI subunits viewed from luminal side. The gray dash lines indicate the hydrogen bonds and ionic interactions formed between lipid head groups and nearby amino acid residues. DGD, LHG, LMG and SQD are three-letter codes for DGDG, PG, MGDG and SQDG, respectively

418), three PG (LHG101/LHG408/LHG409), one MGDG (LMG622), and their symmetry-related molecules flanking on four sides of the PsbTc–PsbL–PsbM–PsbM′–PsbL′–PsbTc′ hexamer (Fig. 3A, B). On one side, two SQDG molecules (SQD621 and SQD623) are sandwiched between the PsbTc–PsbL–PsbM trimer and the first two transmembrane helices of CP47 from the adjacent monomer of PSII core dimer (CP47′), while the third SQDG (SQD418) is located ~18 Å from SQD621 and intercalates at the space between the first transmembrane helix (M1) of D1 and the second transmembrane helix (M2) of CP47′ (Fig. 3C). The head group of SQD621 forms ionic interactions with Arg15 from PsbL and Arg18 from CP47′ and a hydrogen bond with Trp115$_{CP47'}$, while that of SQD623 is hydrogen-bonded to Asn10 and Glu8 from PsbL′ (Fig. 3C). The head group of SQD418 is hydrogen-bonded to the backbone amide group of Arg27 and side chain of Asn26 from D1 subunit on one side, and forms hydrogen bond with Trp113 from CP47′ and ionic interaction with Arg15 from PsbH′ on the other side. The fatty acyl chains of SQD621, SQD623, SQD418 extend from the stromal surface to the middle region and form hydrophobic interactions with non-polar amino acid residues from adjacent protein subunits, β-carotene molecules and the phytyl chains of chlorophyll molecules located nearby. On the other side, three PG molecules (LHG101, LHG408, and LHG409) are located at the cleft between D1 subunit (the fourth and fifth transmembrane helices/M4 and M5 helices) and PsbTc–PsbL–PsbM trimer (Fig. 3D). The head group of LHG101 is hydrogen-bonded to Asn14 and Glu12 of PsbL, Tyr6 and Trp5 from the N-terminal region of CP47, and Asn234 from D1. The adjacent LHG409 forms hydrogen bonds with Ser263, Trp267, and Asn264 of D2 subunit, while LHG408 on the other side forms hydrogen bond with Tyr6$_{CP47}$ and Tyr142$_{D2}$ as well as ionic interactions with Arg7$_{CP47}$ and Arg140$_{D2}$ (Fig. 3D). Such a tightly packed PG trimer also contributes their fatty acyl chains to bridge the D1 subunit with PsbTc–PsbL–PsbM and CP47 through hydrophobic interactions.

Across the membrane, LMG622 (MGDG) has its hydrophobic tails filling in the luminal-side gap between PsbM and the sixth transmembrane helix (M6) of CP47, and its head group forms a hydrogen bond with $Asn332_{CP47}$ and is further bridged to $Asn4_{PsbM}$ through an interstitial water molecule (Fig. 3E).

The remaining two lipid molecules in Cluster 1, namely one DGDG (DGD626) and one MGDG (LMG415), are located on the luminal side in the distal region of the central dimerization interface (Fig. 3F). They are sandwiched between PsbI and M2 of CP47′, and their acyl tails form hydrophobic interactions with the 1-acyl chain of SQD418. The head group of DGDG forms hydrogen bonds with the polar residues from D1 and PsbO simultaneously, and is in van der Waals contact with the side-chain indole ring of Trp75 from CP47′. Therefore, DGD626 is not only involved in dimerization of PSII core, but also has an important role in mediating the assembly of extrinsic protein PsbO with PSII core subunits. The head group of LMG415 is packed closely against that of DGD626, and forms close contact with the N-terminal region of PsbI and Ala100–Val102 region of D1 subunit. On the other side, it faces Thr86–Asn87 region of CP47′ and may be connected to this region through unobserved water molecules.

At the peripheral region of PSII monomer, six lipids molecules (including one SQDG and one PG on the stromal side, two DGDG and two MGDG on the luminal side) are located in the plastoquinone (PQ)–plastoquinol (PQH_2) exchange cavity near PsbK, PsbJ, and Cyt $b559$, and they form Cluster 2 (Fig. 4A, B). The SQDG (SQD412) and PG (LHG410) molecules form a heterodimer and they are surrounded by the fifth transmembrane helix (M5) of D1 and the fourth transmembrane helix (M4) of D2, PsbK as well as the N-terminal region and the sixth transmembrane helix (M6) of CP43 (Fig. 4C). The head group of SQD412 is hydrogen-bonded to Gln28 and Trp36 from CP43, Ser270 from D1 and Asn231 from D2, while the phospho-(1′-sn-glycerol) head group of LHG410 forms ionic

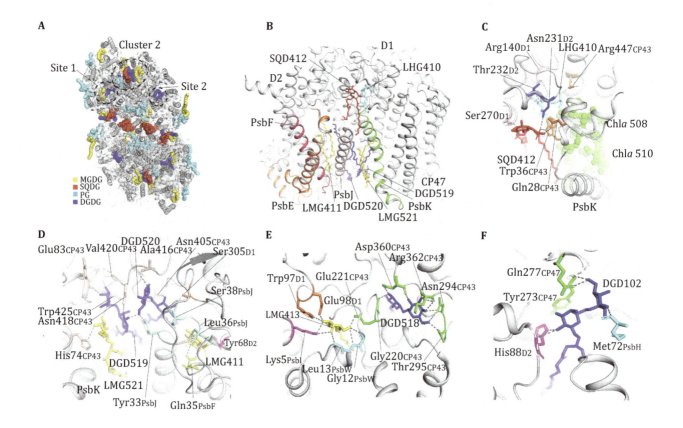

Fig. 4 Lipid molecules involved in the assembly of PsPSII core monomer. **A** The locations of Cluster 2, Site 1 and Site 2 lipid-binding regions within the PsPSII core momomer. **B** Lipid molecules at the peripheral Cluster-2 region of the PSII monomer. **C** The binding sites of two lipid molecules (SGD412 and LHG410) surrounded by D1, D2, PsbK and CP43 subunits at stromal side. **D** The binding sites of four lipid molecules (LMG521, DGD519, DGD520, LMG411) at the luminal side of Cluster 2. **E** Two lipid molecules (LMG413 and DGD518) at Site 1 nearby the interface between CP43 and PsbW. **F** A DGDG molecule (DGD102) on Site 2 surrounded by CP47, D2 and PsbH subunits. The hydrogen bonds and ionic interactions between lipid head groups and nearby amino acid residues are shown as dash lines

interaction with Arg140 from D1 and hydrogen bonds with Arg447$_{CP43}$, Thr232$_{D2}$, and Asn231$_{D2}$. The fatty acyl chains of SQD412 and LHG410 form extensive hydrophobic interactions with non-polar amino acid residues from D1, PsbK, D2 as well as Chl*a* 508 and Chl*a* 510 from CP43 (Fig. 4C). On the luminal side, the four galactolipids fill in the 24-Å wide void space between the first transmembrane helix of D2 subunit and PsbK, and stabilize the local structure by forming hydrophobic interactions with nearby non-polar groups from D1, D2, PsbF, PsbJ, and PsbK. The digalactosyl head group of DGD519 is hydrogen-bonded to the backbone carbonyl groups of Glu83, backbone amide and carbonyl of Val420, and side chains of Asn418 and Trp425 from CP43, while that of DGD520 forms hydrogen bonds with the backbone carbonyl groups of Ala416$_{CP43}$ and Ser38$_{PsbJ}$ as well as the side chains of Tyr33$_{PsbJ}$, Ser305$_{D1}$, and Asn405$_{CP43}$ (Fig. 4D). The head group of LMG411 is located only 4 Å from that of DGD520, and it is bound to the backbone carbonyl groups of Leu36$_{PsbJ}$ and Tyr68$_{D2}$ as well as the side chain of Gln35$_{PsbF}$ through hydrogen bonds. On the other side, LMG521 is hydrogen-bonded to the head group of DGD519 and His74$_{CP43}$ (Fig. 4D). The hydrophobic fatty acyl chains of the six lipid molecules in Cluster 2, the first three transmembrane helices of D2 subunit, M4 of D1 subunit, Cyt *b*559, PsbJ and PsbK outline a hydrophobic cavity measuring ∼22–30 Å wide and 10–22 Å deep. It harbors the PQ molecule on Q$_B$ site on one side and opening to the lipid bilayer on the other side through two lateral portals (one is located between PsbE and PsbJ, and the other lies between PsbF and D2). Such a large cavity may serve as a storage pool for more PQ molecules (or lipid molecules) unobserved in the present structure.

Site 1 contains one DGDG (DGD518) and one MGDG (LMG413) on the luminal side and they are located near the fifth transmembrane helix (M5) of CP43, stabilizing the local structure of CP43 and connecting it with D1 and PsbI subunits (Fig. 4A). The head group of DGD518 is hydrogen-bonded to the side chains of Asn294$_{CP43}$ and Thr295$_{CP43}$, backbone carbonyl and amide of Arg362$_{CP43}$, backbone carbonyl of Asp360$_{CP43}$ and backbone amide of Gly220$_{CP43}$ (Fig. 4E). On the other side, LMG413 interacts with DGD518 through its fatty acyl chains, while its head group binds to the side chains of Glu221$_{CP43}$, Trp97$_{D1}$, Glu98$_{D1}$, and Lys5$_{PsbI}$ as well as the backbone amide groups of Leu13$_{PsbW}$ and Gly12$_{PsbW}$ through hydrogen bonds (Fig. 4E). Thus, these two lipid molecules have crucial roles in mediating the assembly among CP43, D1, PsbI, and PsbW. On Site 2, a DGDG molecule (DGD102) at the luminal side binds to a region near the fifth transmembrane helix (M5) of CP47 and fills in the gap among CP47, D2, and PsbH subunits (Fig. 4F). Its digalactosyl head group forms hydrogen bonds with the side chain of Gln277$_{CP47}$ and the backbone carbonyl of Tyr273$_{CP47}$ as well as the side chains of His88$_{D2}$ and Met72$_{PsbH}$ on the other side, while the two fatty acyl chains extend to the hydrophobic interface between CP47 (M5 and M6) and D2 (M2 and M3). Thereby, DGD102 serves to stabilize the assembly among CP47, D2, and PsbH subunits.

LIPID-BINDING SITES IN PSII FROM DIFFERENT SPECIES

Currently, there are four structures of PSII from different species with lipid-binding sites assigned at resolutions better than 3-Å resolution, namely PsPSII structure as described in this work, red algal PSII structure from *Cyanidium caldarium* (CcPSII) at 2.76-Å resolution (Ago *et al.* 2016) and two cyanobacterial PSII structures (TvPSII and TePSII). As shown in Fig. 5, PsPSII contains evidently more lipid-binding sites than the red algal and cyanobacterial PSII structures. While the peripheral lipid-binding sites in PsPSII are not present in the other three structures, most the internal sites (Clusters 1 and 2, Sites 1 and 2, shown in Fig. 5A) can be found in CcPSII, TvPSII and TePSII. For the Cluster 1 lipids at the dimerization interface of PsPSII, two SQDG binding sites (SQD418 and SQD621) are also conserved in TvPSII and TePSII, but not found in CcPSII. The third SQDG-binding site (SQD623) at the dimerization interface is occupied by MGDG in TePSII, but is vacant in CcPSII and TvPSII. The three PG-binding sites (LHG101, LHG408 and LHG409) in this cluster are found in CcPSII and TvPSII, but were assigned as MGDG in TePSII. In Cluster 2, the SQD412, LHG410, DGD519, DGD520, LMG411, and LMG521 sites are mostly conserved in PSII from all four different species, while the SQD412 site in CcPSII appears to be occupied by SQDG in one monomer and PG in the other monomer. Meanwhile, TvPSII and TePSII have two additional lipid molecules on the stromal side in the PQ—PQH$_2$ exchange cavity area around Cluster 2. The DGD518 and LMG413 of the Site 1 mediate the assembly among CP43, D1, and PsbI in PsPSII, and they are conserved among all four species except that LMG413 site in TePSII is occupied by a DGDG molecule instead of MGDG. The Site-2 lipid DGD102 stabilizes the assembly among CP47, D2, and PsbH, and it is also highly conserved among four species.

In addition to the internal lipid-binding sites, PsPSII contains seven lipid molecules (five PG and two MGDG) at the peripheral regions of CP43 and CP47, namely

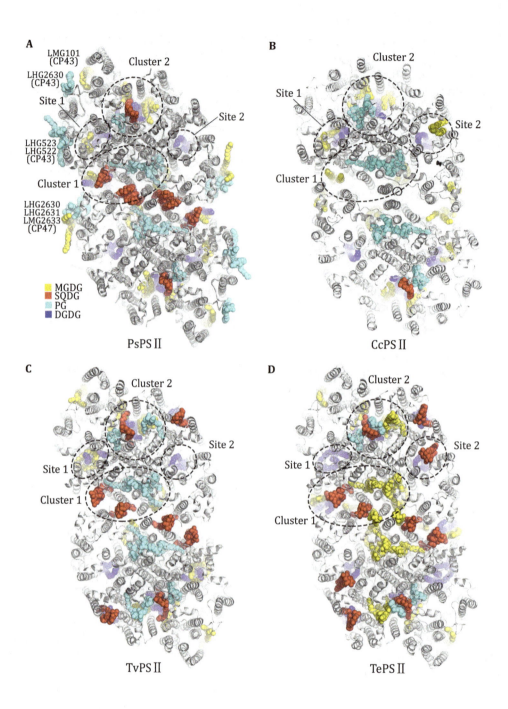

Fig. 5 Comparison of the lipid-binding sites in the structures of PSII from four different species. **A–D** The core region of PsPSII with the outer antennae omitted (**A**), CcPSII (**B**), TvPSII (**C**) and TePSII (**D**) viewed from the stromal side. Lipid molecules are shown as colored sphere models and protein subunits are presented as cartoon models. Pigment molecules and other cofactors are omitted for clarity. PDB codes: PsPSII, 5XNL; CcPSII, 4YUU; TvPSII, 3WU2; TePSII, 4V62

LMG101, LHG2630, LHG522, and LHG523 around CP43 as well as LHG2630, LHG2631, and LMG2633 around CP47 (Fig. 5A). Their roles in mediating the assembly between CP26/LHCII/CP29 and CP43/CP47 will be discussed in detail in the following section. These lipid molecules are absent in CcPSII or the cyanobacterial PSII since they do not bind the peripheral membrane-embedded antenna complexes as those associate with plant PSII. Instead, cyanobacterial PSII (and CcPSII) associates with an extrinsic light-harvesting apparatus named phycobilisome being attached on the outer/stromal surface of PSII (Chang et al. 2015; Zhang et al. 2017).

LOCATIONS AND ROLES OF LIPIDS IN PERIPHERAL ANTENNA COMPLEXES

In the peripheral antenna complexes (LHCII, CP29, CP26, and CP24), one PG molecule per monomer is located on the stromal side within each complex and it is coordinated with the central Mg atom of Chl *a* 611 in all four complexes. In LHCII, CP29 and CP26, the phosphate group of the Chl *a*-ligating PG molecules is further hydrogen-bonded to the side chains of conserved Tyr residues (Tyr44$_{LHCII}$/Tyr32$_{CP29}$/Tyr57$_{CP26}$) and forms ionic interaction with conserved Lys residues (Lys182$_{LHCII}$/Lys200$_{CP29}$/Lys191$_{CP26}$) (Fig. 6A–C). In CP24, Lys182 forms ionic interaction with the PG phosphate group, whereas the site corresponding to the PG-binding Tyr residues in LHCII/CP29/CP26 is occupied by Phe33 (Fig. 6D). The six PG molecules in S-LHCII and M-LHCII trimers are located at the monomer–monomer interfaces, and mediate trimerization of LHCII (Liu *et al.* 2004). Treatment of LHCII trimers with phospholipase A$_2$ cleaved the PG molecules within the complex and led to dissociation of the trimers into monomers (Nussberger *et al.* 1993), demonstrating the important role of PG in stabilizing the LHCII trimer. The PG molecule in CP26 is located at the interface between CP26 and CP43, and mediates the assembly between the two complexes, whereas the PG in CP29 intercalates at the interface between helix A of CP29 and helix C of CP24 and stabilizes the CP29–CP24 heterodimer. The PG molecule in CP24 is located at the peripheral region and may potentially be involved in the assembly of two adjacent C$_2$S$_2$M$_2$ complexes into larger megacomplexes.

LIPIDS AT THE INTERFACES BETWEEN LHCII/CP26/CP29 AND PSII CORE

The assembly between S-LHCII/CP26/CP29 and PSII core involves three small intrinsic subunits, namely PsbW, PsbZ, and PsbH, located at the S-LHCII–CP43, CP26–CP43, and CP29–CP47 interfaces, respectively (Wei *et al.* 2016). In addition to these small subunits,

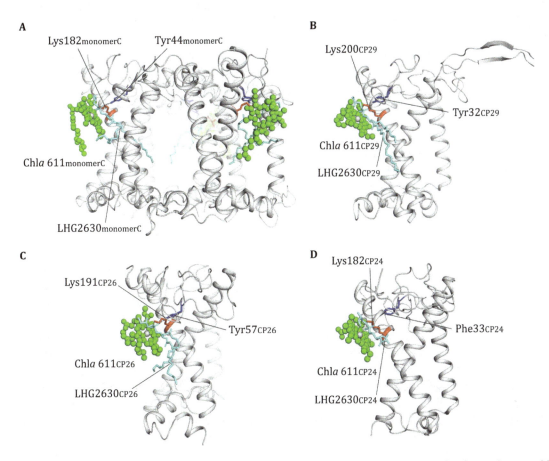

Fig. 6 The locations and roles of PG in the peripheral antenna complexes. **A** The role of three PG molecules in the assembly of LHCII trimer. The structure of S-LHCII from the C$_2$S$_2$M$_2$ supercomplex is shown. **B–D** The binding sites of PG molecules located in CP29 (**B**), CP26 (**C**) and CP24 (**D**). The chlorophyll molecules coordinated by the phosphate group of PG are presented as green sphere models, and the other pigment molecules are omitted for clarity. The amino acid residues involved in binding the head group of PG are highlighted as colored stick models

there are two PG molecules and one MGDG molecule located at the CP26–CP43 interface (Fig. 7A, B), three PG molecules at the LHCII–CP43 interface (Fig. 7C, D) and two PG plus one MGDG molecules at the CP29–CP47 interface (Fig. 7E, F). On the stromal side of the CP26–CP43 interface, LHG2630$_{CP26}$ and LHG2630$_{CP43}$ form a closely packed lipid dimer (Fig. 7B). In addition to the hydrophobic interactions between the 2-acyl groups of these two PG molecules, the head group glycerol of LHG2630$_{CP26}$ is hydrogen-bonded to the phosphate group of LHG2630$_{CP43}$. While LHG2630$_{CP26}$ binds to Lys191 from Helix A as well as Tyr57 and Arg32 from the N-terminal region of CP26 respectively, LHG2630$_{CP43}$ has its head group surrounded by three phenylalanine residues (Phe144, Phe146, Phe147) from M2–M3 loop region of CP43. The four fatty acyl chains of these two PG molecules fill in the stromal-side gap between M3 helix of CP43 and Helix A of CP26 (Fig. 7B). On the luminal side, LMG101 has its head group linked to Asp107 from CP43 and Ser59 from PsbZ through hydrogen bonds on one side, and forms van der Waals contacts with amino acid residues from the C-terminal Helix F of CP26 (Fig. 7B). The two acyl chains of LMG101 insert in the luminal gap between M2 helix of CP43 and Chl a614 of CP26, and connect them through hydrophobic interactions. Thus, these three lipid molecules and PsbZ subunit collectively stabilize the assembly between CP26 and CP43.

At the S-LHCII–CP43 interface, there are three PG molecules (LHG2630$_{S\text{-LHCII/monomer A}}$, LHG523$_{CP43}$, and LHG522 $_{CP43}$) on the stromal side (Fig. 7C). The glycerol head group of LHG523 is hydrogen-bonded to the side chain of Glu48 from the C-terminal region of PsbW as well as the backbone amide of Ala258 and Arg261 from the M4–M5 loop region of CP43 (Fig. 7D). For LHG522, its head group glycerol is in van der Waals contact with the backbones of Ala41–Ser42 on the transmembrane helix of PsbW and the phosphate group forms a strong ionic interaction with Arg262 from CP43. LHG522 and LHG523 form a closed packed dimer, and their fatty acyl chains intercalate between the M4 of CP43 and PsbW, and fill in the space between S-LHCII and CP43. LHG523 is further associated with Chl a611 of S-LHCII (monomer A) through van der Waals contacts and hydrophobic interactions. On the other side, the head group of LHG2630$_{S\text{-LHCII/monomer A}}$ is ligated with the central Mg atom of Chl a 611 and connected to Lys182 and Tyr44 through salt bridge and hydrogen bond respectively. While the 2-acyl chain of LHG2630$_{S\text{-LHCII/monomer A}}$ contributes to the monomer–monomer interface of the LHCII trimer, its 1-acyl chain bends over toward the interfacial region between S-LHCII and CP43 and makes contact with the 2-acyl chain of LHG523 (Fig. 7D). Therefore, these three PG molecules serve to stabilize the assembly among S-LHCII, PsbW, and CP43.

As for the interface between CP29 and CP47, the stromal-side gap between the two complexes is tightly packed by Chl a616 from CP29 and amino acid residues from the N-terminal region of CP29, leaving little space for lipid molecules to bind (Fig. 7E). In addition to direct CP29–CP43 interactions on the stromal side, the N-terminal region of PsbH intercalates between the N-terminal region of CP29 and the stromal surface of CP47 and serves as a bolt securing their interface (Wei et al. 2016). On the luminal side, three lipid molecules (LMG2633$_{CP47}$, LHG2630$_{CP47}$, and LHG2631$_{CP47}$) are sandwiched between CP29 and CP47 (Fig. 7E). The galactosyl head group of LMG2633$_{CP47}$ is hydrogen-bonded to the backbone carbonyl of Thr159 $_{CP47}$, and the head group glycerol of LHG2630$_{CP47}$ binds to the backbone carbonyl groups of Phe162 and Leu161 from the M3–M4 loop region of CP47. The head group of LHG2631$_{CP47}$ is in contact with the Pro88–Trp91 region on the M1–M2 loop of CP47 (Fig. 7F). The six fatty acyl chains of these three lipid molecules fill in the space between the M3 helix of CP47 and Helix C of CP29 and the space between the M4 helix of CP47 and Chl b607$_{CP29}$ near Helix E of CP29. Thereby, the assembly between CP29 and CP47 is stabilized by three interfacial lipid molecules on the luminal side.

At the M-LHCII–CP29, M-LHCII–S-LHCII, S-LHCII–CP26, S-LHCII–CP47, and CP24–PsbH/PsbX interfaces, there are some apparent void areas between the adjacent complexes. The areas may be filled with more lipid molecules, and they are either lost during purification or too disordered to be observed in the structure.

SUMMARY AND PERSPECTIVES

The high-resolution structure of plant $C_2S_2M_2$-type PSII–LHCII supercomplex reveals remarkable structural roles of lipid molecules in stabilizing the PSII core complex. Moreover, they contribute to oligomerization of PSII core dimer and LHCII trimers, and mediate the assembly between PSII and the peripheral antenna complexes including LHCII, CP29, CP26, and CP24. Furthermore, they might influence the biological function of the supercomplex by interacting with the neighboring protein subunits and the function-related cofactors, namely chlorophylls and carotenoids. Curious questions remain open concerning how the interfacial lipid molecules affect the energy transfer and electron transport kinetics as well as the spectroscopic features of the PSII–LHCII supercomplex. In perspective, future investigation on the effect of targeted mutagenesis on

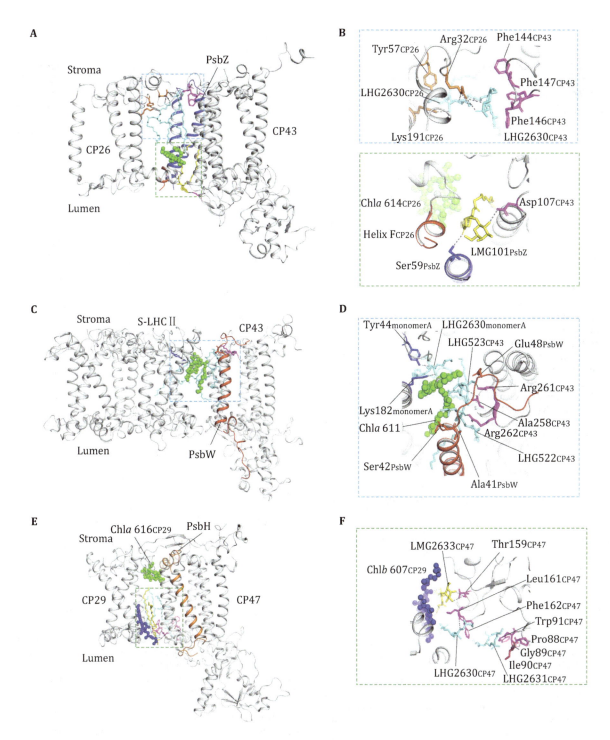

Fig. 7 The lipid molecules at the interfaces between LHCII/CP26/CP29 and PSII core complexes. **A** Side view of the interfacial lipid molecules between CP26 and CP43. **B** The zoom-in views of the lipid-binding sites in the blue and green dashed boxes shown in **A**. In the blue dashed box area (upper part), two PG molecules on the stromal side mediate the interactions between CP26 and CP43. In the green dashed box area (lower part), one MGDG molecule is located on the luminal side of the interface between CP26 and CP43. **C** Side view of the interfacial lipid molecules between S-LHCII and CP43. **D** Zoom-in view of the three PG molecules on the stromal side of the interface between S-LHCII and CP43. **E** Side view of the interfacial lipid molecules between CP29 and CP47. **F** Zoom-in view of the three lipid-binding sites on the luminal side of the interface between CP29 and CP47. The lipid molecules and amino acid residues involved in binding the lipid molecules are highlighted as stick models, while the protein backbones are shown as silver cartoon models. The chlorophyll molecules are shown as sphere models

the above-mentioned lipid-binding sites (through the CRISPR/Cas9 genome editing technique (Bortesi and Fischer 2015), for instance) is highly anticipated. More specifically, it will be interesting if one could select and mutate the amino acid residues involved in binding specific lipid molecules within the PSII–LHCII supercomplex, and examine the functional behaviors of the mutants. The mechanistic insights obtained will improve our understanding on the roles of individual lipid molecules in the assembly and function of PSII–LHCII supercomplex. To this end, the detailed features of the lipid-binding sites described in this review can serve as a guide for designing new experiments to analyze the functional roles of each individual lipid molecules within the supercomplex.

Acknowledgements We thank W. R. Chang for discussion and support, X. Z. Zhang for collaboration on cryo-EM studies of PSII–LHC supercomplexes. We also acknowledge X. P. Wei, X. D. Su and J. Ma for their persistent efforts in solving the structures of plant PSII–LHCII supercomplexes. The work is financially supported by the National Key R&D Program of China (2017YFA0503702), the Strategic Priority Research Program of CAS (XDB08020302), the Key Research Program of Frontier Sciences of CAS (QYZDB-SSW-SMC005), National Natural Science Foundation of China (31670749 and 31570724), and scholarship from the "National Thousand Young Talents Program" from the Office of Global Experts Recruitment in China.

Compliance with Ethical Standards

Conflict of interest All authors declare that they have no conflict of interest.

Human and animal rights and informed consent This article does not contain any studies with human or animal subjects performed by any of the authors.

References

Adir N (1999) Crystallization of the oxygen-evolving reaction centre of photosystem II in nine different detergent mixtures. Acta Crystallogr D Biol Crystallogr 55:891–894

Ago H, Adachi H, Umena Y, Tashiro T, Kawakami K, Kamiya N, Tian L, Han G, Kuang T, Liu Z, Wang F, Zou H, Enami I, Miyano M, Shen JR (2016) Novel features of eukaryotic photosystem II revealed by its crystal structure analysis from a red alga. J Biol Chem 291:5676–5687

Aoki M, Sato N, Meguro A, Tsuzuki M (2004) Differing involvement of sulfoquinovosyl diacylglycerol in photosystem II in two species of unicellular cyanobacteria. Eur J Biochem 271:685–693

Bezouwen LS, Caffarri S, Kale RS, Kouřil R, Awh T, Oostergetel GT, Boekema EJ (2017) Subunit and chlorophyll organization of the plant photosystem II supercomplex. Nat Plants 3:17080

Boekema EJ, Hankamer B, Bald D, Kruip J, Nield J, Boonstra AF, Barber J, Rögner M (1995) Supramolecular structure of the photosystem II complex from green plants and cyanobacteria. Proc Natl Acad Sci USA 92:175

Boekema EJ, Roon HV, Dekker JP (1998) Specific association of photosystem II and light-harvesting complex II in partially solubilized photosystem II membranes. FEBS Lett 424:95–99

Bortesi L, Fischer R (2015) The CRISPR/Cas9 system for plant genome editing and beyond. Biotechnol Adv 33:41–52

Caffarri S, Kouřil R, Kereïche S, Boekema EJ, Croce R (2009) Functional architecture of higher plant photosystem II supercomplexes. EMBO J 28:3052

Chang L, Liu X, Li Y, Liu CC, Yang F, Zhao J, Sui SF (2015) Structural organization of an intact phycobilisome and its association with photosystem II. Cell Res 25:726–737

Essigmann B, Guler S, Narang RA, Linke D, Benning C (1998) Phosphate availability affects the thylakoid lipid composition and the expression of SQD1, a gene required for sulfolipid biosynthesis in Arabidopsis thaliana. Proc Natl Acad Sci USA 95:1950–1955

Frentzen M (2004) Phosphatidylglycerol and sulfoquinovosyldiacylglycerol: anionic membrane lipids and phosphate regulation. Curr Opin Plant Biol 7:270–276

Fujii S, Kobayashi K, Nakamura Y, Wada H (2014) Inducible knockdown of MONOGALACTOSYLDIACYLGLYCEROL SYNTHASE1 reveals roles of galactolipids in organelle differentiation in Arabidopsis cotyledons. Plant Physiol 166:1436–1449

Guskov A, Kern J, Gabdulkhakov A, Broser M, Zouni A, Saenger W (2009) Cyanobacterial photosystem II at 2.9 Å resolution and the role of quinones, lipids, channels and chloride. Nat Struct Mol Biol 16:334–342

Hagio M, Gombos Z, Varkonyi Z, Masamoto K, Sato N, Tsuzuki M, Wada H (2000) Direct evidence for requirement of phosphatidylglycerol in photosystem II of photosynthesis. Plant Physiol 124:795–804

Hankamer B, Morris EP, Barber J (1999) Revealing the structure of the oxygen-evolving core dimer of photosystem II by cryo-electron crystallography. Nat Struct Biol 6:560–564

Hankamer B, Morris E, Nield J, Gerle C, Barber J (2001) Three-dimensional structure of the photosystem II core dimer of higher plants determined by electron microscopy. J Struct Biol 135:262–269

Hasan SS, Cramer WA (2014) Internal lipid architecture of the hetero-oligomeric cytochrome b6f complex. Structure 22:1008–1015

Hashimoto H, Uragami C, Cogdell RJ (2016) Carotenoids and photosynthesis. Subcell Biochem 79:111–139

Holzl G, Dormann P (2007) Structure and function of glycoglycerolipids in plants and bacteria. Prog Lipid Res 46:225–243

Jahns P, Latowski D, Strzalka K (2009) Mechanism and regulation of the violaxanthin cycle: the role of antenna proteins and membrane lipids. Biochim Biophys Acta 1787:3–14

Jarvis P, Dormann P, Peto CA, Lutes J, Benning C, Chory J (2000) Galactolipid deficiency and abnormal chloroplast development in the Arabidopsis MGD synthase 1 mutant. Proc Natl Acad Sci USA 97:8175–8179

Jones MR (2007) Lipids in photosynthetic reaction centres: structural roles and functional holes. Prog Lipid Res 46:56–87

Jordan BR, Chow WS, Baker AJ (1983) The role of phospholipids in the molecular-organization of pea chloroplast membranes—effect of phospholipid depletion on photosynthetic activities. Biochim Biophys Acta 725:77–86

Jordan P, Fromme P, Witt HT, Klukas O, Saenger W, Krauss N (2001) Three-dimensional structure of cyanobacterial photosystem I at 2.5 Å resolution. Nature 411:909–917

Kelly AA, Froehlich JE, Dormann P (2003) Disruption of the two digalactosyldiacylglycerol synthase genes DGD1 and DGD2 in

Arabidopsis reveals the existence of an additional enzyme of galactolipid synthesis. Plant Cell 15:2694–2706

Kern J, Guskov A (2011) Lipids in photosystem I: multifunctional cofactors. J Photochem Photobiol B 104:19–34

Kobayashi K, Narise T, Sonoike K, Hashimoto H, Sato N, Kondo M, Nishimura M, Sato M, Toyooka K, Sugimoto K, Wada H, Masuda T, Ohta H (2013) Role of galactolipid biosynthesis in coordinated development of photosynthetic complexes and thylakoid membranes during chloroplast biogenesis in *Arabidopsis*. Plant J 73:250–261

Kobayashi K, Endo K, Wada H (2016) Roles of lipids in photosynthesis. Subcell Biochem 86:21–49

Krumova SB, Laptenok SP, Kovacs L, Toth T, van Hoek A, Garab G, van Amerongen H (2010) Digalactosyl-diacylglycerol-deficiency lowers the thermal stability of thylakoid membranes. Photosynth Res 105:229–242

Kruse O, Hankamer B, Konczak C, Gerle C, Morris E, Radunz A, Schmid GH, Barber J (2000) Phosphatidylglycerol is involved in the dimerization of photosystem II. J Biol Chem 275:6509–6514

Lee AG (2000) Membrane lipids: it's only a phase. Curr Biol 10:R377–R380

Leng J, Sakurai I, Wada H, Shen JR (2008) Effects of phospholipase and lipase treatments on photosystem II core dimer from a thermophilic cyanobacterium. Photosynth Res 98:469–478

Liu Z, Yan H, Wang K, Kuang T, Zhang J, Gui L, An X, Chang W (2004) Crystal structure of spinach major light-harvesting complex at 2.72 Å resolution. Nature 428:287–292

Mazor Y, Borovikova A, Caspy I, Nelson N (2017) Structure of the plant photosystem I supercomplex at 2.6 Å resolution. Nat Plants 3:17014

Mizusawa N, Wada H (2012) The role of lipids in photosystem II. Biochim Biophys Acta 1817:194–208

Mizusawa N, Sakata S, Sakurai I, Sato N, Wada H (2009a) Involvement of digalactosyldiacylglycerol in cellular thermotolerance in *Synechocystis* sp. PCC 6803. Arch Microbiol 191:595–601

Mizusawa N, Sakurai I, Sato N, Wada H (2009b) Lack of digalactosyldiacylglycerol increases the sensitivity of *Synechocystis* sp. PCC 6803 to high light stress. FEBS Lett 583:718–722

Nelson N, Ben-Shem A (2004) The complex architecture of oxygenic photosynthesis. Nat Rev Mol Cell Biol 5:971

Nield J, Barber J (2006) Refinement of the structural model for the photosystem II supercomplex of higher plants. Biochim Biophys Acta 1757:353–361

Nield J, Orlova EV, Morris EP, Gowen B, Van HM, Barber J (2000) 3D map of the plant photosystem II supercomplex obtained by cryoelectron microscopy and single particle analysis. Nat Struct Biol 7:44

Nussberger S, Dorr K, Wang DN, Kuhlbrandt W (1993) Lipid-protein interactions in crystals of plant light-harvesting complex. J Mol Biol 234:347–356

Reifarth F, Christen G, Seeliger AG, Dormann P, Benning C, Renger G (1997) Modification of the water oxidizing complex in leaves of the *dgd1* mutant of *Arabidopsis thaliana* deficient in the galactolipid digalactosyldiacylglycerol. Biochemistry 36:11769–11776

Rhee KH, Morris EP, Zheleva D, Hankamer B, Kuhlbrandt W, Barber J (1997) Two-dimensional structure of plant photosystem II at 8 Å resolution. Nature 389:522–526

Rhee KH, Morris EP, Barber J, Kühlbrandt W (1998) Three-dimensional structure of the plant photosystem II reaction centre at 8 Å resolution. Nature 396:283–286

Sakurai I, Hagio M, Gombos Z, Tyystjarvi T, Paakkarinen V, Aro EM, Wada H (2003) Requirement of phosphatidylglycerol for maintenance of photosynthetic machinery. Plant Physiol 133:1376–1384

Sakurai I, Mizusawa N, Wada H, Sato N (2007) Digalactosyldiacylglycerol is required for stabilization of the oxygen-evolving complex in photosystem II. Plant Physiol 145:1361–1370

Sato N (2004) Roles of the acidic lipids sulfoquinovosyl diacylglycerol and phosphatidylglycerol in photosynthesis: their specificity and evolution. J Plant Res 117:495–505

Sato N, Sonoike K, Tsuzuki M, Kawaguchi A (1995) Impaired photosystem II in a mutant of *Chlamydomonas reinhardtii* defective in sulfoquinovosyl diacylglycerol. Eur J Biochem 234:16–23

Sato N, Aoki M, Maru Y, Sonoike K, Minoda A, Tsuzuki M (2003) Involvement of sulfoquinovosyl diacylglycerol in the structural integrity and heat-tolerance of photosystem II. Planta 217:245–251

Siegenthaler P-A (1998) Molecular organization of acyl lipids in photosynthetic membranes of higher plants. In: Paul-André S, Norio M (eds) Lipids in photosynthesis: structure, function and genetics. Dordrecht, Netherland: Springer, p 119–144

Su X, Ma J, Wei X, Cao P, Zhu D, Chang W, Liu Z, Zhang X, Li M (2017) Structure and assembly mechanism of plant $C_2S_2M_2$-type PSII–LHCII supercomplex. Science 357:815–820

Umena Y, Kawakami K, Shen JR, Kamiya N (2011) Crystal structure of oxygen-evolving photosystem II at a resolution of 1.9 Å. Nature 473:55–60

Wada H, Murata N (2007) The essential role of phosphatidylglycerol in photosynthesis. Photosynth Res 92:205–215

Wei X, Su X, Cao P, Liu X, Chang W, Li M, Zhang X, Liu Z (2016) Structure of spinach photosystem II–LHCII supercomplex at 3.2 Å resolution. Nature 534:69–74

Zhang J, Ma J, Liu D, Qin S, Sun S, Zhao J, Sui SF (2017) Structure of phycobilisome from the red alga *Griffithsia pacifica*. Nature 551:57–63

Zhou F, Liu S, Hu Z, Kuang T, Paulsen H, Yang C (2009) Effect of monogalactosyldiacylglycerol on the interaction between photosystem II core complex and its antenna complexes in liposomes of thylakoid lipids. Photosynth Res 99:185–193

Dissection of structural dynamics of chromatin fibers by single-molecule magnetic tweezers

Xue Xiao[1,3], Liping Dong[2,3], Yi-Zhou Wang[1,3], Peng-Ye Wang[1,3], Ming Li[1,3], Guohong Li[2,3], Ping Chen[2✉], Wei Li[1✉]

[1] National Laboratory for Condensed Matter Physics and Key Laboratory of Soft Matter Physics, Institute of Physics, Chinese Academy of Sciences, Beijing 100190, China
[2] National Laboratory of Biomacromolecules, CAS Center for Excellence in Biomacromolecules, Institute of Biophysics, Chinese Academy of Sciences, Beijing 100101, China
[3] University of Chinese Academy of Sciences, Beijing 100049, China

Abstract The accessibility of genomic DNA, as a key determinant of gene-related processes, is dependent on the packing density and structural dynamics of chromatin fiber. However, due to the highly dynamic and heterogeneous properties of chromatin fiber, it is technically challenging to study these properties of chromatin. Here, we report a strategy for dissecting the dynamics of chromatin fibers based on single-molecule magnetic tweezers. Using magnetic tweezers, we can manipulate the chromatin fiber and trace its extension during the folding and unfolding process under tension to investigate the dynamic structural transitions at single-molecule level. The highly accurate and reliable *in vitro* single-molecule strategy provides a new research platform to dissect the structural dynamics of chromatin fiber and its regulation by different epigenetic factors during gene expression.

Keywords Magnetic tweezers, Chromatin fiber, Dynamics, Single molecule

INTRODUCTION

In eukaryotic cells, genomic DNA is wrapped on histones to form the nucleosome (Richmond and Davey 2003), which is the basic repeating unit of chromatin, and then further folds to condensed chromatin fibers (Bickmore and van Steensel 2013). But the transcription and replication of DNA require the accessibility of DNA double-helix structure by transcription factors and polymerases. The packing density and structural dynamics of chromatin fibers are regulated by many epigenetic factors, including chromatin modifications, histone chaperones, histone variants, chromatin remodelers, and chromatin architectural proteins. For the genomic regions where gene expression are active, a series of epigenetic modifications (Zentner and Henikoff 2013), enzymes (Maier *et al.* 2000), and transcription factors (Li *et al.* 2016b; Pavri *et al.* 2006) assist in unfolding of chromatin fibers to expose DNA, allowing the transcription and replication to proceed. When the transcription or duplication is completed, the chromatin fiber needs to be reassembled (Fleming *et al.* 2008). Therefore, deciphering the structural dynamics and its epigenetic regulation of chromatin fibers is the basis for understanding the epigenetic regulation of gene-related biological processes including transcription and DNA replication.

Traditional biochemical methods have shown some difficulty in studying the dynamic structural changes of chromatin fibers (Fleming *et al.* 2008; Georgel *et al.* 2003). For example, the properties of chromatin fibers

---

Xue Xiao, Liping Dong and Yi-Zhou Wang have contributed equally to this work.

✉ Correspondence: chenping@moon.ibp.ac.cn (P. Chen), weili007@iphy.ac.cn (W. Li)

obtained in gel electrophoresis and circular dichroism are the sum of a huge number of molecules involved in the reaction; thus it is impossible to characterize the dynamic process of structural changes of chromatin fibers. Although high-resolution imaging techniques, such as X-ray and Cryo-electron microscopy (Cryo-EM), have successfully determined the high-resolution structures of the nucleosome (Luger et al. 1997) and the 30-nm chromatin fiber (Song et al. 2014), these techniques are only suitable for studying static structures and cannot monitor the dynamic structural transitions of the nucleosome and chromatin fibers. Fluorescence resonance transfer (FRET) technique can track the dynamic changes of chromatin structure at single-molecule level (Chen et al. 2016; Fierz et al. 2011; Ngo and Ha 2015). However, in most time, fluorescence probe quenches in a few minutes; thus it is difficult to monitor the structural transitions in such a short-time window.

Single-molecule force spectroscopy is an ideal technique to study the structural change of biological macromolecules, especially the nucleosome and chromatin fibers. By applying different external forces on the single chromatin fiber or mononucleosome, we can manipulate conformational changes of the molecule, which could be monitored by tracking the extension of the chromatin fiber or mononucleosome. We can also deduce the stability and assembly pattern of the chromatin fiber or mononucleosome by analyzing the step size and the transition force of structural intermediates. The main techniques of single-molecular force spectroscopy include atomic force microscopy (AFM; Piontek and Roos 2018), optical tweezers (Neuman and Block 2004), and magnetic tweezers (Gosse and Croquette 2002). AFM captures biological macromolecules by a chemically modified probe and applies a pulling force, but the pulling force is usually very large and it is difficult to ensure only one single-molecule is captured. The optical tweezers require the two ends of molecule binding to two polystyrene beads respectively, and then the optical tweezers grab the polystyrene beads through the optical traps constructed by the laser, and then move the laser to apply the pulling force. Optical tweezers can achieve relatively high resolution (Abbondanzieri et al. 2005; Mahamdeh and Schäffer 2009), but it is difficult to keep constant force when tracking the conformational changes. Besides that, optical tweezers are not suitable for long time measurement, because high-intensity laser may damage the sample (Simpson et al. 1998). The magnetic tweezers connect the two ends of biological macromolecules to surface and paramagnetic bead respectively, and then apply the pulling force with an external magnetic field. Magnetic tweezers can track many biological macromolecules at the same time, but the resolution is slightly lower than optical tweezers. In this paper, we report a strategy to study the structural dynamics of chromatin fibers based on single-molecule magnetic tweezers. We modified the magnetic tweezers by adopting solid-microscope optical system, large-field objective lens, and high-speed sampling CCD camera to improve the time resolution of magnetic tweezers and the number of magnetic beads tracked at the same time. We also developed a home-built control software of magnetic tweezers based on LabVIEW to achieve more advanced bead tracking and data analysis algorithm, which further improves the resolution of magnetic beads position and pulling force. The chromatin fiber labeled with either biotin or digoxigenin at the two ends of DNA was reconstituted in vitro and attached to the modified magnetic tweezers. The extensions of chromatin fibers under different tension were recorded at single-molecule level to trace the dynamic structural transition. This strategy provides a new research platform to dissect the structural dynamics of chromatin fiber and its regulation by different chromatin factors.

EXPERIMENTAL SECTION

Preparation of chromatin fibers

Recombinant histones and DNA templates of 24 tandem 177 bp repeats of the 601 sequence (24 × 177 bp) were cloned and purified as previously described (Li et al. 2010). The 24 × 177 bp 601 DNA was subcloned into a modified pWM530 plasmid containing two BseYI sites flanking multiple cloning sites. The pWM530-BseYI-177-24 plasmid was digested by BseYI enzyme, with the 24 × 177 DNA templates purified by gel extraction. The two single-stranded ends of 24 × 177 DNA templates were filled with either dUTP-digoxigenin or dATP-biotin by Klenow reaction at 30 °C for 3 h. Remaining enzymes were removed by phenolic chloroform extraction and ethanol precipitation. For histone purification, pET-histone expression plasmids were transformed into BL21 (DE3) pLysS cells individually and then a single colony each was inoculated from a freshly streaked agar plate into 5 ml of LB medium containing 100 μg/mL ampicillin and 25 μg/mL chloramphenicol. 100 ml overnight culture from optimal clones was inoculated into 5 l LB medium and overexpression of histones were induced at $OD_{600\ nm}$ of 0.4–0.6 with 0.5 mmol/L IPTG. After growing for another 3–4 h, cells were harvested by centrifugation and resuspended with buffer A (50 mmol/L Tris–HCl, pH 7.5, 100 mmol/L NaCl, 1 mmol/L EDTA, 1 mmol/L bezamidine). After sonication and centrifugation, the pellets containing inclusion bodies of the corresponding histones were washed three times by buffer A plus 1% Triton X-100 and later washed with buffer A twice to get rid of Triton X-100. Then pellets were dissolved by gently stirring for 1 h at room temperature with buffer B (7 mol/L guanidinium HCl, 20 mmol/L Tris–HCl, pH 7.5, 10 mmol/L DTT). With another centrifugation (23,000 g for 10 min) to remove undissolved pellets, supernatants containing unfolded histones were collected and analyzed by 15% SDS-PAGE.

The respective histone octamers were reconstituted as previously described (Dyer et al. 2004). Equimolar amounts of individual histones in unfolding buffer B (7 mol/L guanidinium HCl, 20 mmol/L Tris–HCl, pH 7.5, 10 mmol/L DTT) were dialyzed into refolding buffer (2 mol/L NaCl, 10 mmol/L Tris–HCl, pH 7.5, 1 mmol/L EDTA, 5 mmol/L 2-mercaptoethanol), and purified through a Superdex S200 column. Chromatin samples were assembled using the salt-dialysis method as previously described (Song et al. 2014). The reconstitution reaction mixture with histone octamers and DNA templates in TEN buffers (10 mmol/L Tris–HCl, pH 8.0, 1 mmol/L EDTA, 2 mol/L NaCl) were dialyzed for 16 h at 4 °C in TEN buffer, which was continuously diluted by slowly pumping in TE buffer (10 mmol/L Tris–HCl, pH 8.0, 1 mmol/L EDTA) to a lower concentration of NaCl from 2 to 0.6 mol/L. For nucleosomal arrays, samples were collected after final dialysis in measurement HE buffer (10 mmol/L HEPES, pH 8.0, 0.1 mmol/L EDTA) for 4 h. For histone H1 incorporation, an equal molar amount of histone H1 (relative to mono-nucleosomes) was added before the final dialysis step and further dialyzed in TE buffer with 0.6 mol/L NaCl for 3 h. The stoichiometry of histone octamer binding to the DNA template was determined by EM investigation (Song et al. 2014).

Stoichiometry of histone octamer to DNA template and histone H1 to nucleosome were evaluated by EM analysis as described previously (Chen et al. 2013). The samples were fixed with 0.4% glutaraldehyde (Fluka) in HE buffer on ice for 30 min. For metal shadowing experiment, chromatin samples were prepared in HE buffer with DNA concentrations of 5 μg/mL. 2 mmol/L spermidine was added into the sample solution to enhance the absorption of chromatin to the grids. Samples were applied to the glow-discharged carbon-coated EM grids and incubated for 2 min and then blotted. Grids were washed stepwise in 20 ml baths of 0%, 25%, 50%, 75%, and 100% ethanol solution for 4 min, each at room temperature, air dried and then shadowed with tungsten at an angle of 10° with rotation. For negative staining studies, chromatin samples were prepared in HE buffer with DNA concentrations of 20 μg/mL. Chromatin samples in fixative solution were incubated on glow-discharged carbon-coated EM grids for 1 min. The excess sample solution was removed using filter papers. The grid was stained in 2% uranylacetate for 2×30 s, blotted with filter papers and allowed to air-dry for several minutes. Samples were examined using a FEI Tecnai G2 Spirit 120 kV transmission electron microscope.

Setup of magnetic tweezers

As shown in Fig. 1, the setup of magnetic tweezers is mainly composed of light source (625 nm red LED, THORLABS), two NdFeB magnets with 0.5 mm, flow cell, 60× oil immersed objective (UPLSAPO60XO, NA 1.35, Olympus), and 1280×2024 CCD camera (MC1362, Mikrotron). The flow cell consists of a bottom coverslip, a double-sided tap with a rectangular channel (5×50 mm^2), and an upper coverslip with an inlet and outer at each end. In the flow cell, one end of the chromatin fiber is anchored to the streptavidin coated magnetic beads (M280 Invitrogen Norway), and the other end is anchored to the surface coated with anti-digoxigenin. The two magnets are fixed on a high-resolution translation stage (M-126, PI) and a rotation stage (C-150, PI). The exerted tension (0–100 pN) on chromatin fibers is controlled by tuning the distance of the magnets in z direction (Dammer et al. 1996; Lee et al. 1994; Merkel et al. 1999; Moy et al. 1994; Pincet and Husson 2005). At a given magnet position, the tension exerted on the chromatin samples is constant. The twist can be regulated in the twist-constrained assays by rotating the magnets. The images of beads would go through the objective and then into the CCD camera. The thermal drift is the barrier for the image tracing in a long time up to several hours. In our magnetic tweezers, four Peltier modules are applied to control the sample temperature with the resolution of 0.1 °C.

Coverslip cleaning and surface functionalization

To anchor the digoxin and biotin labeled chromatin fiber on the surface, the bottom coverslip needs to be cleaned and functionalized. Firstly, several pieces of coverslips were put in a staining jar filled with deionized water and detergent, and then the staining jar was placed in an ultrasonic cleaning bath for 30 min. After that, the coverslips were rinsed with deionized water for 5–10 times to remove the detergent. The washing process was repeated using acetone and methanol, respectively

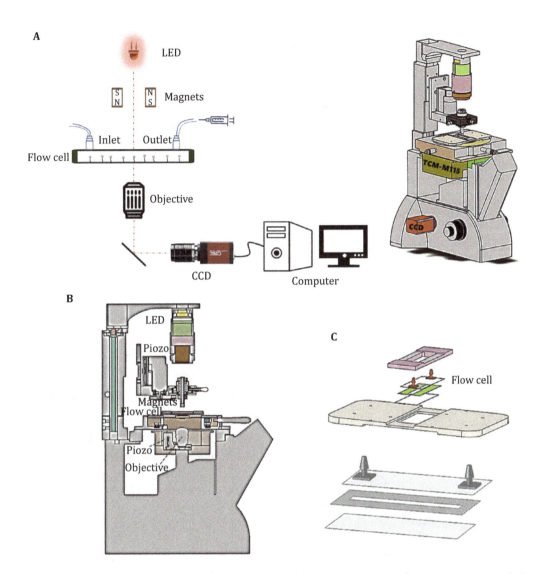

Fig. 1 Setup of magnetic tweezers. **A** The schematic (left) and assembly (right) diagram of magnetic tweezers. **B** Sectional view of magnetic tweezers. Magnetic tweezers are consisted of LED light source, a pair of magnets, a flow cell, a flow cell and a CCD camera. **C** The configuration of the flow cell. The flow cell is formed by two coverslips with a two-sided tap in the center

to rinse some organic impurities. To functionalize the surface of coverslip, the nitrocellulose solution (0.4%–0.8% w/v) was coated on the surface and heated at 120 °C for 3 min, then the surface could bind with anti-digoxin for tethering chromatin fibers (Cnossen *et al.* 2014).

To control the thermal drift, furthermore, the diluted polystyrene beads solution was dropped on the surface and heated at ~150 °C for 5 min to melt on the cover-slip surface. The fixed beads were tracked simultaneously and served as the indicator of the drift. We subtracted these drift signals in our LabVIEW software.

Result

Preparation of chromatin fibers

The nucleosomal arrays and chromatin fibers used for the single-molecule magnetic tweezers investigation were reconstituted *in vitro* by purified histones and DNA template (Fig. 2A, B). The DNA template containing 24 tandem repeats of 177-bp Widom 601 nucleosome positioning sequence (24 × 177 bp) was labeled with either biotin or digoxigenin at the two ends. In the absence of linker histone, the nucleosomal arrays reconstituted on the 24 × 177 bp DNA templates adopted an extended beads-on-a-string conformation, as shown by EM analysis (Fig. 2C). In the presence of H1,

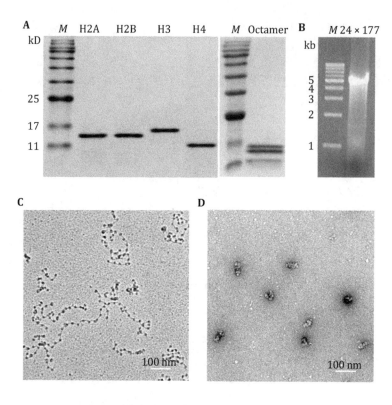

Fig. 2 Preparation of chromatin samples. **A** SDS-PAGE analysis of the purified four histones and the reconstituted histone octamers. **B** Agarose gel analysis of the purified 24 × 177 bp DNA template for magnetic tweezers. **C** EM (metal shadowing) images of the nucleosomal arrays without H1. **D** EM (negatively stained) images of the compact chromatin fibers with H1

the arrays were further condensed into compact chromatin fibers (Fig. 2D).

The chromatin fiber labeled with either biotin or digoxigenin at the two ends was attached to the modified magnetic tweezers. To avoid non-specific binding of chromatin fibers to the surface and magnetic beads, the flow cell requires passivation by incubating with PBS passivation buffer (10 mg/mL BSA, 1 mmol/L EDTA, 10 mmol/L PB, 10 mg/mL Pluronic F127 surfactant (Sigma-Aldrich), 3 mmol/L NaN$_3$, pH 7.8) for 2 h (Fulconis et al. 2006; Koster et al. 2005). To tether the chromatin fiber to the surface of super paramagnetic beads (M280 Invitrogen Norway), the diluted chromatin fiber solution (∼10 pmol/L) was first mixed with M280 beads for 30 min at room temperature on the Hula Mixer (ThermoFisher) at 1 turns/min. Secondly, passivation buffer in the flow cell was rinsed by HE buffer (10 mmol/L HEPES, 1 mmol/L EDTA) using a pump (500 μl/min). Thirdly, the mixture of chromatin fibers and magnetic beads was incubated in the flow cell for 10 min to anchor the other end of the chromatin fiber to the surface of the functionalized coverslip. At last, the unbound beads and chromatin fibers in the flow cell were subsequently washed out with 500 μl HE buffer.

To identify the beads that only bind with single chromatin fiber, we pull the beads with a relatively large force (2–3 pN) and rotate the permanent magnets ∼50 turns. If the bead binds more than one chromatin fiber, the chromatin fibers would be braided and the distance of the bead to surface would decrease. On the other hand, the beads that only bind single chromatin fiber have no changes in distance to surface, which could be tracked in further stretching experiments.

Tracking the magnetic beads in three-dimensions

In stretching experiments, the extension of chromatin fibers is determined by tracking the three-dimensional (3D) position of the magnetic beads. The positions of a magnetic bead in the horizontal direction (X, Y) can be determined by calculating the center of its diffraction pattern. A common bead tracking method is one-dimensional (1D) cross-correlation algorithm (Gosse and Croquette 2002), but the accuracy is relatively low. Our magnetic tweezers use the quadrant interpolation algorithm developed by Cees Dekker (van Loenhout et al. 2012), which is an improved version of conventional 1D cross-correlation algorithm. In brief, we calculate the estimated center (X_0, Y_0) of a bead using a center-of-mass computation, which is shown as follow:

$$X_0 = \frac{\sum_{i=1}^{N}\sum_{j=1}^{N}(iI_{ij})}{\sum_{i=1}^{N}\sum_{j=1}^{N}I_{ij}}, \quad Y_0 = \frac{\sum_{i=1}^{N}\sum_{j=1}^{N}(jI_{ij})}{\sum_{i=1}^{N}\sum_{j=1}^{N}I_{ij}}, \tag{1}$$

where i and j are pixel subscripts of extracted bead image and I_{ij} represents the light intensity of the jth pixel point of the ith line. Based on the estimated center, the image is divided into four quadrants to sample radial profiles at polar coordinates, respectively. When calculating the X position of a bead, the sum of right radial profiles is concatenated with the left profiles, thus creating an intensity profile $I_x(r)$ along the x direction. $I_x(r)$ is cross-correlated with its mirror profile $I_x(-r)$:

$$C_{xx} = \text{IFFT}(\text{FFT}(I_x(r)) * \overline{\text{FFT}(I_x(-r))}), \quad (2)$$

where FFT is fast Fourier transform and IFFT is inverse fast Fourier transform, C_{xx} is resulting cross correlative curve. The X position of bead can be derived from the peak position d_r of resulting cross correlative curve C_{xx} as follow:

$$X = X_0 + \frac{2d_r}{\pi}, \quad (3)$$

where X_0 is X coordinate of estimated center, π is circumference rate. By creating the intensity profile along the Y direction, we could calculate the Y position accordingly.

Tracking the bead position in vertical direction (z direction) requires a prerequisite calibration that building a relationship between defocused distance and diffraction pattern of beads (Gosse and Croquette 2002). Firstly, by moving magnets toward flow cell, a relatively large force (~ 2–3 pN) is applied on beads to reduce the fluctuation of beads. Then the inversed objective is moving at ~ 100 nm step size, while the CCD is recording the diffraction pattern at every step (Fig. 3A–D). The relationship between diffraction patterns and defocused distance can create a two-dimensional (2D) look-up table (LUT) through bi-linear interpolation (Fig. 3E). To determine the Z position of a given image, the χ^2 value of the radial profile $I(r)$ with LUT is calculated:

$$\chi^2 = \sum_{r=0}^{R} (I(r) - I_{\text{LUT}}(r))^2. \quad (4)$$

Thus, the minimum χ^2 corresponds to the Z position of the image.

Calculation of applied force

To calculate force applied on samples, a common used method is treating the bead–chromatin system as an inversed pendulum with small displacements in the vicinity equilibrium position (Klaue and Seidel 2009; Strick et al. 1998). In this model, the magnetic bead is attached to a rigid rod with length l, and the force F_{mag} exerts on it. When the pendulum has a small displacement a, restoring force is giving by $F_a = -\frac{F_{\text{mag}}}{l} \times a$. This equation is equivalent to the rule of a Hookean spring with the spring constant $k_a = \frac{F_{\text{mag}}}{l}$, according the equipartition theorem,

$$E_{\text{therm}} = \frac{1}{2} k_a \langle a^2 \rangle = \frac{1}{2} k_B T, \quad (5)$$

where k_B is Boltzmann constant and T is temperature. Then we get the force F_{mag}:

$$F_{\text{mag}} = \frac{k_B T}{\langle a^2 \rangle} l, \quad (6)$$

where $\langle a^2 \rangle$ is the variance of bead in lateral fluctuation, and l is effective pendulum length, namely the sum of radius of bead and extension of chromatin fiber. Therefore, the applied force can be determined from the lateral fluctuation of the bead (Fig. 4A, B). However, in practice, at forces greater than ~ 10 pN, the measurement of $\langle a^2 \rangle$ in real space appears systematic error because of the finite camera acquisition frequency. Therefore, power spectral density (PSD) analysis is typically used to analyze the in-plane fluctuations.

Considering the viscous drag coefficient γ_a, the PSD can be expressed:

$$S_a(f) = \frac{4 k_B T \gamma_a}{k_a^2} \frac{1}{1 + (f/f_c)^2}, \quad (7)$$

where f is the frequency and $f_c = \frac{k_a}{2\pi \gamma_a}$ is characteristic cutoff frequency. Due to the limiting of acquisition frequency of camera f_s, low-pass filtering and aliasing should be considered (Klaue and Seidel 2009; Lansdorp and Saleh 2012; te Velthuis et al. 2010) as follows,

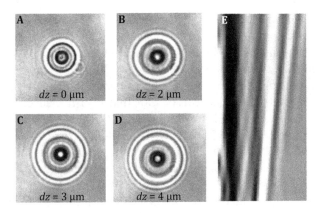

Fig. 3 Generation of a calibration profile in the z direction for a bead. **A–D** The diffraction patterns of the bead with various distances to the focal plane. **E** By recording the size and density of the diffraction rings at different distance to the focal plane, a generated calibration profile is used to measure vertical relative displacements of the bead

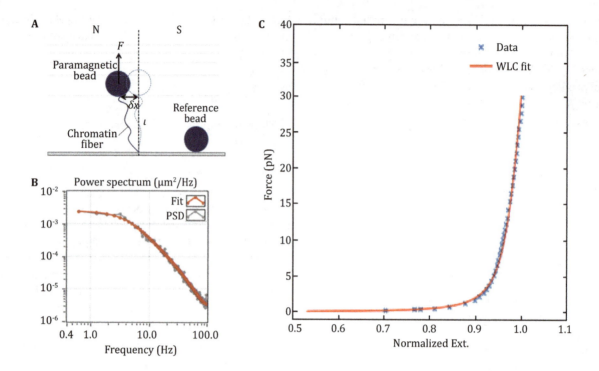

Fig. 4 The force calculation for the DNA tether. **A** A schematic representation of a tethered bead with a reference bead fixed to the bottom coverslip. The exerted tension F on the paramagnetic bead arises from the gradient of the magnetic field. **B** The spectrum and the fitted curve of the one-dimensional fluctuations of the paramagnetic bead under a constant tension. **C** For a 10 kb dsDNA, the relationship between the measured force and extension is consistent with the WLC (worm-like chain) model

$$S_a^{\text{corr}}(f) = \sum_{n=-\infty}^{\infty} \frac{4k_B T \gamma_a}{k_a^2} \frac{1}{1+(|f+nf_s|/f_c)^2} \frac{\sin^2(\pi \tau_e |f+nf_s|)}{(\pi \tau_e |f+nf_s|)^2}, \quad (8)$$

where τ_e is the exposure time of the camera. This formula can be used to fit the PSD, obtaining the k_a, then the F_{mag}. In our experiment, we used an improved PSD fitting method developed by Ralf Seidel, which takes into account the exposure time, direction of magnet field and rotation of bead (Daldrop et al. 2015).

To calibrate the force derived from our magnetic tweezers, we stretched a dsDNA with a length of 10 kb, then we derived the applied force from lateral fluctuation of beads and fitted with an extensible WLC model (Odijk 1995):

$$\frac{z}{L} = 1 - \frac{1}{2}\left(\frac{k_B T}{F L_P}\right)^{1/2} + \frac{F}{S}, \quad (9)$$

where z and L are extension and the contour length of molecule, L_P is persistence length that characterize the rigidity of molecule, and S is stretch modulus of the molecule. As shown in Fig. 4C, our experiment result agrees well with the WLC model, indicating a good performance of our magnetic tweezers.

Structural dynamics of chromatin fibers revealed by magnetic tweezers

Using our improved magnetic tweezers, we investigated the hierarchical organization and structural dynamics of chromatin fibers reconstituted *in vitro* in the presence of H1 (Fig. 5A), whose 3D Cryo-EM structures have been resolved recently (Li et al. 2016c; Song et al. 2014). To trace the structural transition, we applied a continuously increasing force on a single chromatin fiber by moving the magnets in z direction at 2 μm/s. The force–extension curve of a chromatin fiber containing H1 is shown by the blue curve, with that of a nucleosomal array without H1 for comparison (Fig. 5B). At force above 5 pN, similar stepwise unfolding events were observed for both extended nucleosomal arrays and compacted chromatin fibers, which correspond to the rupture states of nucleosomes in the chromatin fiber (Bintu et al. 2012; Brower-Toland et al. 2002; Hall et al. 2009). At low-force region (<5 pN), the curve of the chromatin fiber begins to deviate from that of the nucleosomal array at ~1 pN, and a distinct force plateau spanning a range of ~100 nm near 3 pN is clearly identified (Fig. 5B). The structural transition near 3 pN should be attributed to the disruption of

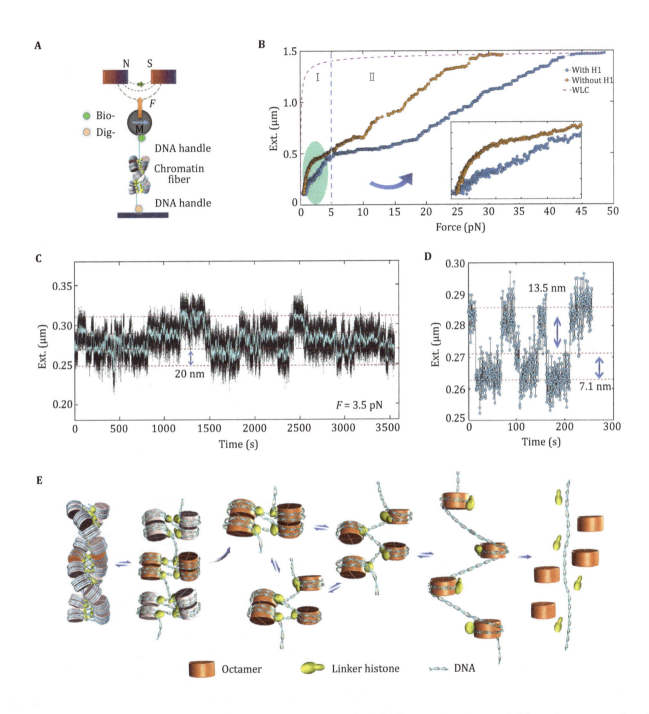

Fig. 5 Investigation of structural dynamics of chromatin fibers by single-molecule magnetic tweezers. **A** Schematic representation of chromatin fiber studied by magnetic tweezers. **B** Comparison of two typical force–extension curves of a chromatin fiber with H1 (*blue curve*) and without H1 (*orange curve*) in HE buffer, with two major stages at high force (>5 pN) or low force (<5 pN) recognized. The inset shows the details of stage at low force (<5 pN). **C** Stepwise folding and unfolding dynamics of tetranucleosomal units at 3.5 pN for the chromatin fiber with H1. **D** Stepwise folding and unfolding dynamics of each tetranucleosomal unit with two alternative pathways at 3.5 pN for the chromatin fiber with H1. **E** Model for the dynamic organization of chromatin fiber. The left-handed double-helix chromatin fiber unfolds to a "tetranucleosomes-on-a-string" extended structure, and then unfolds to a complete open nucleosomal array in one or two steps

nucleosome–nucleosome interactions in the higher-order chromatins, as the nucleosomes remain intact at such low forces.

In order to gain more insights into the nature of structural transition near 3 pN, we performed the force-clamp measurements of chromatin fibers. Interestingly, a reversible folding and unfolding dynamical behavior was observed at 3.5 pN (Fig. 5C), with the obvious transition step size of ~20 nm. The details of the folding and unfolding dynamics of chromatin fiber at 3.5 pN indicate that the structure transition with step of 20 nm is always accomplished within two steps, which corresponds to the two-step disruption of tetra-nucleosome structural unit (Fig. 5D). Importantly, similar dynamic processes can be observed in the chromatin fibers assembled not only on regular tandem repeat of 601 DNA sequence but also on the scrambled (non-repetitive) DNA sequence, suggesting that the existence of tetra-nucleosome units is not dependent on DNA sequence. By using the modified single-molecule magnetic tweezers method, we show that the tetra-nucleosome unit exists as a stable structural intermediate of the 30-nm chromatin fiber and further unfolds to a more extended "beads-on-a-string" conformation by disrupting the nucleosome–nucleosome interactions within tetranucleosomal unit (Fig. 5E) (Li et al. 2016b).

DISCUSSION

In eukaryotic gene activation, the compact chromatin fibers must unfold for transcription and other DNA-related biological processes. Chromatin dynamics is critical to regulate the accessibility of transcription factors via dynamic transitions between the compact chromatin fiber and the more accessible nucleosomal array in vivo (Li et al. 2010; Li and Reinberg 2011). Therefore, it is of great importance to study how the dynamic organization of the chromatin fiber is regulated. However, there are various technical challenges to probe the detailed structure and the dynamics of chromatin fiber using traditional biochemical assays (Vilfan et al. 2009). For example, static conformations obtained by Cryo-EM structural study in vitro cannot provide much detailed information for dynamic processes. As a single-molecule method, magnetic tweezers is a powerful tool to study the dynamic organization of chromatin fibers by tracing the real-time folding and unfolding of single chromatin fiber (Bintu et al. 2012; Brower-Toland et al. 2002; Hall et al. 2009; Kruithof et al. 2009; Meng et al. 2015; Yan et al. 2007).

In this paper, we present a platform for dissecting dynamics of chromatin fibers using single-molecule magnetic tweezers. With more advanced tracking algorithms and higher resolution, we can observe the unfolding details of chromatin fibers and monitor the folding dynamics of chromatin fibers by careful tension adjustment. With magnetic tweezers, we have revealed that tetra-nucleosome unit exists as a stable structural intermediate during the formation of 30-nm chromatin fiber. These results are the complement of traditional biochemical experiments and high-resolution static imaging techniques such as EM, depicting the dynamics of single chromatin fiber, and providing a new perspective for understanding the dynamic organization of chromatin fibers and its regulation by chromatin factors.

Magnetic tweezers are widely used in the study of biological macromolecules because of their simple principle, low cost, high parallelism and single-molecule tracking. In addition to studying the mechanical properties of DNA and its higher-order structures, magnetic tweezers can also be used to study the folding kinetics of various proteins, such as actin (Yuan et al. 2017) and trans-membrane proteins (Min et al. 2015). The constant force measurement, which is easy to be realized by magnetic tweezers, is also suitable for studying the working mechanism of DNA-related enzymes, such as condensin (Eeftens et al. 2017), helicase (Li et al. 2016a; Seol et al. 2016), and polymerase (Maier et al. 2000).

In the last 20 years, the technology and analytical theory of magnetic tweezers have been continuously improved. According to the needs of experiments, a series of improvements on magnetic tweezers have been made. For example, Yan Jie's group developed the transverse magnetic tweezers, which can greatly improve the force applied by the magnetic beads (Yan et al. 2004). They also developed a non-disturbing buffer exchange technology to replace the buffer environment in the measurement process (Le et al. 2015). In order to improve the resolution of magnetic tweezers, Ralf Seidel et al. used CMOS camera with ultra-high sampling rate and image processing technology base on GPU, improving the resolution of magnetic tweezers to angstrom level (Huhle et al. 2015). On the other hand, Terence R. Strick et al. used single-molecule FRET technology to combine with magnetic tweezers (Graves et al. 2015), which not only can improve the resolution of length change, but also can be used to reveal whether the protein falls off in the process of a structural transformation. Further development of these technologies may provide new platform to solve the hard questions in life science.

Acknowledgements This work was supported by Grants from the National Natural Science Foundation of China (11474346, 31471218, 31630041, 31525013, 31521002 and 61275192), the Key Research Program of Frontier Sciences, CAS (QYZDB-SSW-SLH045, QYZDY-SSW-SMC020), the Youth Innovation Promotion Association CAS (2015071), the Ministry of Science and Technology of China (2017YFA0504200, 2015CB856200), (HHMI International Research Scholar Grant (55008737),and the National Key Research and Development Program (2016YFA0301500).

Compliance with Ethical Standards

Conflict of interest Xue Xiao, Liping Dong, Yi-Zhou Wang, Peng-Ye Wang, Ming Li, Guohong Li, Ping Chen and Wei Li declare that they have no conflict of interest.

Human and animal rights and informed consent This article does not contain any studies with human or animal subjects performed by any of the authors.

References

Abbondanzieri EA, Greenleaf WJ, Shaevitz JW, Landick R, Block SM (2005) Direct observation of base-pair stepping by RNA polymerase. Nature 438:460–465

Bickmore WA, van Steensel B (2013) Genome architecture: domain organization of interphase chromosomes. Cell 152:1270–1284

Bintu L, Ishibashi T, Dangkulwanich M, Wu Y-Y, Lubkowska L, Kashlev M, Bustamante C (2012) Nucleosomal elements that control the topography of the barrier to transcription. Cell 151:738–749

Brower-Toland BD, Smith CL, Yeh RC, Lis JT, Peterson CL, Wang MD (2002) Mechanical disruption of individual nucleosomes reveals a reversible multistage release of DNA. Proc Natl Acad Sci USA 99:1960

Chen P, Zhao J, Wang Y, Wang M, Long H, Liang D, Huang L, Wen Z, Li W, Li X, Feng H, Zhao H, Zhu P, Li M, Wang QF, Li G (2013) H3.3 actively marks enhancers and primes gene transcription via opening higher-ordered chromatin. Genes Dev 27:2109–2124

Chen Y, Tokuda JM, Topping T, Meisburger SP, Pabit SA, Gloss LM, Pollack L (2016) Asymmetric unwrapping of nucleosomal DNA propagates asymmetric opening and dissociation of the histone core. Proc Natl Acad Sci USA. https://doi.org/10.1073/pnas.1611118114

Cnossen JP, Dulin D, Dekker NH (2014) An optimized software framework for real-time, high-throughput tracking of spherical beads. Rev Sci Instrum. https://doi.org/10.1063/1.4898178

Daldrop P, Brutzer H, Huhle A, Kauert DJ, Seidel R (2015) Extending the range for force calibration in magnetic tweezers. Biophys J 108:2550–2561

Dammer U, Hegner M, Anselmetti D, Wagner P, Dreier M, Huber W, Güntherodt HJ (1996) Specific antigen/antibody interactions measured by force microscopy. Biophys J 70:2437–2441

Dyer PN, Edayathumangalam RS, White CL, Bao Y, Chakravarthy S, Muthurajan UM, Luger K (2004) Reconstitution of nucleosome core particles from recombinant histones and DNA. Methods Enzymol 375:23–44

Eeftens JM, Bisht S, Kerssemakers J, Kschonsak M, Haering CH, Dekker C (2017) Real-time detection of condensin-driven DNA compaction reveals a multistep binding mechanism. EMBO J. https://doi.org/10.15252/embj.201797596

Fierz B, Chatterjee C, McGinty RK, Bar-Dagan M, Raleigh DP, Muir TW (2011) Histone H2B ubiquitylation disrupts local and higher-order chromatin compaction. Nat Chem Biol 7:113–119

Fleming AB, Kao C-F, Hillyer C, Pikaart M, Osley M (2008) H2B ubiquitylation plays a role in nucleosome dynamics during transcription elongation. Mol Cell 31:57–66

Fulconis R, Mine J, Bancaud A, Dutreix M, Viovy J-L (2006) Mechanism of RecA-mediated homologous recombination revisited by single molecule nanomanipulation. EMBO J 25:4293–4304

Georgel PT, Horowitz-Scherer RA, Adkins N, Woodcock CL, Wade PA, Hansen JC (2003) Chromatin compaction by human MeCP2. Assembly of novel secondary chromatin structures in the absence of DNA methylation. J Biol Chem 278:32181–32188

Gosse C, Croquette V (2002) Magnetic tweezers: micromanipulation and force measurement at the molecular level. Biophys J 82:3314–3329

Graves ET, Duboc C, Fan J, Stransky F, Leroux-Coyau M, Strick TR (2015) A dynamic DNA–repair complex observed by correlative single-molecule nanomanipulation and fluorescence. Nat Struct Mol Biol 22:452–457

Hall MA, Shundrovsky A, Bai L, Fulbright RM, Lis JT, Wang MD (2009) High-resolution dynamic mapping of histone–DNA interactions in a nucleosome. Nat Struct Mol Biol 16:124

Huhle A, Klaue D, Brutzer H, Daldrop P, Joo S, Otto O, Keyser UF, Seidel R (2015) Camera-based three-dimensional real-time particle tracking at kHz rates and Ångström accuracy. Nat Commun 6:5885

Klaue D, Seidel R (2009) Torsional stiffness of single superparamagnetic microspheres in an external magnetic field. Phys Rev Lett 102:028302

Koster DA, Croquette V, Dekker C, Shuman S, Dekker NH (2005) Friction and torque govern the relaxation of DNA supercoils by eukaryotic topoisomerase IB. Nature 434:671

Kruithof M, Chien F-T, Routh A, Logie C, Rhodes D, van Noort J (2009) Single-molecule force spectroscopy reveals a highly compliant helical folding for the 30-nm chromatin fiber. Nat Struct Mol Biol 16:534

Lansdorp BM, Saleh OA (2012) Power spectrum and Allan variance methods for calibrating single-molecule video-tracking instruments. Rev Sci Instrum 83:025115

Le S, Yao M, Chen J, Efremov AK, Azimi S, Yan J (2015) Disturbance-free rapid solution exchange for magnetic tweezers single-molecule studies. Nucleic Acids Res. https://doi.org/10.1093/nar/gkv554

Lee GU, Kidwell DA, Colton RJ (1994) Sensing discrete streptavidin-biotin interactions with atomic force microscopy. Langmuir 10:354–357

Li G, Reinberg D (2011) Chromatin higher-order structures and gene regulation. Curr Opin Genet Dev 21:175–186

Li G, Margueron R, Hu G, Stokes D, Wang Y-H, Reinberg D (2010) Highly compacted chromatin formed *in vitro* reflects the dynamics of transcription activation *in vivo*. Mol Cell 38:41–53

Li JH, Lin WX, Zhang B, Nong DG, Ju HP, Ma JB, Xu CH, Ye FF, Xi XG, Li M, Lu Y, Dou SX (2016a) Pif1 is a force-regulated helicase. Nucleic Acids Res. https://doi.org/10.1093/nar/gkw295

Li W, Chen P, Yu J, Dong L, Liang D, Feng J, Yan J, Wang P-YY, Li Q, Zhang Z, Li M, Li G (2016b) FACT remodels the tetranucleosomal unit of chromatin fibers for gene transcription. Mol Cell 64:120–133

Li X, Liu H, Cheng L (2016c) Symmetry-mismatch reconstruction of genomes and associated proteins within icosahedral viruses using cryo-EM. Biophys Rep 2:25–32

Luger K, Mäder AW, Richmond RK, Sargent DF, Richmond TJ (1997) Crystal structure of the nucleosome core particle at 2.8 Å resolution. Nature 389:251–260

Mahamdeh M, Schäffer E (2009) Optical tweezers with millikelvin precision of temperature-controlled objectives and base-pair resolution. Opt Express 17:17190–17199

Maier B, Bensimon D, Croquette V (2000) Replication by a single DNA polymerase of a stretched single-stranded DNA. Proc Natl Acad Sci USA 97:12002–12007

Meng H, Andresen K, van Noort J (2015) Quantitative analysis of single-molecule force spectroscopy on folded chromatin fibers. Nucleic Acids Res 43:3578–3590

Merkel R, Nassoy P, Leung A, Ritchie K, Evans E (1999) Energy landscapes of receptor–ligand bonds explored with dynamic force spectroscopy. Nature 397:50

Min D, Jefferson RE, Bowie JU, Yoon T-YY (2015) Mapping the energy landscape for second-stage folding of a single membrane protein. Nat Chem Biol. https://doi.org/10.1038/nchembio.1939

Moy VT, Florin EL, Gaub HE (1994) Intermolecular forces and energies between ligands and receptors. Science 266:257

Neuman KC, Block SM (2004) Optical trapping. Rev Sci Instrum 75:2787–2809

Ngo TTM, Ha T (2015) Nucleosomes undergo slow spontaneous gaping. Nucleic Acids Res. https://doi.org/10.1093/nar/gkv276

Odijk T (1995) Stiff chains and filaments under tension. Macromolecules 28:7016–7018

Pavri R, Zhu B, Li G, Trojer P, Mandal S, Shilatifard A, Reinberg D (2006) Histone H2B monoubiquitination functions cooperatively with FACT to regulate elongation by RNA polymerase II. Cell 125:703–717

Pincet F, Husson J (2005) The solution to the streptavidin–biotin paradox: the influence of history on the strength of single molecular bonds. Biophys J 89:4374–4381

Piontek MC, Roos WH (2018) Atomic force microscopy: an introduction. Methods Mol Biol (Clifton NJ) 1665:243–258

Richmond TJ, Davey CA (2003) The structure of DNA in the nucleosome core. Nature 423:145–150

Seol Y, Strub M-PP, Neuman KC (2016) Single molecule measurements of DNA helicase activity with magnetic tweezers and *t*-test based step-finding analysis. Methods (S Diego Calif) 105:119–127

Simpson NB, McGloin D, Dholakia K, Allen L, Padgett MJ (1998) Optical tweezers with increased axial trapping efficiency. J Mod Opt 45:1943–1949

Song F, Chen P, Sun D, Wang M, Dong L, Liang D, Xu R-MM, Zhu P, Li G (2014) Cryo-EM study of the chromatin fiber reveals a double helix twisted by tetranucleosomal units. Science (NY NY) 344:376–380

Strick TR, Croquette V, Bensimon D (1998) Homologous pairing in stretched supercoiled DNA. Proc Natl Acad Sci USA 95:10579–10583

te Velthuis AJW, Kerssemakers JWJ, Lipfert J, Dekker NH (2010) Quantitative guidelines for force calibration through spectral analysis of magnetic tweezers data. Biophys J 99:1292–1302

van Loenhout MT, Kerssemakers JW, De Vlaminck I, Dekker C (2012) Non-bias-limited tracking of spherical particles, enabling nanometer resolution at low magnification. Biophys J 102:2362–2371

Vilfan ID, Lipfert J, Koster DA, Lemay SG, Dekker NH (2009) Magnetic tweezers for single-molecule experiments. In: Hinterdorfer P, Oijen A (eds) Handbook of single-molecule biophysics. Springer, New York, pp 371–395

Yan J, Skoko D, Marko JF (2004) Near-field-magnetic-tweezer manipulation of single DNA molecules. Phys Rev E 70:011905

Yan J, Maresca TJ, Skoko D, Adams CD, Xiao B, Christensen MO, Heald R, Marko JF (2007) Micromanipulation studies of chromatin fibers in *Xenopus* egg extracts reveal ATP-dependent chromatin assembly dynamics. Mol Biol Cell 18:464–474

Yuan G, Le S, Yao M, Qian H, Zhou X, Yan J, Chen H (2017) Elasticity of the transition state leading to an unexpected mechanical stabilization of titin immunoglobulin domains. Angew Chem Int Ed 56:5490–5493

Zentner GE, Henikoff S (2013) Regulation of nucleosome dynamics by histone modifications. Nat Struct Mol Biol 20:259–266

Identification of small ORF-encoded peptides in mouse serum

Yaqin Deng[1,2], Adekunle Toyin Bamigbade[1], Mirza Ahmed Hammad[1], Shimeng Xu[1], Pingsheng Liu[1,2]

[1] National Laboratory of Biomacromolecules, CAS Center for Excellence in Biomacromolecules, Institute of Biophysics, Chinese Academy of Sciences, Beijing 100101, China
[2] University of Chinese Academy of Sciences, Beijing 100049, China

Abstract Identification of the coding elements in the genome is fundamental to interpret the development of living systems and species diversity. Small peptides (length < 100 amino acids) have played an important role in regulating the biological metabolism, but their identification has been limited by their size and abundance. Serum is the most important body fluid and is full of small peptides. In this study, we have established a small ORF-encoded peptides (SEPs) database from mouse GENCODE release. This database provides about half a million putative translated SEPs in mouse. We also extract serum proteins from wild type and *ob/ob* mice, and collect the low molecular weight proteins for mass spectrometric analysis. More than 50 novel SEPs have been discovered. Several SEPs are further verified by biochemical method with newly raised antibodies. These novel SEPs enhance the knowledge about the complexity of serum and provide new clues for the annotation and functional analysis of genes, especially the noncoding elements in the genome.

Keywords Small ORF-encoded peptides (SEPs), Serum, Database, Mass spectrometric analysis, *ob/ob* mice

INTRODUCTION

While biologists generally focus on the protein-coding open reading frame (ORF) of mRNA, it is now emerging that many mRNAs, even noncoding RNAs, also possess small ORF (sORF), and have significant roles in different organisms (Chu *et al.* 2015). It is a consensus that the ORFs in the transcribed mRNA will be translated into corresponding proteins due to in-frame codons defined by start and end codons. However, it still remains a big challenge in the field of gene annotation to distinguish the bona fide proteins from the translation noise. Moreover, most ORF-finding algorithms have historically set 300 nucleotides as the minimum ORF size for gene annotation, which incorrectly classifies genuine proteins corresponded RNA into noncoding RNAs (ncRNAs). On account to the great development of bioinformatics and biotechnology, numerous large-scale genomic studies have identified many nonclassical protein-coding genes, previous thought to be noncoding (Aramayo and Polymenis 2017; Bazzini *et al.* 2014; Chew *et al.* 2013; Derrien *et al.* 2012; Ingolia *et al.* 2011, 2014; Makarewich and Olson 2017; Tautz 2009; Ulitsky *et al.* 2011). More studies find that sORFs in ncRNA can encode small peptides, often referred as small ORF-encoded peptides (SEPs) that play important roles in the fundamental biological processes and in the maintenance of cellular homeostasis in different organisms, such as yeast, plant, zebra fish, *Drosophila*, and mammals (Anderson *et al.* 2015, 2016; Bazzini *et al.* 2014; Cohen 2014; Hanada *et al.* 2013; Ingolia *et al.* 2011; Ji *et al.* 2015; Lee *et al.*

✉ Correspondence: xushimeng@ibp.ac.cn (S. Xu), pliu@ibp.ac.cn (P. Liu)

2015; Magny *et al.* 2013; Matsumoto *et al.* 2017; Nelson *et al.* 2016; Smith *et al.* 2014).

Serum is the most important body fluid in mammals and possesses many important but low abundant small molecular proteins, such as peptide hormones, growth factors, lymphokines, and cytokines. However, few studies have revealed the existence and bioactivity of SEPs in serum. The major challenge in serum SEPs discovery arises from its extraordinary complexity in protein composition with the addition of post-translational modifications (PTMs) and protein variability, as well as the great concentration range (more than ten orders of magnitude) (Anderson and Anderson 2002; Omenn 2007).

To characterize the existence and bioactivity of SEPs in serum, we first established a mouse SEP database. This SEP database was then merged with mouse Uniprot database and Contamination database to form Mouse Merged database (MMD) for mass spectrometry (MS) data mining in this study. On the other hand, we extracted proteins with small molecular weight in different mouse sera and subjected to Q Exactive MS detection. After data mining, we discovered 54 novel SEPs in 15 serum samples. Furthermore, we raised four antibodies for four typical SEPs and finally confirmed the existence of two SEPs at the biochemical level.

RESULTS

Construction and verification of Mouse Merged database

To characterize the existence of SEPs in serum, a novel mouse SEP database was constructed according to the RNA transcripts released from Gencode (vM4). This database provided about half a million putative translated SEPs in mouse. This database was then combined with mouse Uniprot database and Contamination database, forming MMD (Fig. 1). In order to verify the quality of the MMD, several recently identified functional SEPs were chosen and blasted within MMD. All of them could match one list in the MMD (Table 1). For example, MOTS-c is derived from a sORF in mitochondrial DNA and regulates insulin sensitivity and metabolic homeostasis (Lee *et al.* 2015). MLN, a conserved skeletal muscle-specific micropeptide, is derived from a sORF in a putative long noncoding RNA (lncRNA) and regulates skeletal muscle physiology (Anderson *et al.* 2015). SPAR is derived from a sORF in an lncRNA and inhibits muscle regeneration (Matsumoto *et al.* 2017). NoBody, a novel component of the mRNA decapping complex, is derived from a sORF in an lncRNA (D'Lima

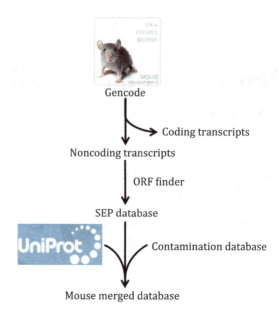

Fig. 1 Construction of Mouse Merged database. Around 110 thousand transcripts of mouse were released from Gencode (vM4). All transcripts, except the known coding transcripts, were translated to SEPs (length between 8 and 100 a.a.) by the ORF Finder program and an in-house program. The SEP database was then merged with mouse Uniprot database and Contamination database to form Mouse Merged database for mass spectrometry data mining in this study

et al. 2017). Together, we have successfully constructed a high-quality MMD for the following MS data mining.

ob/ob mice show severe impaired glucose tolerance

As the species and the concentration of serum proteins show vast variability, wild-type (WT) and *ob/ob* mice were chosen for the serum protein preparation. We thought that some serum SEPs might show different expression patterns between WT and pathological mouse model. The body weight of *ob/ob* mice was significant higher than that of WT as indicated by previous studies (Fig. 2A). We then verified the glucose metabolism states of these two mouse models. *ob/ob* mice showed severely damaged glucose tolerance (Fig. 2B, C). These results show that these two typical mouse models possess dramatically different metabolic states.

We next extracted the serum proteins of WT and *ob/ob* mice according to the workflow showed in Fig. 3A. Since the putative serum SEPs might show very low abundance and high dynamics, 11 WT mouse samples and four *ob/ob* mouse samples were chosen for the following MS detection, respectively (Fig. 3B, C). As shown in Fig. 3B and C, all the serum samples showed clear protein staining signal in the low molecular weight range. We then sliced the gel area below 14 kDa for the

Table 1 Verification of the MMD

Accession	Sequence	Pub. SEPs	Reference
ENSMUSG00000064337.1	MKWEEMGYIFL	MOTS-c	Lee et al. (2015)
ENSMUSG00000019933.3	MSGKSWVLISTTSPQSLEDEILGRLLKILFVLFVDLMSIMYVVITS	MLN	Anderson et al. (2015)
ENSMUSG00000028475.8	METAVIGMVAVLFVITMAITCILCYFSYDSHTQDPERSSRRSFTVATFHQEASLFTGPALQSRPLPRPQNFWTVV	SPAR	Matsumoto et al. (2017)
ENSMUSG00000086316.3	MGDQPCASGRSTLPPGNTREPKPPKKRCVLAPRWDYPEGTPSGGSSTLPSAPPPASAGLKSHPPPEK	NoBody	D'Lima et al. (2017)

following sample preparation and MS (the area between two red lines in every lane).

MS detection discovers novel serum SEPs

The sliced gels were further processed for MS detection according to the workflow in Fig. 3A. 54 novel SEPs were detected in total from the 15 samples (Table 2). Eight SEPs were detected in more than one sample. 38 SEPs were only detected in WT mouse serum and 12 SEPs were only detected in *ob/ob* mouse serum (Table 2). We sequentially named the SEPs from SEP1 to SEP54 (Table 2). SEP3, SEP12, SEP33, and SEP54 were chosen for further study to confirm the accuracy of Q Exactive MS results and to verify the existence of the SEPs in serum. The MS/MS spectrums of these four SEPs were presented in Fig. 4. SEP3 was detected in Sample 1 and Sample 2, and was encoded from a sORF in processed transcript of *Epha7* gene (Table 2, Fig. 4A). Besides, SEP3 was conserved in mammals (Fig. 5A). SEP12 was detected in five samples and was encoded from a sORF in processed_transcript of *Ufsp2* gene (Table 2, Fig. 4B). SEP12 was also conserved in mammals (Fig. 5A). SEP33 was detected in four samples and was encoded from a sORF in retained_intron of *Tnnt2* gene (Table 2, Fig. 4C). SEP54 was detected only once with high X correlation score in Sample 15 and was encoded from a sORF in lncRNA Gm2670 (Table 2, Fig. 4D). All of the four primary MS results strongly suggested the detection of targeted peptides. Taken together, these lines of evidence suggest that SEPs widely exist in the serum and might show wide individual differences.

Western blot results confirm the existence of serum SEPs

In order to further confirm the existence of the SEPs, four antibodies were raised to against SEP3, SEP12, SEP33, and SEP54. The antigens were designed as indicated in materials and methods. The sera from the immunized rabbits were used as the antibodies to detect the corresponding SEPs in mouse sera by Western blot with human serum as control. Consistent with the MS results, SEP3 antibody recognized an 8-kDa protein in WT mouse serum (Fig. 6A), rather than that of human and *ob/ob* mouse. Similarly, SEP54 antibody recognized a 10-kDa protein in mouse serum (Fig. 6B), rather than that of human. Furthermore, consistent with the MS result, SEP54 showed higher concentration in *ob/ob* mouse. However, SEP12 and SEP33 antibodies failed to recognize any specific band in all of the serum samples (data not shown). Altogether, these results

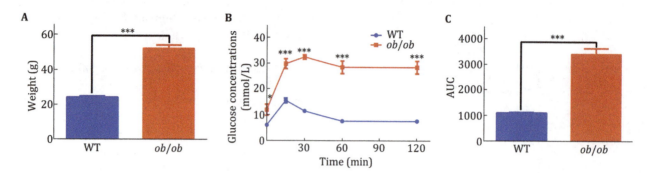

Fig. 2 Verification of *ob/ob* mice. Twelve-week-old male mice were chosen for experiments in this study. **A** The weight of WT and *ob/ob* mice. **B** The IPGTT test for the WT and *ob/ob* mice. All the mice were fasting for 18 h before IPGTT. 2 g/kg glucose was injected for the IPGTT. **C** Area under the curve was calculated for the IPGTT. WT mice, $n = 5$; *ob/ob* mice, $n = 6$. Data were analyzed by Student t tests and presented as mean ± SEM. Significance, $*p < 0.05$; $***p < 0.001$

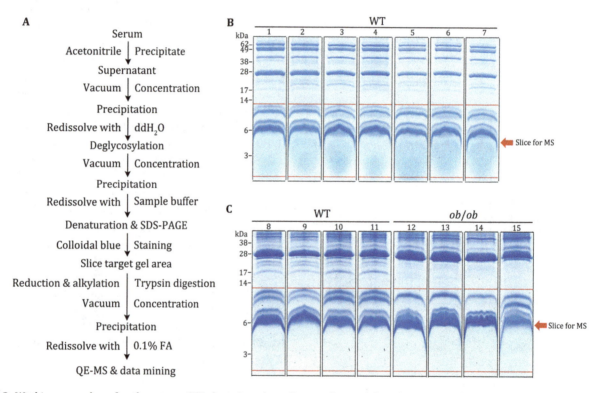

Fig. 3 Working procedure for the serum SEP detection. According to the workflow for the enrichment and identification of low abundance mouse serum proteins (**A**), two rounds of mouse serum proteins were separated by SDS-PAGE and stained by Colloidal blue. **B** The first round, seven WT mice. **C** The second round, four WT mice and four *ob/ob* mice. The proteins below 14 kDa (the proteins between two *red lines* in every lane) were sliced for mass spectrometric analysis

further demonstrate the existence of SEPs in serum with different expression levels.

DISCUSSION

In this study, we constructed a novel SEP database and discovered some SEPs in serum by MS. Our data provided two key insights into the genome-wide expression of SEPs in mammals. First, SEPs were widely distributed and translated from a large body of transcripts. We annotated hundreds of thousands of SEPs (length ranging from 8 to 100 a.a.) according to the noncoding transcripts in mouse GENCODE (vM4), and validated 54 novel SEPs in the mouse serum (Table 2). This was the first systematic study to explore the existence of SEPs in serum. Previous studies have successfully used computational approach and ribosome profiling to define the transcripts in translating ribosome (Bazzini et al. 2014; Chew et al. 2013; Ingolia et al. 2009, 2011; Menschaert et al. 2013) and identified some SEPs in specific tissues and cell lines. However, it has been

Table 2 MS identification of sliced bands

Band NO.	SEP NO.	Accession	Description	Detected peptide	XCorr	M.W. (kDa)
1	1	ENSMUSG00000032985.11	5730522E02Rik|processed_transcript	LPQLAAAEPNRPR	2.22	10.6
	2	ENSMUSG00000085596.1	Gm11476| processed_transcript	QLPLYQIEILVcNlTAmTHPSNFSIESNQcRLSPRSQQLEcPK	2.11	8.8
	3	ENSMUSG00000028289.8	Epha7| processed_transcript	MHLQSRLSAKR	1.65	8.0
	4	ENSMUSG00000029049.10	Morn1	processed_transcript	NGTGcVSPLELR	1.09	6.4
	5	ENSMUSG00000021177.11	Tdp1|retained_intron	MVcLSFTTK	0.75	1.4
	6	ENSMUSG00000042105.14	Inpp5f|retained_intron	MLVLPVnLPR	1.73	1.3
	7	ENSMUSG00000029647.11	Pan3|nonsense_mediated_decay	MLLIcNQKQMHLPSSWPTSSDR	1.47	2.7
	8	ENSMUSG00000081985.1	lincRNA 1700047M11Rik|lincRNA	SNRSLTLR	1.05	1.4
	9	ENSMUSG00000081985.1	Gng2-ps1|processed_pseudogeiie	KLlELLKmEAnIDR	1.05	6.0
	10	ENSMUSG00000006423.11	C330007P06Rik|retained_intron	HPPVIFVTYTHMQANTHAHKIR	1.36	6.3
	11	ENSMUSG00000059974.6	Ntm|processed_transcript	TTQAKMHnSISWAIFTGLAALcLFQGK	1.49	3.2
2	3	ENSMUSG00000028289.8	Epha7|processed_transcript	MHLQSRLSAKR	1.65	8.0
	12	ENSMUSG00000031634.8	Ufsp2|processed_transcript	mISSKPIER	1.66	2.8
3	1	ENSMUSG00000032985.11	5730522E02Rik|procssed_transcript	LPQLAAAEPNRPR	1.60	10.6
	13	ENSMUSG00000054693.10	Adam10|nonsense_mediated_decay	KEAlVmGLSLMEDLKVSSR	0.90	9.6
	14	ENSMUSG00000035953.9	Tmem55b|retained_intron	HFPRSLRDIQPccLER	1.28	3.4
	15	ENSMUSG00000103242.1	RP23-132G24.3|processed_pseudogene	MAEAIYIIEVKEWGK	1.33	2.5
4	16	ENSMUSG00000085553.1	antisense Gm14808|antisense	QnNHGGWLWVPKEScALGR	1.65	7.4
	12	ENSMUSG00000031634.8	Ufsp2|processed_transcript	mISSKPIER	1.65	2.8
	17	ENSMUSG00000048215.10	lincRNA A630023P12Rik|lincRNA	SQNFSWlmLLcPSQM	0.47	4.9
	18	ENSMUSG00000086528.1	antisense Gm15731|antisense	TIQKAPPHYmSIELR	1.60	4.1
	19	ENSMUSG00000102503.1	TEC RP23-388I22.1|TEC	mYYLVKmScYmKcLR	0.32	2.7
5	20	ENSMUSG00000085865.1	lincRNA Gm15966|lincRNA	SASSWNQPLPGPSGFGLEEVSREGGWR	3.36	5.8
	21	ENSMUSG00000081123.1	Gm11469|processed_pseudogene	EGVNIAEAIER	1.60	10.1
	4	ENSMUSG00000029049.10	Morn1|processed_transcript	NGTGcVSPLELR	1.45	6.4
	22	ENSMUSG00000053199.9	Arhgap20|retained_intron	QSTVKcWRPFQmSHmQTFmK	1.15	7.7
	23	ENSMUSG00000029464.6	Gpn3|retained_intron	PGGAERnSR	0.78	2.4
	24	ENSMUSG00000098033.1	Gm9381|processed_pseudogene	IVSnAScTTnclVLLAKVIFGmTTLALER	1.69	9.3
	25	ENSMUSG00000102240.1	TEC RP23-242B14.1|TEC	mNLKILTYVcFASQRQTIYLENR	2.21	5.5
	26	ENSMUSG00000020063.12	Sirt1|nonsense_mediated_decay	mVFHTFLFVTLnSLK	1.06	3.1
	11	ENSMUSG00000059974.6	Ntm|processed_transcript	TIQAKMHnSISWAIFTGLAALcLFQGK	2.30	3.2
6	27	ENSMUSG00000024073.10	Birc6|retained_intron	QLFLVEnKNLNIIIPmFYcFFPIR	1.22	11.3
	28	ENSMUSG00000057406.12	Whsc1|nonsense_mediated_decay	SLPSQKcSPKYSENEAR	0.83	3.9
	29	ENSMUSG00000031559.10	4930555F03Rik|processed_transcript	mLHcVHSSLIYnSQTLER	1.58	4.9

Table 2 continued

Band NO.	SEP NO.	Accession	Description	Detected peptide	XCorr	M.W. (kDa)
7	30	ENSMUSG00000072929.6	Gm15109|unprocessed_pseudogene	DTMVQEEEMDQGMHHHQDLSQK	0.24	3.9
	31	ENSMUSG00000090699.1	Gm9071|unprocessed_pseudogene	mKEKEVMSFLHNLEMEYIEAR	1.51	6.2
	32	ENSMUSG00000255495.10	Ptdss2|nonsense_mediated_decay	NPSGYSLQHQERYcGQYFGFLMFWSHT	1.01	6.9
8	33	ENSMUSG00000026414.9	Tnnt2|retained_intron	DAILEALR	1.67	4.8
	34	ENSMUSG00000025089.11	Gfra1|nonsense_mediated_decay	FPHTFYHRVLIcSTAWDPNK	1.12	7.0
9	35	ENSMUSG00000034285.11	Nipsnap1|nonsense_mediated_decay	IEVLGSLFR	1.97	6.6
	33	ENSMUSG00000026414.9	Tnnt2|retained_intron	DAILEALR	1.61	4.8
	36	ENSMUSG00000084274.2	Gm12504|processed_transcript	MNYFcFHImWcYVLSFmAR	0.33	3.7
	15	ENSMUSG00000103242.1	RP23-132G24.3|processed_pseudogene	MAEAIYIIEVKEWGK	1.06	2.5
	37	ENSMUSG00000102415.1	TEC RP23-284P20.1|TEC	LTKTYQHVYcMLK	0.91	3.2
10	33	ENSMUSG00000031626.12	Tnnt2|retained_intron	DAILEALR	1.72	4.8
	38	ENSMUSG00000031626.12	Pros1|retained_intron	EnmDSnHKKTVFSILLEMR	0.23	4.8
	39	ENSMUSG00000031626.12	Gm29365|unprocessed_pseudogene	SVTDmDTIEKSNLnRQFLFcPWDVTK	0.64	8.5
11	1	ENSMUSG00000032985.11	5730522E02Rik|processed_transcript	LPQLAAAEPNRPR	2.16	10.6
	40	ENSMUSG00000092054.3	Kif4-ps|transcribed_processed_pseudogene	mLTELEK	1.09	5.7
	41	ENSMUSG00000090109.1	Ear-ps10|unprocessed_pseudogene	TTVAMKSYTVAcNPR	1.50	7.9
	42	ENSMUSG00000099956.1	Gm29365|unprocessed_pseudogene	DPAFYAYQLLDDYKEGnLHMIPDTPPAEERSGDDSDVLIGn	0.61	6.0
12	43	ENSMUSG00000031626.12	Sorbs2|processed_transcript	YQIFnFnR	1.70	2.4
	44	ENSMUSG00000022686.10	B3gnt5|processed_transcript	FVLETFPPGLLGGQRTSGTFK	1.10	4.5
	45	ENSMUSG00000083128.1	Gm12723|processed_pseudogene	TEAIEALVK	1.18	5.7
	12	ENSMUSG00000031634.8	Ufsp2|processed_transcript	mISSKPIER	1.82	2.8
	46	ENSMUSG00000020361.9	Hspa4|processed_transcript	TQYVDHAGLELKGSHQPLPPK	1.03	5.2
	47	ENSMUSG00000091078.1	antisense Gm17218|antisense	MASVSPEIKR	1.38	1.6
13	12	ENSMUSG00000031634.8	Ufsp2|processed_transcript	mISSKPIER	1.88	2.8
	48	ENSMUSG00000103862.1	TEC RP23-198F7.2|TEC	mRNWLVSPmnSK	1.24	4.1
14	6	ENSMUSG00000042105.14	Inpp5f|retained_intron	MLVLPVnLPR	2.34	1.3
	1	ENSMUSG00000032985.11	5730522E02Rik|processed_transcript	LPQLAAAEPNRPR	1.85	10.6
	49	ENSMUSG00000042688.12	Mapk6|retained_intron	FLFTnR	1.65	1.7
	12	ENSMUSG00000031634.8	Ufsp2|processed_transcript	mISSKPIER	1.58	2.8
	50	ENSMUSG00000076594.1	Ikgv6-13|IG_LV_gene	ASQn	1.30	10.1
	51	ENSMUSG00000005360.10	Slca3|retained_intron	VWEAPRYnK	1.40	6.2
	52	ENSMUSG00000103591.1	TEC RP24-369B15.2|TEC	mMLKTIcRIINVFLILLnEDDAK	1.26	5.4
	53	ENSMUSG00000006010.10	BC003331 |retained_intron	ELSWIIWmKNGPQNMPAR	1.48	3.0
15	54	ENSMUSG00000097002.1	lincRNA Gm2670|lincRNA	KVnLFQAK	1.98	3.5
	33	ENSMUSG00000026414.9	Tnnt2|retained_intron	DAILEALR	1.93	4.8

Mouse Merged database was used for mass spectrometry database search

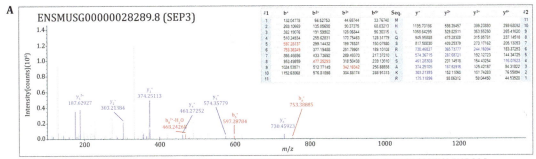

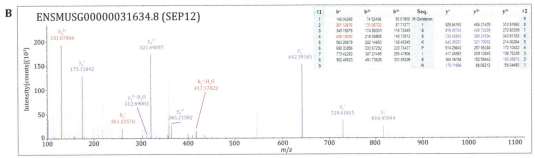

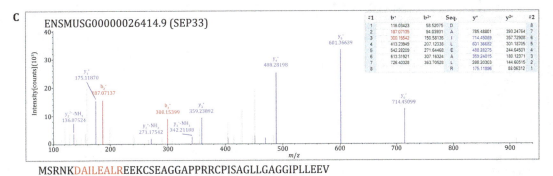

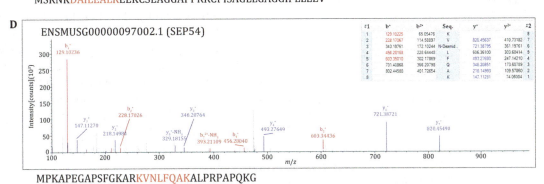

Fig. 4 MS/MS spectrum of the four example peptides. The matched fragment ions of precursor ions were listed in the right of MS/MS spectra. All the matched ions were labeled with different colors, b-ions were labeled with *red color*, y-ions were labeled with *blue color*. The sequences below the spectra were the corresponding full length SEPs according to the Mouse Merged database. *Red highlights* represent the detected peptide fragments. **A** The spectrum result of SEP3. **B** The spectrum result of SEP12. **C** The spectrum result of SEP33. **D** The spectrum result of SEP54

strongly argued whether the RNA fragments protected by the ribosome always reflect the actively translated transcripts. The RNA bound to RNA-binding proteins will also be improperly classified as coding sequence, as well as the ribosome randomly bound RNA. In consistent with the known serum small peptides, the putative serum SEPs might also be low abundant, highly dynamic, and low molecular weight. Therefore, a

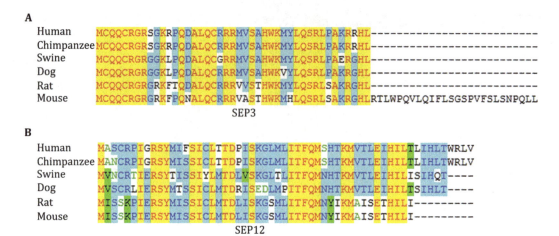

Fig. 5 SEP3 and SEP12 are conserved in mammals. Conservation analysis of SEP3 (**A**) and SEP12 (**B**) with clustal multiple alignment in six species

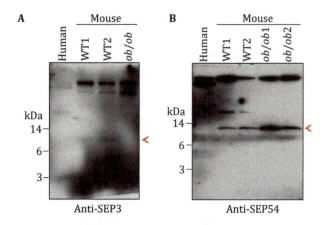

Fig. 6 WB verification of SEPs in mouse serum. Polyclonal antibodies for four SEPs were raised in rabbits. Two antibodies showed specific bands in the low molecular weight area of mouse serum samples. **A** Anti-SEP3 antibody recognized a target protein in around 8 kDa, indicated by the *red arrow*. **B** Anti-SEP54 antibody recognized a target protein in around 10 kDa, indicated by the *red arrow*

high-precision mass spectrometer and a homemade database were chosen to verify the existence of SEPs in the serum.

Second, the serum SEPs were very low abundant and highly dynamic among individuals. For example, ApoA2, a well-known high abundant serum peptide (Uniprot: P09813; length: 102 a.a.), was detected more than ten fragments in every sample of all of the 15 samples (data not shown). However, all of the 54 SEPs detected in this study matched only one fragment in the corresponding sample (Table 2), and only eight SEPs were repeatedly detected in more than one sample (Table 2). Besides, the low abundance of those SEPs might also be one of the reasons for the weak Western blot signal of SEP3 and SEP54 (Fig. 6). On the other hand, insulin, a protein existing in serum with nanogram level, was not detected in the 15 samples (data not shown), which suggests that the abundance of the above-detected SEPs might be higher than that of insulin and further proved the existence of those SEPs in serum. Besides, the low repeatability detection of the above 54 SEPs among 15 samples from 11 WT mice and four *ob/ob* mice implied, to some degree, the high dynamics of serum SEPs among individuals (with different metabolic states). These MS results were further verified by Western bolt analysis. Consistently, SEP3 antibody detected stronger signal in WT mouse serum than that in *ob/ob* mouse serum, in agreement with the result that SEP3 was only detected in WT mouse samples in MS results. Similarly, SEP54 antibody detected stronger signal in *ob/ob* mouse serum samples than that in WT mouse sample. These high dynamics of SEPs were similar with that of known small peptides in serum. For example, serum insulin level increases after feeding and serum irisin level increases after exercise (Jedrychowski *et al.* 2015). Besides, signal peptide prediction showed only three of the detected SEPs had the secretion signal peptide (see Supplemental table), which indicated that most SEPs tended to be secreted by uncanonical pathway, or released from broken cells.

Several approaches have been used to validate the putative SEPs (Housman and Ulitsky 2016). Ideally, the generation of antibodies against target SEPs is the most effective method (Anderson *et al.* 2015). However, the optimal antigen designing to the SEPs is challenging for their small size. This may be the reason why antibodies raised by SEP12 and SEP33 could not recognize specific bands in serum samples. As antigen peptides for SEP3 possessed 4 a.a. difference between human and mouse, SEP3 antibody could not recognize its ortholog in human SEP3 (Figs. 5A, 6A). Another concern for the usage of antibody is that even the highest-affinity antibody may not be sufficient to produce a strong enough signal for the detection of low abundance SEPs. Alternatively, clustered regularly interspaced short palindromic repeats (CRISPR)-CRISPR associated protein 9 (Cas9) mediated gene-editing for the target SEPs in vitro or in vivo could also provide direct evidence for the existence of the SEPs and substantially support the functional study of the SEPs. Besides, for the biological function study of serum SEPs, such as SEP3 and SEP54, the mouse tail vein injection of artificially synthesized full length SEPs will be a high-efficiency approach.

It still remains unclear how many SEPs exist in serum and what the biological functions of the serum SEPs are. New methods and new ideas are still needed to further study the SEPs in serum. Together, our study opens a new avenue for the identification of small peptides in serum, and provides an entry point to investigate their function in vivo.

MATERIALS AND METHODS

Animals

Twelve-week-old male WT (C57BL/6J) and ob/ob mice were housed in our animal facility on a 12-h light/dark cycle with ad libitum access to water and food. All animal protocols were approved by the Animal Care and Use Committee of the Institute of Biophysics, Chinese Academy of Sciences, SYXK (SPF 2011-0029).

IPGTT

Tail blood samples were collected from 12-week-old mice that had been fasted for 18 h, before and at 15, 30, 60, and 120 min after i.p. injection of glucose (2 g/kg). Glucose levels were measured at prespecified times. Blood glucose was measured using glucometer (ACCU-CHEK, Roche).

Mouse Merged database construction

For the construction of SEP database, both the ORF Finder program and an in-house program were used to identify ORFs from noncoding transcripts in mouse GENCODE (vM4). Ensembl transcripts (release 73) were downloaded from the Ensembl FTP repository to annotate the noncoding transcripts in mouse GENCODE. The peptide sequences associated with predicted ORFs of noncoding RNAs, ranging from 8 to 100 a.a., were selected to construct the SEP database. The SEP database was then merged with mouse Uniprot database and Contamination database to form Mouse Merged database (Supplemental data).

Serum sample preparation

The workflow for the preparation of serum samples was shown in Fig. 3A. Serum samples were collected from 12-week-old WT and ob/ob mice by removalling eyeballs. After clotting, serum was separated by centrifugation at 3000 g for 10 min at 4 °C. The low molecular weight and low abundance serum proteins were enriched with 60% acetonitrile as previous reported (Echan et al. 2005; Kay et al. 2008; Wu et al. 2010). Briefly, 100 μl serum was mixed with 300 μl H_2O and 600 μl acetonitrile and placed for 30 min at 4 °C. After centrifuged at 12,000 g for 30 min at 4 °C, the supernatant was concentrated by vacuum centrifugation. The precipitate was redissolved with 100 μl H_2O and processed to deglycosylation according to the instruction (NEB, USA). The deglycosylated proteins were redissolved with Sample buffer (125 mmol/L Tris Base, 20% glycerol, 4% SDS, 4% β-mercaptoethanol, and 0.04% bromophenol blue) with EDTA-free protease and phosphatase inhibitors (Thermo, USA). The protein samples were further denatured at 95 °C for 5 min.

Colloidal blue staining and mass spectrometry detection

Serum protein samples were separated on 10% Tricine-gels and subjected to Colloidal blue staining (Life Technologies, USA) (Schagger 2006). The indicated bands were cut into slices for MS detection (Fig. 3B, C). In-gel digestion of every slice was performed as previously described (Chen et al. 2016). The resulting peptide mixtures were dried and stored at −80 °C until further LC–MS/MS analysis.

LC–MS/MS analysis of serum peptide mixtures was performed on a Q Exactive mass spectrometer with a nano-electrospray ion source (Thermo, USA) coupled with an EasyLC nano HPLC system. The digested peptides were then loaded onto a C18 trap column with an autosampler, eluted onto a C18 column (100 μm × 15 cm) packed with ReproSil-Pur 130 C18-AQ 3 μm particles (Dr. Maisch HPLC GmbH, Germany).

All MS/MS spectra were acquired in a data-dependent scan mode, where one full-MS scan was followed with ten MS/MS scans. The full-scan MS spectra (300–1600 m/z) were acquired with a resolution of 60,000 at m/z 400 after accumulation to a target value of 3e6. The 20 most abundant ions found in MS1 were selected for fragmentation at a normalized collision energy of 27% (Chen et al. 2016).

The LC–MS/MS data were searched against the homemade MMD using the Proteome Discoverer 1.4 with SEQUEST as search engine (Thermo, USA). Search parameters were set as follows: enzyme: trypsin; precursor ion mass tolerance: 10 ppm; fragment ion mass tolerance: 0.02 Da. The maximum number of misscleavages by trypsin was set as two for peptides. The variable modification was set to oxidation of methionine. The fixed modification was set to carboxyamidomethylation of cysteine.

Signal peptide prediction

According to websites "http://phobius.sbc.su.se" and "http://www.cbs.dtu.dk/services/TargetP/," Phobius and TargetP were used for the prediction of SEP signal peptide. The output format was based on TargetP. The SEPs were listed positively in the supplemental table only when both methods returned a positive signal peptide prediction. The final prediction was based on the scores on mTP, SP, and another. mTP was a mitochondrial targeting peptide. SP was a signal peptide for secretory pathway, which was shown as "S" in the supplemental table, and "–" was any other location. Reliability class (RC) contains five classes, in which "1" means the strongest prediction. TPlen showed the predicted presequence length.

Conservation analyses

The corresponding nucleotide sequences for SEP3, SEP12, SEP33, and SEP54 ORFs were obtained from NCBI database (https://www.ncbi.nlm.nih.gov/), respectively, as reported previously (Lee *et al.* 2015). BLAST search was processed to ensure correct extraction of the nucleotide sequences. The protein sequences of six species, human (*Homo sapiens*), chimpanzee (*Pan troglodytes*), swine (*Homo sapiens*), dog (*Canis lupus familiaris*), rat (*Rattus norvegicus*), and mouse (*Mus musculus*), were aligned using Clustal Multiple Alignment.

Immunoassay

The corresponding antigen peptide for SEPs was conjugated to Keyhole Limpet Hemocyanin (KLH) and injected into rabbits. The antigen information was listed here: RGRKFPQNAL for SEP3, SSKPIERSYMI for SEP12, RNKDAILEALRE for SEP33, and KAPEGAPSFGKA for SEP54. IgG purified sera were used for the detection of serum SEPs by Western blot.

For Western blot, serum protein samples were prepared in Sample buffer with EDTA-free protease and phosphatase inhibitors (Thermo, USA), heated at 95 °C for 5 min, ran on a 10% Tricine-gels and transferred to 0.4 μm PVDF membranes (Merck, Germany) at 100 mA for 30 min. Membranes were blocked with 5% nonfat dry milk for 1 h at room temperature (RT) and incubated with primary antibody (1:500–1:2,000 dilution) overnight at 4 °C, followed by secondary HRP-conjugated antibodies (1:10,000) for 1 h at RT. Chemiluminescence was detected and imaged using ECL (PerkinElmer Life Sciences, Waltham, MA).

Statistical analyses

Data were presented as mean ± SEM unless specifically indicated. The statistical analyses were performed using GraphPad Prism 6. Comparisons of significance between groups were performed using Student t tests as indicated.

Acknowledgements The authors thank Dr. Jiao Yuan for her useful suggestions in the construction of Mouse Merged database. The authors also thank Dr. Xiulan Chen for her useful suggestions in the analysis of MS results. This work was supported by the Ministry of Science and Technology of China (2016YFA0500100), National Natural Science Foundation of China (U1402225, 31571388, 31671402, 31671233, 31701018, and 81471082). This work was also supported by the "Personalized Medicines–Molecular Signature-based Drug Discovery and Development," Strategic Priority Research Program of the Chinese Academy of Sciences (XDA12030201). This work was also supported by the CAS-Croucher Joint Laboratory Project (CAS16SC01). Bamigbade Adekunle Toyin sincerely acknowledged the CAS-TWAS President's Fellowship.

Compliance with Ethics Standards

Conflict of interest Yaqin Deng, Adekunle Toyin Bamigbade, Mirza Ahmed Hammad, Shimeng Xu, and Pingsheng Liu declare that they have no conflict of interests.

Human and animal rights and informed consent All animal protocols were approved by the Animal Care and Use Committee of the Institute of Biophysics, Chinese Academy of Sciences. All institutional and national guidelines for the care and use of laboratory animals were followed.

References

Anderson NL, Anderson NG (2002) The human plasma proteome: history, character, and diagnostic prospects. Mol Cell Proteomics 1:845–867

Anderson DM, Anderson KM, Chang CL, Makarewich CA, Nelson BR, McAnally JR, Kasaragod P, Shelton JM, Liou J, Bassel-Duby R, Olson EN (2015) A micropeptide encoded by a putative long noncoding RNA regulates muscle performance. Cell 160:595-606

Anderson DM, Makarewich CA, Anderson KM, Shelton JM, Bezprozvannaya S, Bassel-Duby R, Olson EN (2016) Widespread control of calcium signaling by a family of SERCA-inhibiting micropeptides. Sci Signal 9:119

Aramayo R, Polymenis M (2017) Ribosome profiling the cell cycle: lessons and challenges. Curr Genet 63(6):959-964

Bazzini AA, Johnstone TG, Christiano R, Mackowiak SD, Obermayer B, Fleming ES, Vejnar CE, Lee MT, Rajewsky N, Walther TC, Giraldez AJ (2014) Identification of small ORFs in vertebrates using ribosome footprinting and evolutionary conservation. EMBO J 33:981-993

Chen X, Xu S, Wei S, Deng Y, Li Y, Yang F, Liu P (2016) Comparative proteomic study of fatty acid-treated myoblasts reveals role of Cox-2 in palmitate-induced insulin resistance. Sci Rep 6:21454

Chew GL, Pauli A, Rinn JL, Regev A, Schier AF, Valen E (2013) Ribosome profiling reveals resemblance between long noncoding RNAs and 5′ leaders of coding RNAs. Development 140:2828-2834

Chu Q, Ma J, Saghatelian A (2015) Identification and characterization of sORF-encoded polypeptides. Crit Rev Biochem Mol Biol 50:134-141

Cohen SM (2014) Everything old is new again: (linc)RNAs make proteins! EMBO J 33:937-938

Derrien T, Johnson R, Bussotti G, Tanzer A, Djebali S, Tilgner H, Guernec G, Martin D, Merkel A, Knowles DG, Lagarde J, Veeravalli L, Ruan X, Ruan Y, Lassmann T, Carninci P, Brown JB, Lipovich L, Gonzalez JM, Thomas M, Davis CA, Shiekhattar R, Gingeras TR, Hubbard TJ, Notredame C, Harrow J, Guigo R (2012) The GENCODE v7 catalog of human long noncoding RNAs: analysis of their gene structure, evolution, and expression. Genome Res 22:1775-1789

D'Lima NG, Ma J, Winkler L, Chu Q, Loh KH, Corpuz EO, Budnik BA, Lykke-Andersen J, Saghatelian A, Slavoff SA (2017) A human microprotein that interacts with the mRNA decapping complex. Nat Chem Biol 13:174-180

Echan LA, Tang HY, Ali-Khan N, Lee K, Speicher DW (2005) Depletion of multiple high-abundance proteins improves protein profiling capacities of human serum and plasma. Proteomics 5:3292-3303

Hanada K, Higuchi-Takeuchi M, Okamoto M, Yoshizumi T, Shimizu M, Nakaminami K, Nishi R, Ohashi C, Iida K, Tanaka M, Horii Y, Kawashima M, Matsui K, Toyoda T, Shinozaki K, Seki M, Matsui M (2013) Small open reading frames associated with morphogenesis are hidden in plant genomes. Proc Natl Acad Sci USA 110:2395-2400

Housman G, Ulitsky I (2016) Methods for distinguishing between protein-coding and long noncoding RNAs and the elusive biological purpose of translation of long noncoding RNAs. Biochem Biophys Acta 1859:31-40

Ingolia NT, Ghaemmaghami S, Newman JR, Weissman JS (2009) Genome-wide analysis in vivo of translation with nucleotide resolution using ribosome profiling. Science 324:218-223

Ingolia NT, Lareau LF, Weissman JS (2011) Ribosome profiling of mouse embryonic stem cells reveals the complexity and dynamics of mammalian proteomes. Cell 147:789-802

Ingolia NT, Brar GA, Stern-Ginossar N, Harris MS, Talhouarne GJ, Jackson SE, Wills MR, Weissman JS (2014) Ribosome profiling reveals pervasive translation outside of annotated protein-coding genes. Cell Rep 8:1365-1379

Jedrychowski MP, Wrann CD, Paulo JA, Gerber KK, Szpyt J, Robinson MM, Nair KS, Gygi SP, Spiegelman BM (2015) Detection and quantitation of circulating human irisin by tandem mass spectrometry. Cell Metab 22(4):734-740

Ji Z, Song R, Regev A, Struhl K (2015) Many lncRNAs, 5′UTRs, and pseudogenes are translated and some are likely to express functional proteins. eLife 4:e08890

Kay R, Barton C, Ratcliffe L, Matharoo-Ball B, Brown P, Roberts J, Teale P, Creaser C (2008) Enrichment of low molecular weight serum proteins using acetonitrile precipitation for mass spectrometry based proteomic analysis. Rapid Commun Mass Spectrom 22:3255-3260

Lee C, Zeng J, Drew BG, Sallam T, Martin-Montalvo A, Wan J, Kim SJ, Mehta H, Hevener AL, de Cabo R, Cohen P (2015) The mitochondrial-derived peptide MOTS-c promotes metabolic homeostasis and reduces obesity and insulin resistance. Cell Metab 21:443-454

Magny EG, Pueyo JI, Pearl FM, Cespedes MA, Niven JE, Bishop SA, Couso JP (2013) Conserved regulation of cardiac calcium uptake by peptides encoded in small open reading frames. Science 341:1116-1120

Makarewich CA, Olson EN (2017) Mining for micropeptides. Trends Cell Biol 27:685-696

Matsumoto A, Pasut A, Matsumoto M, Yamashita R, Fung J, Monteleone E, Saghatelian A, Nakayama KI, Clohessy JG, Pandolfi PP (2017) mTORC1 and muscle regeneration are regulated by the LINC00961-encoded SPAR polypeptide. Nature 541:228-232

Menschaert G, Van Criekinge W, Notelaers T, Koch A, Crappe J, Gevaert K, Van Damme P (2013) Deep proteome coverage based on ribosome profiling aids mass spectrometry-based protein and peptide discovery and provides evidence of alternative translation products and near-cognate translation initiation events. Mol Cell Proteomics 12:1780-1790

Nelson BR, Makarewich CA, Anderson DM, Winders BR, Troupes CD, Wu F, Reese AL, McAnally JR, Chen X, Kavalali ET, Cannon SC, Houser SR, Bassel-Duby R, Olson EN (2016) A peptide encoded by a transcript annotated as long noncoding RNA enhances SERCA activity in muscle. Science 351:271-275

Omenn GS (2007) The HUPO human plasma proteome project. Proteomics Clin Appl 1:769-779

Schagger H (2006) Tricine-SDS-PAGE. Nat Protoc 1:16-22

Smith JE, Alvarez-Dominguez JR, Kline N, Huynh NJ, Geisler S, Hu W, Coller J, Baker KE (2014) Translation of small open reading frames within unannotated RNA transcripts in *Saccharomyces cerevisiae*. Cell Rep 7:1858-1866

Tautz D (2009) Polycistronic peptide coding genes in eukaryotes—how widespread are they? Brief Funct Genomics Proteomics 8:68-74

Ulitsky I, Shkumatava A, Jan CH, Sive H, Bartel DP (2011) Conserved function of lincRNAs in vertebrate embryonic development despite rapid sequence evolution. Cell 147:1537-1550

Wu J, An Y, Pu H, Shan Y, Ren X, An M, Wang Q, Wei S, Ji J (2010) Enrichment of serum low-molecular-weight proteins using C18 absorbent under urea/dithiothreitol denatured environment. Anal Biochem 398:34-44

Class C G protein-coupled receptors: reviving old couples with new partners

Thor C. Møller[1], David Moreno-Delgado[1], Jean-Philippe Pin[1 ✉], Julie Kniazeff[1]

[1] Institut de Génomique Fonctionnelle (IGF), CNRS, INSERM, Univ. Montpellier, 34094 Montpellier, France

Abstract G protein-coupled receptors (GPCRs) are key players in cell communication and are encoded by the largest family in our genome. As such, GPCRs represent the main targets in drug development programs. Sequence analysis revealed several classes of GPCRs: the class A rhodopsin-like receptors represent the majority, the class B includes the secretin-like and adhesion GPCRs, the class F includes the frizzled receptors, and the class C includes receptors for the main neurotransmitters, glutamate and GABA, and those for sweet and umami taste and calcium receptors. Class C receptors are far more complex than other GPCRs, being mandatory dimers, with each subunit being composed of several domains. In this review, we summarize our actual knowledge regarding the activation mechanism and subunit organization of class C GPCRs, and how this brings information for many other GPCRs.

Keywords G protein-coupled receptors, Dimerization, Activation mechanism, Glutamate, GABA

INTRODUCTION: CLASS C GPCRS AND TOPOLOGY

The class C of G protein-coupled receptors (GPCRs) contains 22 members, including the eight metabotropic glutamate receptors (mGlu$_{1-8}$), the GABA$_B$ receptors (GABA$_{B1}$ and GABA$_{B2}$), the calcium-sensing receptor (CaS) and the taste receptors (T1R1–T1R3) (Fredriksson et al. 2003). The mGlu receptors have been further classified into three different groups depending on their similarities in sequence, pharmacology, signalling and localization: Group I includes mGlu$_1$ and mGlu$_5$, Group II mGlu$_2$ and mGlu$_3$ and Group III mGlu$_4$, mGlu$_6$, mGlu$_7$ and mGlu$_8$.

Structurally, most class C GPCRs contain an extracellular so-called Venus flytrap (VFT) domain, a bilobed structure with a crevice between the two lobes that encloses the orthosteric binding site (Fig. 1). Agonist binding stabilizes a conformation with a shorter distance between the two lobes termed the closed conformation (Kunishima et al. 2000). The VFT domain is connected to the seven-transmembrane (7TM) domain through a cysteine-rich domain (CRD), which is notably absent in the GABA$_B$ receptor (Kunishima et al. 2000; Muto et al. 2007). The 7TM domain shares similarities with other class C GPCRs both in topology and in activating similar G proteins. In addition, class C GPCRs are either homodimers (e.g. mGlu receptors) or heterodimers (e.g. GABA$_B$) (Fig. 1). Here, we summarize the insights into the activation mechanism of this class of dimeric receptors gained in particular from structural and mutagenesis studies, and then we review the emerging evidence for new types of class C GPCR heterodimers or higher order oligomers.

DIMERIZATION OF CLASS C GPCRS: A NECESSITY FOR SIGNAL TRANSDUCTION

Insights from homodimers: mGlu and CaS receptors

mGlu and CaS receptors are prototypical homodimers that are stabilized by an inter-protomer disulphide bond, polar contacts between VFT domains and interactions between 7TM domains. The dimerization of

---

Thor C. Møller and David Moreno-Delgado have contributed equally to the preparation of this review article.

✉ Correspondence: Jean-philippe.pin@igf.cnrs.fr (J.-P. Pin)

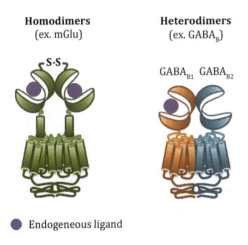

Fig. 1 Structural model and schematic representation of class C GPCRs. Class C GPCRs are composed of a Venus flytrap (VFT) domain, a cysteine-rich domain (CRD) and a transmembrane (7TM) domain. This class of receptors forms obligatory dimers, either homodimers (*e.g.* mGlu) or heterodimers (*e.g.* GABA_B)

these receptors is critical for promoting the activation mechanism, leading from agonist binding to G protein activation. Indeed, different studies indicate that the conformation of one protomer is relative to the other changes upon activation, and this has been observed in all the structural domains found in class C GPCRs (Pin and Bettler 2016).

The dimer of VFT domains is in equilibrium between a resting and an active orientation, and agonist binding displaces the equilibrium towards the active state (Fig. 2). This reorientation is directly linked to G protein activation and, as a consequence, has been used to design a FRET-based sensor to monitor the receptor activation (Doumazane *et al.* 2013). Recently, single-molecule analyses using either the isolated dimer of VFT domains or the full-length mGlu$_2$ receptor dimer have confirmed that the VFT domains oscillate rapidly between the resting and active orientations (Olofsson *et al.* 2014; Vafabakhsh *et al.* 2015). In addition, the studies revealed that agonists with different efficacies diverge in their ability to shift the conformational equilibrium towards the fully active state, rather than stabilizing intermediate conformations.

Because reorientation of the VFT domain dimer is tightly linked to G protein activation, it is implied that this conformational change is somehow transmitted to the 7TM domains. The CRD has been shown to play a critical role in this process for mGlu receptors. This domain is highly rigid due to four intramolecular disulphide bonds (Muto *et al.* 2007), and disrupting them by mutagenesis impairs the capacity of orthosteric agonists to activate G proteins (Huang *et al.* 2011). In addition, crosslinking the two CRDs in a dimer results in a constitutively active receptor (Huang *et al.* 2011). Altogether, this indicates that the transmission from VFT to 7TM domain is mediated by the rigid CRDs coming into close proximity (Fig. 2).

At the 7TM domain level, activation of the receptor requires a rearrangement of the interface between the

Fig. 2 Mechanism of activation of homodimers and heterodimers. Both homodimers and heterodimers undergo conformational changes upon activation. The relative orientation of the VFT dimer is changed upon agonist binding; the CRDs (not in GABA$_B$) are getting closer and the 7TM dimer changes conformation such that a single 7TM is in the active state

7TM domains in the dimer. Actually, it has been shown for the mGlu$_2$ receptor that this interface in the inactive state is formed by transmembrane helix 4 (TM4) and TM5 in each protomer, while the two TM6 s are facing each other in the active state (Xue et al. 2015). This major change in the dimer interface is required for receptor activity, demonstrated by locking the TM4–TM5 interface, which prevents activation by agonist, and locking the TM6 interface, which leads to a constitutively active receptor (Xue et al. 2015). However, the crystal structure of the mGlu$_1$ receptor 7TM domain in the presence of a negative allosteric modulator (NAM) suggested an alternative dimerization interface involving TM1 (Wu et al. 2014). While this difference might be attributed to crystal packing or lack of the VFT domain, further studies are required to determine whether a common mechanism describing the movement of the 7TM domain dimer can be defined for all class C GPCRs.

The data mentioned above suggest that the activation of mGlu receptors relies on the conformational changes of one protomer relative to the other one, which underlies the strict requirement of the mGlu dimerization for glutamate to activate G proteins (Fig. 2). This is further confirmed by experiments showing that glutamate fails to activate an isolated mGlu monomer reconstituted in nanodiscs whereas it activates an mGlu dimer (El Moustaine et al. 2012).

Several studies indicate that a single 7TM domain reaches the active conformation in an mGlu homodimer (Goudet et al. 2005; Hlavackova et al. 2005). In addition, an isolated mGlu monomer purified and reconstituted in nanodiscs activates G proteins when stimulated by a positive allosteric modulator (PAM) (El Moustaine et al. 2012). Hence, in the context of a class C GPCR homodimer, G proteins might be activated through the ligand-bound subunit (cis-activation) and/or through the other subunit (trans-activation). It has been shown for the mGlu$_5$ receptor that Gq protein can be activated either by cis- or trans-activation (Brock et al. 2007). However, it is also possible that in some cases, depending on whether the receptor is cis- or trans-activated, the pathways activated may differ. For instance, it has been observed using a combination of glutamate binding-deficient and G protein coupling-deficient receptors that the mGlu$_1$ receptor triggers Gq-coupled signalling through cis- and trans-activation, while Gi/o and Gs are exclusively activated through cis-activation (Tateyama and Kubo 2011).

Another point to acknowledge when considering class C GPCR homodimers is cooperativity. It has been observed that although glutamate binding to one protomer could induce receptor activation, binding to both protomers was required for full activity (Kniazeff et al. 2004). In addition, binding to one protomer can induce negative cooperativity to the second protomer (Suzuki et al. 2004), which suggests additional complexity in mGlu receptor pharmacology.

Altogether, in class C GPCR homodimers, G protein activation can be achieved upon binding of a single agonist and by a single protomer with an active 7TM domain. However, the dimeric structure is an absolute prerequisite for the conformational transitions from endogenous agonist binding to G protein activation and thus for physiological receptor function.

Insights from heterodimers: GABA$_B$ and T1Rs

The GABA$_B$, sweet taste and umami taste receptors are prototypical class C heterodimers: two different subunits are required to activate G proteins upon agonist binding, confirmed in vivo for the GABA$_B$ receptor by the disappearance of all physiological responses attributed to the heterodimer when either of the two subunits is knocked out (Prosser et al. 2001; Schuler et al. 2001; Queva et al. 2003; Zhao et al. 2003; Gassmann et al. 2004). The GABA$_B$ receptor is a non-covalently linked obligatory heterodimer composed of the subunits GABA$_{B1}$ and GABA$_{B2}$ (Jones et al. 1998; Kaupmann et al. 1998; White et al. 1998), while the taste receptors are composed of T1R3 and either T1R1 or T1R2 resulting in umami or sweet taste receptors, respectively (Nelson et al. 2001, 2002) (Fig. 1). For the GABA$_B$ receptor, the GABA$_{B1}$ subunit contains the binding site for orthosteric ligands (Galvez et al. 1999, 2000), while the GABA$_{B2}$ subunit is necessary for G protein activation (Margeta-Mitrovic et al. 2001a; Robbins et al. 2001; Duthey et al. 2002), confirming the absolute necessity of heterodimer formation. In contrast, the T1R2 subunit seems to be responsible for binding of most ligands and for G protein activation in the sweet taste receptor (Xu et al. 2004). Interestingly, the attempt to create a homomeric receptor by fusing the ligand binding and the G protein coupling domains from GABA$_B$ resulted in a non-functional receptor (Galvez et al. 2001), indicating a unique activation mechanism for these heterodimeric receptors.

In the GABA$_B$ receptor, the correct assembly of the heterodimer is ensured by the C-terminal tail: when expressed alone, GABA$_{B1}$ is retained in the ER due to a RSRR retention motif located in its C-terminal tail (Couve et al. 1998; Margeta-Mitrovic et al. 2000; Pagano et al. 2001). In the presence of GABA$_{B2}$, the retention motif is masked by a coiled-coil interaction between the subunits, thus ensuring that only correctly assembled heterodimers are trafficked to the cell surface (Margeta-Mitrovic et al. 2000; Pagano et al. 2001). However, functional GABA$_B$ receptors can assemble on the cell surface independently of the coiled-coil domains (Pagano *et al.* 2001). In addition to the coiled-coil domain interaction, the $GABA_B$ subunits form interactions between the VFT domains (Geng *et al.* 2013) and most likely also the 7TM domains. Crystal structures of the $GABA_B$ VFT dimer show that in the resting state, they interact exclusively *via* a tight interface involving only one lobe, whereas in the active state a reorientation of the $GABA_{B1}$ VFT domain facilitates an additional, looser interaction between the second lobes (Geng *et al.* 2013).

During activation, agonists bind to the VFT domain of one protomer and promote active conformation of the 7TM domain of the other protomer (Galvez *et al.* 2001; Margeta-Mitrovic *et al.* 2001b), implying that *trans*-activation is the main process for G protein activation in these heterodimers (Fig. 2). Compared to homodimers, this results from a slightly different mechanism. For example, in the $GABA_B$ receptor, the $GABA_{B2}$ VFT domain is unable to bind ligands (Kniazeff *et al.* 2002) and its closed conformation is not necessary for full activation (Geng *et al.* 2012). Total deletion of the $GABA_{B2}$ VFT domain results in a functional receptor suggesting that the signal can proceed from the $GABA_{B1}$ VFT domain to the $GABA_{B2}$ 7TM domain through the $GABA_{B1}$ 7TM domain (Monnier *et al.* 2011). On the other hand, replacing the $GABA_{B1}$ 7TM domain by a single transmembrane helix also produced a functional receptor, suggesting that the signal may also be transmitted through the $GABA_{B2}$ VFT domain (Monnier *et al.* 2011). Altogether, these data propose that two ways of activation exist in the $GABA_B$ receptor.

Cooperativity between protomers exists also in class C GPCR heterodimers. It has been shown in the $GABA_B$ receptor that although the $GABA_{B2}$ VFT and the $GABA_{B1}$ 7TM domains are not directly involved in ligand binding and G protein activation, they play a key role in defining the activation potency. Indeed, when expressed alone, $GABA_{B1}$ exhibits low-affinity agonists binding; however, when co-expressed with $GABA_{B2}$, the interaction with the $GABA_{B2}$ VFT domain increases agonist affinity tenfold (Kaupmann *et al.* 1998). Along the same lines, the $GABA_{B1}$ 7TM domain improves the G protein coupling efficiency of $GABA_{B2}$ (Galvez *et al.* 2001).

Altogether, in class C GPCR heterodimers, such as the $GABA_B$ receptor, one subunit contains the ligand binding domain, but the other subunit is critical for high-affinity agonist binding and functional responses. For the $GABA_B$ receptor, the necessity of dimerization and allosteric transition is ensured by specific targeting of the heterodimer to the cell surface.

NEW FOLKS IN CLASS C GPCR OLIGOMERS

Heterodimers of mGlu receptors

In addition to homodimers, mGlu receptors have recently been reported to possibly form eleven different heterodimers in heterologous systems: the $mGlu_1$ and $mGlu_5$ receptors can only heteromerize between them, whereas all combinations are possible among five other mGlu receptors ($mGlu_2$, $mGlu_3$, $mGlu_4$, $mGlu_7$ and $mGlu_8$) (Doumazane *et al.* 2011). These various combinations are likely strict heterodimers and not the result of association of two homodimers as well demonstrated for the $mGlu_2$–$mGlu_4$ combination (Yin *et al.* 2014; Niswender *et al.* 2016). Interestingly, all possible combinations were found between receptors that share neuronal localization and G protein coupling, which suggests that heterodimer formation is not an artefact of receptor co-expression, but a specific process controlled by structural and functional properties of the receptors.

The study of mGlu heterodimers is a difficult issue to address due to two main points: the lack of specific ligands (especially for receptors from the same group) and the presence of both homodimers and heterodimers in cells co-expressing two mGlu receptors (Doumazane *et al.* 2011). However, some studies have tried to address the topic focusing on heterodimers between $mGlu_2$ and $mGlu_4$ receptors. Although the precise function and pharmacological properties of these heterodimers in native tissues remain open questions, $mGlu_2$ and $mGlu_4$ receptors were found to co-immunoprecipitate in rat dorsal striatum and medial prefrontal cortex (Yin *et al.* 2014). Regarding orthosteric agonist activation, partial agonists such as DCG-IV seem to have a reduced effect in heterodimer activation in comparison with $mGlu_2$ homodimers (Kammermeier 2012), and full agonists such as LY379268 seem to be less potent in activating $mGlu_2$–$mGlu_4$ heterodimers and exhibit dose–response curves with reduced slope (Yin *et al.* 2014). Regarding heterodimer activation by PAMs, $mGlu_2$ PAMs do not activate the heterodimer (Kammermeier 2012), whereas the effect of $mGlu_4$ PAMs depends strongly on their scaffold: VU0155041 seems to activate the heterodimer, but not PHCCC (Yin *et al.* 2014) or VU0418506 (Niswender *et al.* 2016). Further research in the binding site of these PAMs could lead to compounds activating only $mGlu_{2/4}$ heterodimers, which could help to understand their physiological role.

Higher order $GABA_B$ oligomers

The existence of higher order oligomers of GPCRs is still a topic open for discussion, especially because most of the observations have been done in heterologous cells and not validated in native tissues (Vischer et al. 2015). However, increasing experimental evidence suggests that the $GABA_B$ receptor forms oligomers larger than heterodimers. First, in time-resolved FRET experiments, a strong FRET signal was measured between $GABA_{B1}$ subunits, whereas the signal between $GABA_{B2}$ subunits was weak. This led to the proposal that the $GABA_B$ receptor forms at least dimers of heterodimers associated through the $GABA_{B1}$ subunits (Maurel et al. 2008; Comps-Agrar et al. 2011, 2012). Notably, the FRET/receptor ratio was constant over a wide range of receptor densities, including expression levels similar to the endogenous levels in the brain (Maurel et al. 2008). In addition, the existence of oligomers larger than tetramers was suggested by single-molecule microscopy experiments in CHO cells, which showed that at low densities, the majority of $GABA_B$ receptors were dimers with a smaller population ($\sim$30%) of tetramers, whereas at higher densities the dimer population disappeared and complexes larger than tetramers appeared, representing $\sim$60% at the highest density (Calebiro et al. 2013).

In native tissues, the existence of $GABA_B$ oligomers is more complex to prove, but some evidence supports the proposal. Indeed, a FRET signal between anti-$GABA_{B1a}$ antibodies was detected in brain membrane from wild-type animals, but not from $GABA_{B1a}$ knockout animals (Comps-Agrar et al. 2011), and the migration of $GABA_B$ receptors from brain membranes on native gels is consistent with complexes larger than dimers (Schwenk et al. 2010).

A possible function of the oligomerization of the $GABA_B$ heterodimer is modulation of receptor signalling. Indeed, it was found that inhibiting $GABA_{B1}$–$GABA_{B1}$ interactions using either a non-functional $GABA_{B1}$ subunit as competitor or introducing a mutation in the $GABA_{B1}$ VFT domain increased signalling efficacy by approximately 50% (Maurel et al. 2008; Comps-Agrar et al. 2011). It was further shown that one ligand or one G protein per oligomer was sufficient to achieve full activation, suggesting negative cooperativity between heterodimers (Comps-Agrar et al. 2011).

CONCLUSIONS

Class C GPCRs are acknowledged to be dimeric. Over the last two decades, an increasing number of studies have shed light on the necessity of this dimerization for their mechanism of activation. These studies also proposed general concepts for the activation of GPCR dimers. In recent years, new combinations of class C GPCRs with specific pharmacological properties have been reported in heterologous systems and may reveal an even higher complexity of the glutamatergic and GABAergic modulation of the synaptic activity.

Abbreviations

7TM domain	Seven-transmembrane domain
CaS receptor	Calcium-sensing receptor
CRD	Cysteine-rich domain
GABA	γ-Aminobutyric acid
GPCR	G protein-coupled receptor
mGlu receptor	Metabotropic glutamate receptor
NAM	Negative allosteric modulator
PAM	Positive allosteric modulator
TM	Transmembrane
VFT	Venus flytrap

Acknowledgements This work was supported by the CNRS, INSERM and Univ Montpellier, and by grants from the Agence National de la Recherche (ANR-12-BSV2-0015; ANR-13-RPIB-0009), the Fondation Recherche Médicale (FRM DEQ 20130326522). TCM was supported by a funding from the European Union's Seventh Framework Programme for research, technological development and demonstration under grant agreement no 627227.

Compliance with ethical standards

Conflict of interest Thor C. Møller, David Moreno-Delgado, Jean-Philippe Pin and Julie Kniazeff declare that they have no conflict of interest.

Human and animal rights and informed consent This review article does not contain any studies with human or animal subjects performed by any of the authors.

References

Brock C, Oueslati N, Soler S, Boudier L, Rondard P, Pin JP (2007) Activation of a dimeric metabotropic glutamate receptor by intersubunit rearrangement. J Biol Chem 282:33000–33008

Calebiro D, Rieken F, Wagner J, Sungkaworn T, Zabel U, Borzi A, Cocucci E, Zurn A, Lohse MJ (2013) Single-molecule analysis of fluorescently labeled G-protein-coupled receptors reveals complexes with distinct dynamics and organization. Proc Natl Acad Sci USA 110:743–748

Comps-Agrar L, Kniazeff J, Nørskov-Lauritsen L, Maurel D, Gassmann M, Gregor N, Prézeau L, Bettler B, Durroux T,

Trinquet E, Pin JP (2011) The oligomeric state sets GABA$_B$ receptor signalling efficacy. EMBO J 30:2336–2349

Comps-Agrar L, Kniazeff J, Brock C, Trinquet E, Pin JP (2012) Stability of GABA$_B$ receptor oligomers revealed by dual TR-FRET and drug-induced cell surface targeting. FASEB J 26:3430–3439

Couve A, Filippov AK, Connolly CN, Bettler B, Brown DA, Moss SJ (1998) Intracellular retention of recombinant GABA$_B$ receptors. J Biol Chem 273:26361–26367

Doumazane E, Scholler P, Zwier JM, Trinquet E, Rondard P, Pin JP (2011) A new approach to analyze cell surface protein complexes reveals specific heterodimeric metabotropic glutamate receptors. FASEB J 25:66–77

Doumazane E, Scholler P, Fabre L, Zwier JM, Trinquet E, Pin JP, Rondard P (2013) Illuminating the activation mechanisms and allosteric properties of metabotropic glutamate receptors. Proc Natl Acad Sci USA 110:E1416–E1425

Duthey B, Caudron S, Perroy J, Bettler B, Fagni L, Pin JP, Prezeau L (2002) A single subunit (GB2) is required for G-protein activation by the heterodimeric GABA$_B$ receptor. J Biol Chem 277:3236–3241

El Moustaine D, Granier S, Doumazane E, Scholler P, Rahmeh R, Bron P, Mouillac B, Baneres JL, Rondard P, Pin JP (2012) Distinct roles of metabotropic glutamate receptor dimerization in agonist activation and G-protein coupling. Proc Natl Acad Sci USA 109:16342–16347

Fredriksson R, Lagerstrom MC, Lundin LG, Schioth HB (2003) The G-protein-coupled receptors in the human genome form five main families. Phylogenetic analysis, paralogon groups, and fingerprints. Mol Pharmacol 63:1256–1272

Galvez T, Parmentier ML, Joly C, Malitschek B, Kaupmann K, Kuhn R, Bittiger H, Froestl W, Bettler B, Pin JP (1999) Mutagenesis and modeling of the GABA$_B$ receptor extracellular domain support a venus flytrap mechanism for ligand binding. J Biol Chem 274:13362–13369

Galvez T, Prézeau L, Milioti G, Franek M, Joly C, Froestl W, Bettler B, Bertrand HO, Blahos J, Pin JP (2000) Mapping the agonist-binding site of GABA$_B$ type 1 subunit sheds light on the activation process of GABA$_B$ receptors. J Biol Chem 275:41166–41174

Galvez T, Duthey B, Kniazeff J, Blahos J, Rovelli G, Bettler B, Prézeau L, Pin JP (2001) Allosteric interactions between GB1 and GB2 subunits are required for optimal GABA$_B$ receptor function. EMBO J 20:2152–2159

Gassmann M, Shaban H, Vigot R, Sansig G, Haller C, Barbieri S, Humeau Y, Schuler V, Muller M, Kinzel B, Klebs K, Schmutz M, Froestl W, Heid J, Kelly PH, Gentry C, Jaton AL, Van der Putten H, Mombereau C, Lecourtier L, Mosbacher J, Cryan JF, Fritschy JM, Lüthi A, Kaupmann K, Bettler B (2004) Redistribution of GABA$_{B(1)}$ protein and atypical GABA$_B$ responses in GABA$_{B(2)}$-deficient mice. J Neurosci 24:6086–6097

Geng Y, Xiong D, Mosyak L, Malito DL, Kniazeff J, Chen Y, Burmakina S, Quick M, Bush M, Javitch JA, Pin JP, Fan QR (2012) Structure and functional interaction of the extracellular domain of human GABA$_B$ receptor GBR2. Nat Neurosci 15:970–978

Geng Y, Bush M, Mosyak L, Wang F, Fan QR (2013) Structural mechanism of ligand activation in human GABA$_B$ receptor. Nature 504:254–259

Goudet C, Kniazeff J, Hlavackova V, Malhaire F, Maurel D, Acher F, Blahos J, Prezeau L, Pin JP (2005) Asymmetric functioning of dimeric metabotropic glutamate receptors disclosed by positive allosteric modulators. J Biol Chem 280:24380–24385

Hlavackova V, Goudet C, Kniazeff J, Zikova A, Maurel D, Vol C, Trojanova J, Prézeau L, Pin JP, Blahos J (2005) Evidence for a single heptahelical domain being turned on upon activation of a dimeric GPCR. EMBO J 24:499–509

Huang S, Cao J, Jiang M, Labesse G, Liu J, Pin JP, Rondard P (2011) Interdomain movements in metabotropic glutamate receptor activation. Proc Natl Acad Sci USA 108:15480–15485

Jones KA, Borowsky B, Tamm JA, Craig DA, Durkin MM, Dai M, Yao WJ, Johnson M, Gunwaldsen C, Huang LY, Tang C, Shen Q, Salon JA, Morse K, Laz T, Smith KE, Nagarathnam D, Noble SA, Branchek TA, Gerald C (1998) GABA$_B$ receptors function as a heteromeric assembly of the subunits GABA$_B$R1 and GABA$_B$R2. Nature 396:674–679

Kammermeier PJ (2012) Functional and pharmacological characteristics of metabotropic glutamate receptors 2/4 heterodimers. Mol Pharmacol 82:438–447

Kaupmann K, Malitschek B, Schuler V, Heid J, Froestl W, Beck P, Mosbacher J, Bischoff S, Kulik A, Shigemoto R, Karschin A, Bettler B (1998) GABA$_B$-receptor subtypes assemble into functional heteromeric complexes. Nature 396:683–687

Kniazeff J, Galvez T, Labesse G, Pin JP (2002) No ligand binding in the GB2 subunit of the GABA$_B$ receptor is required for activation and allosteric interaction between the subunits. J Neurosci 22:7352–7361

Kniazeff J, Bessis AS, Maurel D, Ansanay H, Prézeau L, Pin JP (2004) Closed state of both binding domains of homodimeric mGlu receptors is required for full activity. Nat Struct Mol Biol 11:706–713

Kunishima N, Shimada Y, Tsuji Y, Sato T, Yamamoto M, Kumasaka T, Nakanishi S, Jingami H, Morikawa K (2000) Structural basis of glutamate recognition by a dimeric metabotropic glutamate receptor. Nature 407:971–977

Margeta-Mitrovic M, Jan YN, Jan LY (2000) A trafficking checkpoint controls GABA$_B$ receptor heterodimerization. Neuron 27:97–106

Margeta-Mitrovic M, Jan YN, Jan LY (2001a) Function of GB1 and GB2 subunits in G protein coupling of GABA$_B$ receptors. Proc Natl Acad Sci USA 98:14649–14654

Margeta-Mitrovic M, Jan YN, Jan LY (2001b) Ligand-induced signal transduction within heterodimeric GABA$_B$ receptor. Proc Natl Acad Sci USA 98:14643–14648

Maurel D, Comps-Agrar L, Brock C, Rives ML, Bourrier E, Ayoub MA, Bazin H, Tinel N, Durroux T, Prézeau L, Trinquet E, Pin JP (2008) Cell-surface protein–protein interaction analysis with time-resolved FRET and snap-tag technologies: application to GPCR oligomerization. Nat Methods 5:561–567

Monnier C, Tu H, Bourrier E, Vol C, Lamarque L, Trinquet E, Pin JP, Rondard P (2011) Trans-activation between 7TM domains: implication in heterodimeric GABA$_B$ receptor activation. EMBO J 30:32–42

Muto T, Tsuchiya D, Morikawa K, Jingami H (2007) Structures of the extracellular regions of the group II/III metabotropic glutamate receptors. Proc Natl Acad Sci USA 104:3759–3764

Nelson G, Hoon MA, Chandrashekar J, Zhang Y, Ryba NJ, Zuker CS (2001) Mammalian sweet taste receptors. Cell 106:381–390

Nelson G, Chandrashekar J, Hoon MA, Feng L, Zhao G, Ryba NJ, Zuker CS (2002) An amino-acid taste receptor. Nature 416:199–202

Niswender CM, Jones CK, Lin X, Bubser M, Thompson Gray A, Blobaum AL, Engers DW, Rodriguez AL, Loch MT, Daniels JS, Lindsley CW, Hopkins CR, Javitch JA, Conn PJ (2016) Development and antiparkinsonian activity of VU0418506, a selective positive allosteric modulator of metabotropic glutamate receptor 4 homomers without activity at mGlu$_{2/4}$ heteromers. ACS Chem Neurosci 7:1201–1211

Olofsson L, Felekyan S, Doumazane E, Scholler P, Fabre L, Zwier JM, Rondard P, Seidel CA, Pin JP, Margeat E (2014) Fine tuning of sub-millisecond conformational dynamics controls metabotropic glutamate receptors agonist efficacy. Nat Commun 5:5206

Pagano A, Rovelli G, Mosbacher J, Lohmann T, Duthey B, Stauffer D, Ristig D, Schuler V, Meigel I, Lampert C, Stein T, Prezeau L, Blahos J, Pin J, Froestl W, Kuhn R, Heid J, Kaupmann K, Bettler B (2001) C-terminal interaction is essential for surface trafficking but not for heteromeric assembly of $GABA_B$ receptors. J Neurosci 21:1189–1202

Pin JP, Bettler B (2016) Organization and functions of mGlu and $GABA_B$ receptor complexes. Nature 540:60–68

Prosser HM, Gill CH, Hirst WD, Grau E, Robbins M, Calver A, Soffin EM, Farmer CE, Lanneau C, Gray J, Schenck E, Warmerdam BS, Clapham C, Reavill C, Rogers DC, Stean T, Upton N, Humphreys K, Randall A, Geppert M, Davies CH, Pangalos MN (2001) Epileptogenesis and enhanced prepulse inhibition in $GABA_{B1}$-deficient mice. Mol Cell Neurosci 17:1059–1070

Queva C, Bremner-Danielsen M, Edlund A, Ekstrand AJ, Elg S, Erickson S, Johansson T, Lehmann A, Mattsson JP (2003) Effects of GABA agonists on body temperature regulation in $GABA_{B(1)}^{-/-}$ mice. Br J Pharmacol 140:315–322

Robbins MJ, Calver AR, Filippov AK, Hirst WD, Russell RB, Wood MD, Nasir S, Couve A, Brown DA, Moss SJ, Pangalos MN (2001) $GABA_{B2}$ is essential for g-protein coupling of the $GABA_B$ receptor heterodimer. J Neurosci 21:8043–8052

Schuler V, Luscher C, Blanchet C, Klix N, Sansig G, Klebs K, Schmutz M, Heid J, Gentry C, Urban L, Fox A, Spooren W, Jaton AL, Vigouret J, Pozza M, Kelly PH, Mosbacher J, Froestl W, Kaslin E, Korn R, Bischoff S, Kaupmann K, van der Putten H, Bettler B (2001) Epilepsy, hyperalgesia, impaired memory, and loss of pre- and postsynaptic $GABA_B$ responses in mice lacking $GABA_{B(1)}$. Neuron 31:47–58

Schwenk J, Metz M, Zolles G, Turecek R, Fritzius T, Bildl W, Tarusawa E, Kulik A, Unger A, Ivankova K, Seddik R, Tiao JY, Rajalu M, Trojanova J, Rohde V, Gassmann M, Schulte U, Fakler B, Bettler B (2010) Native $GABA_B$ receptors are heteromultimers with a family of auxiliary subunits. Nature 465:231–235

Suzuki Y, Moriyoshi E, Tsuchiya D, Jingami H (2004) Negative cooperativity of glutamate binding in the dimeric metabotropic glutamate receptor subtype 1. J Biol Chem 279:35526–35534

Tateyama M, Kubo Y (2011) The intra-molecular activation mechanisms of the dimeric metabotropic glutamate receptor 1 differ depending on the type of G proteins. Neuropharmacology 61:832–841

Vafabakhsh R, Levitz J, Isacoff EY (2015) Conformational dynamics of a class C G-protein-coupled receptor. Nature 524:497–501

Vischer HF, Castro M, Pin JP (2015) G protein-coupled receptor multimers: a question still open despite the use of novel approaches. Mol Pharmacol 88:561–571

White JH, Wise A, Main MJ, Green A, Fraser NJ, Disney GH, Barnes AA, Emson P, Foord SM, Marshall FH (1998) Heterodimerization is required for the formation of a functional GABAB receptor. Nature 396:679–682

Wu H, Wang C, Gregory KJ, Han GW, Cho HP, Xia Y, Niswender CM, Katritch V, Meiler J, Cherezov V, Conn PJ, Stevens RC (2014) Structure of a class C GPCR metabotropic glutamate receptor 1 bound to an allosteric modulator. Science 344:58–64

Xu H, Staszewski L, Tang H, Adler E, Zoller M, Li X (2004) Different functional roles of T1R subunits in the heteromeric taste receptors. Proc Natl Acad Sci USA 101:14258–14263

Xue L, Rovira X, Scholler P, Zhao H, Liu J, Pin JP, Rondard P (2015) Major ligand-induced rearrangement of the heptahelical domain interface in a GPCR dimer. Nat Chem Biol 11:134–140

Yin S, Noetzel MJ, Johnson KA, Zamorano R, Jalan-Sakrikar N, Gregory KJ, Conn PJ, Niswender CM (2014) Selective actions of novel allosteric modulators reveal functional heteromers of metabotropic glutamate receptors in the CNS. J Neurosci 34:79–94

Zhao GQ, Zhang Y, Hoon MA, Chandrashekar J, Erlenbach I, Ryba NJ, Zuker CS (2003) The receptors for mammalian sweet and umami taste. Cell 115:255–266

Significant expansion and red-shifting of fluorescent protein chromophore determined through computational design and genetic code expansion

Li Wang[1,2], Xian Chen[3], Xuzhen Guo[1,2], Jiasong Li[1], Qi Liu[1], Fuying Kang[1,2], Xudong Wang[1], Cheng Hu[1,2], Haiping Liu[1], Weimin Gong[1], Wei Zhuang[4]✉, Xiaohong Liu[1]✉, Jiangyun Wang[1,2]✉

[1] Institute of Biophysics, Chinese Academy of Sciences, Beijing 100101, China
[2] College of Life Sciences, University of Chinese Academy of Sciences, Beijing 100049, China
[3] Key Laboratory of Physics and Technology for Advanced Batteries (Ministry of Education), Department of Physics, Jilin University, Changchun 130012, China
[4] State Key Laboratory of Structural Chemistry, Fujian Institute of Research on the Structure of Matter, Chinese Academy of Sciences, Fuzhou 350002, China

Graphical abstract

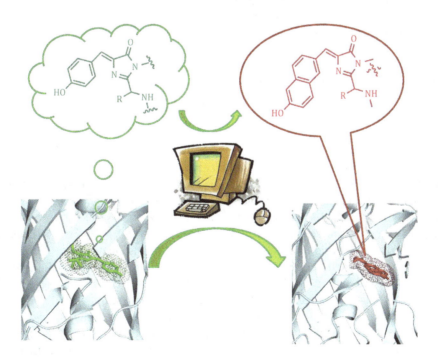

Li Wang, Xian Chen and Xuzhen Guo have contributed equally to this work.

✉ Correspondence: wzhuang@fjirsm.ac.cn (W. Zhuang), liuxh@moon.ibp.ac.cn (X. Liu), ang@ibp.ac.cn (J. Wang)

Abstract Fluorescent proteins (FPs) with emission wavelengths in the far-red and infrared regions of the spectrum provide powerful tools for deep-tissue and super-resolution imaging. The development of red-shifted FPs has evoked widespread interest and continuous engineering efforts. In this article, based on a computational design and genetic code expansion, we report a rational approach to significantly expand and red-shift the chromophore of green fluorescent protein (GFP). We applied computational calculations to predict the excitation and emission wavelengths of a FP chromophore harboring unnatural amino acids (UAA) and identify in silico an appropriate UAA, 2-amino-3-(6-hydroxynaphthalen-2-yl)propanoic acid (naphthol-Ala). Our methodology allowed us to formulate a GFP variant (cpsfGFP-66-Naphthol-Ala) with red-shifted absorbance and emission spectral maxima exceeding 60 and 130 nm, respectively, compared to those of GFP. The GFP chromophore is formed through autocatalytic post-translational modification to generate a planar 4-(p-hydroxybenzylidene)-5-imidazolinone chromophore. We solved the crystal structure of cpsfGFP-66-naphthol-Ala at 1.3 Å resolution and demonstrated the formation of a much larger conjugated π-system when the phenol group is replaced by naphthol. These results explain the significant red-shifting of the excitation and emission spectra of cpsfGFP-66-naphthol-Ala.

Keywords Green fluorescent protein, Red-shift, Unnatural amino acids, Computational design

INTRODUCTION

Fluorescent proteins (FPs) with far-red and infrared-region emission wavelengths are powerful tools for tracking protein localization in cells, cell migration, and deep-tissue imaging (Enterina et al. 2015; Nienhaus and Nienhaus 2014; Niu and Guo 2013; Sengupta et al. 2014; Shaner et al. 2011). Methods to achieve red-shifted FPs have evoked widespread interest and continuous engineering efforts (Chang et al. 2012; Enterina et al. 2015; Grimm et al. 2015; Newman et al. 2011; Niu and Guo 2013; Shu et al. 2009; Subach et al. 2011; Subach and Verkhusha 2012). Despite recent results, however, the design of these mutated FPs remains mostly dependent on high-throughput screening, which is a costly and tedious task. The development of modern computational chemistry allows a better understanding of the transition-state structure of FPs through theoretical calculation (Chica et al. 2010; Mou et al. 2015). Through this technology, key factors can be determined for the design of FPs with significantly red-shifted excitation and emission spectra. When the 66Tyr residue of the chromophore in various FPs is substituted with HqAla, excitation and emission spectra red-shifted by 30 nm are achieved (Liu et al. 2013). In the present work, a computation-based design and genetic code expansion are applied to rationally design an appropriate unnatural amino acid (UAA), 2-amino-3-(6-hydroxynaphthalen-2-yl)propanoic acid (naphthol-Ala), and genetically incorporate the same to the chromophore of cpsfGFP. cpsfGFP-66-naphthol-Ala has significantly red-shifted absorbance and emission spectral maxima compared to those of wild-type green FP (GFP). Structural analysis reveals the formation of a novel naphthol-imidazolinone (NapI) chromophore with a significantly larger conjugated π system compared to that of the original 4-(p-hydroxybenzylidene)-5-imidazolinone (HBI) chromophore in GFP. Furthermore, as a noteworthy and promising technology in super-resolution imaging, GFP and some of its mutants undergo efficient photoconversion into a red fluorescent state (Bogdanov et al. 2009; Elowitz et al. 1997; Saha et al. 2013). The oxidation red fluorescence of cpsfGFP-66-Naphthol-Ala revealed remarkably red-shifted excitation and emission spectra. This oxidation redding phenomenon was observed in vivo through confocal and fluorescence microscopy.

RESULTS AND DISCUSSION

Calculation and rational design of red-shift FP chromophore

When the sfGFP 66Tyr site is substituted with a heterocyclic amino acid, the resulting excitation and emission spectra are red-shifted by approximately 30 nm (Liu et al. 2013). The fluorescence frequencies of the red-shifted chromophores (Fig. 1, ST-1–ST-8) were first computed by the time-dependent density functional theory (TDDFT) method with the long-range-corrected (LRC) exchange–correlation functional (Laurent and Jacquemin 2013). The gas-phase vertical absorption and emission energies of the eight chromophores (ST-1–ST-8 in Fig. 1), including their neutral and anionic states, are presented in Table 1. A trend of spectral shifting trend was well represented by the calculations. From left to right in each row of Table 1, the absorption and emission energies of either neutral or anionic chromophores are red-shifted compared to those of their left-hand neighbors. In addition, the absorption and emission energies of neutral and anionic chromophores in lower rows are red-shifted compared to the corresponding values indicated in upper rows. All frequency-change features observed were consistent with the experiment values.

The transition dipole moments of the chromophores also revealed a regular pattern. Table 1 shows that, in general, larger transition dipole moments corresponded to more red-shifted excitation and emission spectra. The transition dipole moment of emission was consistently larger than that of absorption. Intramolecular charge transfer significantly affects the spectra of many conjugated chromophores or polymers. Population analysis indicated that, besides large intramolecular charge transfers, a free electron flow channel can also decrease the transition (absorption and emission) energy.

The calculations provide a reliable guide for the design of red-shifted FPs. First, large conjugated π electrons afford a spacious and free electron flow place. Second, an ideal side chain should provide efficient electron-withdrawing and -donating effects. "Left-ring" type (HBI and HQI) charge transfers affect the anionic "right-ring to left-ring" charge transfer process more extensively than the neutral charge transfer process, and the neutral "left-ring to right-ring" charge transfer process is more influenced by the side chain of the "right-ring" charge transfer process than the "left-ring" transfer process. Third, in many conjugated polymer materials, N is a typical electron-withdrawing atom. In the case of HQI, elimination of N from HQI may promote the anionic "right-ring to left-ring" charge transfer process.

We predicted the structures of the heterocycle-containing chromophores ST-9–ST-12 (Fig. 1). Using the same calculation, the fluorescence frequencies of these chromophores were computed to design more red-shifted FPs. The gas-phase vertical absorption and emission energies of these chromophores (ST-9–ST-12 in Fig. 1) in their neutral and anionic states are presented in Table 1. Compared with the absorption and

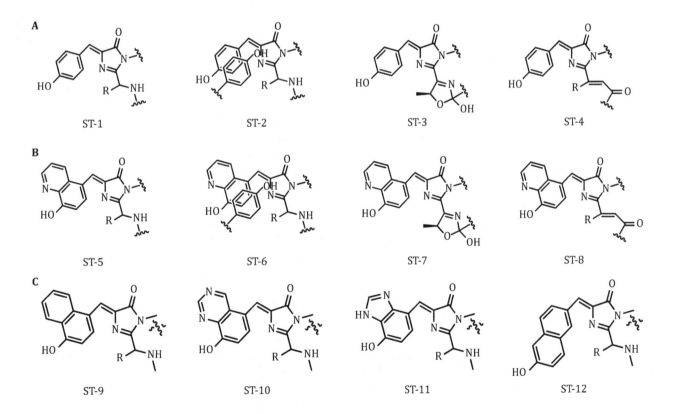

Fig. 1 A ST-1, ST-2, ST-3, and ST-4, respectively, representing the chemical structures of sfGFP, sfYFP, PsmOrange, and eqFP650 containing the 4-(p-hydroxybenzylidene)-5-imidazolinone chromophore. **B** ST-5, ST-5, ST-7, and ST-8, respectively, representing the chemical structures of sfGFP-66-HqAla, sfYFP-66-HqAla, PsmOrange-72-HqAla, and eqFP650-67-HqAla containing the 8-hydroxyquinolin-imidazolinone chromophore. **C** Chemical structures of the predicted chromophores ST-9–ST-12

Table 1 Calculated absorption and emission energies (eV) of chromophore models ST-1–ST-8 and their corresponding dipole moments (a.u.)

	ST-1		ST-2		ST-3		ST-4	
	$E\ (\Delta E)$	Dip.	$E\ (\Delta E)$	Dip.	$E\ (\Delta E)$	Dip.	$E\ (\Delta E)$	Dip.
Neutral								
$S_0 \rightarrow S_1$	3.87	8.04	3.78 (− 0.09)	6.60	3.50 (− 0.37)	9.59	3.16 (− 0.71)	10.47
$S_1 \rightarrow S_0$	3.27	9.28	3.16 (− 0.11)	8.01	2.93 (− 0.34)	11.49	2.69 (− 0.58)	12.98
Anionic								
$S_0 \rightarrow S_1$	3.23	12.84	3.16 (− 0.07)	11.08	2.75 (− 0.48)	15.03	2.56 (− 0.67)	18.41
$S_1 \rightarrow S_0$	3.04	12.76	2.95 (− 0.09)	10.67	2.59 (− 0.45)	15.41	2.43 (− 0.61)	18.80
Exp.			Abs. (− 0.15)		Abs. (− 0.29)		Abs. (− 0.46)	
			Emi. (− 0.08)		Emi. (− 0.24)		Emi. (− 0.52)	
	ST-5		ST-6		ST-7		ST-8	
	$E\ (\Delta E)$	Dip.	$E\ (\Delta E)$	Dip.	$E\ (\Delta E)$	Dip.	$E\ (\Delta E)$	Dip.
Neutral								
$S_0 \rightarrow S_1$	3.55 (− 0.32)	8.12	3.50 (− 0.37)	6.62	3.21 (− 0.66)	10.41	3.00 (− 0.87)	11.36
$S_1 \rightarrow S_0$	3.03 (− 0.24)	9.75	2.94 (− 0.33)	8.08	2.72 (− 0.55)	12.80	2.49 (− 0.78)	14.42
Anionic								
$S_0 \rightarrow S_1$	3.12 (− 0.11)	11.67	3.01 (− 0.22)	9.12	2.67 (− 0.56)	16.12	2.46 (− 0.77)	19.45
$S_1 \rightarrow S_0$	2.94 (− 0.10)	11.19	2.83 (− 0.21)	9.17	2.52 (− 0.52)	16.75	2.34 (− 0.70)	20.07
Exp.	Abs. (− 0.25)		Abs. (− 0.28)		Abs. (− 0.42)		Abs. (− 0.56)	
	Emi. (− 0.15)		Emi. (− 0.19)		Emi. (− 0.34)		Emi. (− 0.61)	

"ΔE" represents the energy red-shift of models ST-2–ST-8 compared to that of the ST-1 model. The experimental absorption and emission energy red-shifts of these models relative to those of the ST-1 model are also shown

emission energies of the ST-1 chromophore in Table 2, those of the anionic ST-12 chromophore showed larger red-shifts.

Synthesis and genetic incorporation of 2-amino-3-(6-hydroxynaphthalen-2-yl) propanoic acid (hereafter termed naphthol-Ala)

We synthesized the UAA naphthol-Ala from the best candidate chromophore (ST-12) and genetically encoded it into cpsfGFP. Naphthol-Ala can be synthesized through six steps (Fig. 2A). In brief, 6-hydroxy-2-naphthoic acid is first methylated to yield methyl ester and then reduced to obtain a hydroxyl group through halogenation by sulfoxide chloride. The corresponding chloride is reacted with diethyl acetamidomalonate in alkaline conditions to form the ester precursor of the target UAA, which is then deprotected to form the hydrobromide salt of naphthol-Ala.

To deliver naphthol-Ala to the defined protein sites in *E. coli* through the TAG codon, a mutant *M. jannaschii* tyrosyl amber suppressor tRNA (*Mj*tRNA$_{CUA}^{Tyr}$)/tyrosyl-tRNA synthetase (*Mj*TyrRS) pair must be constructed. The desired *Mj*TyrRS clone was obtained after three rounds of positive selection and two rounds of negative selection from the *Mj*TyrRS library. This clone can resist 120 μg/mL chloramphenicol due to the expression of the chloramphenicol acetyl transferase (CAT) gene in the presence of 1 mmol/L naphthol-Ala but can only resist 20 μg/mL chloramphenicol in the absence of the UAA. The desired UAA RS was named naphthol-Ala RS. Sequencing of this clone revealed the following mutations (Table 3): Y32R, L65H, H70G, F108N, Q109C, D158N, and L162S. The Y32R and H70G mutations create additional space to accommodate the bulky naphthol group. L65, F108, and D158 are mutated to hydrophilic amino acids, thus creating a large binding pocket while providing additional hydrogen bonding interactions to stabilize the naphthol ring. Naphthol-Ala can also be incorporated into proteins with the following mutants (Chen and Tsao 2013): Y32E, L65T, D158S, I159A, H160P, Y161T, L162Q, A167W, and D286R in NpOH-RS1; and Y32E, L65V, K90E, I159A, H160W, Y161G, L162Q, A167I, and D286R in NpOH-RS2. The yield of the Z-domain mutant with NpOH-RS1 was roughly 7 mg/L culture in the presence of naphthol-Ala. These results demonstrate that the active site of tRNA (*Mj*tRNA$_{CUA}^{Tyr}$)/tyrosyl-tRNA synthetase has high capacity

Table 2 Calculated absorption and emission energies (eV) of the chromophore models ST-9–ST-12

	ST-9 $E\,(\Delta E)$	ST-10 $E\,(\Delta E)$	ST-11 $E\,(\Delta E)$	ST-12 $E\,(\Delta E)$
Neutral				
$S_0 \to S_1$	3.51 (− 0.36)	3.56 (− 0.31)	3.72 (− 0.15)	3.60 (− 0.27)
$S_1 \to S_0$	3.03 (− 0.24)	3.00 (− 0.27)	3.19 (− 0.08)	3.08 (− 0.19)
Anionic				
$S_0 \to S_1$	3.09 (− 0.14)	3.06 (− 0.17)	3.21 (− 0.02)	2.72 (− 0.51)
$S_1 \to S_0$	2.94 (− 0.10)	2.58 (− 0.46)	2.97 (− 0.07)	2.56 (− 0.48)
Exp.				Abs. (− 0.27)
				Emi. (− 0.49)

"ΔE" refers to the energy red-shift of the four predicted chromophore models compared to that of the ST-1 chromophore model

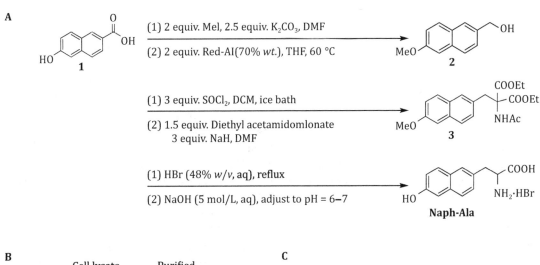

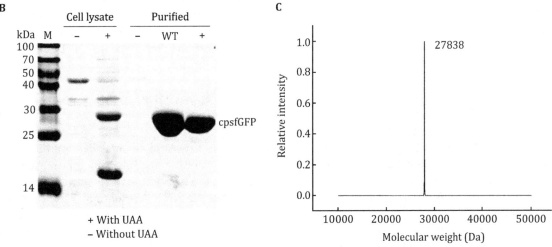

Fig. 2 **A** Synthesis of 2-amino-3-(6-hydroxynaphthalen-2-yl) propanoic acid (naphthol-Ala). **B** Coomassie-stained SDS-PAGE gel indicating the expression of the cpsfGFP-66-naphthol-Ala mutant in the presence and absence of 1 mmol/L naphthol-Ala. **C** ESI–MS spectra of the TAG66 mutant of cpsfGFP, expected mass: 27873 Da, found: 27873.27 Da

and variability for incorporating even the same UAA. To enhance the yield of the mutant proteins, naphthol-Ala RS was constructed on the pEVOL system.

To determine the efficiency and fidelity of naphthol-Ala incorporation, an amber stop code was substituted for the Tyr66 site in cpsfGFP containing a C-terminal

His6 tag. Protein expression was carried out in *E. coli* in the presence of the selected synthetase (naphthol-Ala RS), $MjtRNA_{CUA}^{Tyr}$ with 0.2% arabinose, 1 mmol/L IPTG, and 1 mmol/L UAA. The protein expressed in the absence of the UAA was taken as the negative control. Analysis of the cell lysate and purified protein by Coomassie-stained SDS-PAGE showed that the full-length cpsfGFP mutant is expressed efficiently only in the presence of naphthol-Ala, thereby indicating the specific activity of naphthol-Ala RS for naphthol-Ala only (Fig. 2B). The yield of the Tyr66 mutant cpsfGFP (termed cpsfGFP-66-Naphthol-Ala) was 30 mg/L, whereas that of sfGFP was 100 mg/L. ESI–MS revealed that cpsfGFP-66-naphthol-Ala has an average mass of 27,873.27 Da (Fig. 2C), consistent with the calculated mass of 27,873 Da for the Y66 → naphthol-Ala cpsfGFP mutant, where the chromophore of the cpsfGFP mutant is fully matured after self-catalyzed dehydration and oxidation.

Fluorescence characterization of cpsfGFP-66-naphthol-Ala

We characterized the fluorescent property of cpsfGFP-66-naphtol-Ala, which shows red-shifted anionic state absorbance and emission spectral maxima at 545 and 640 nm, respectively (Fig. 3A, Table 4); these values are higher than those of sfGFP by 60 and 130 nm, respectively. The excitation and emission spectra of naphthol-Ala do not show any significant overlap with those of cpsfGFP-66-naphthol-Ala (Fig. 3B), thus indicating that the observed red-shift is not due to UAA small molecules.

We performed UV–Vis titrations of cpsfGFP-66-naphthol-Ala at various pH. Figure 4A shows that neutral cpsfGFP-66-naphthol-Ala displays a strong absorption peak at 450 nm and that new peaks corresponding to the anionic state appear at 545 nm with the increasing pH (pKa = 7.7). The CD spectra of cpsfGFP-66-naphthol-Ala from pH 5–10 show that the secondary β-sheet structure does not change with pH fluctuations, thereby indicating that the protein remains correctly folded throughout pH titration (Fig. 4B). These results are consistent with the results predicted by the computational calculations. The detected emission red-shift energy (− 0.49 eV) of cpsfGFP-66-naphthol-Ala is comparable with the calculated emission red-shift energy of the ST-12 chromophore, which has a ΔE value of − 0.47 eV (Table 2). The CD spectra of cpsfGFP-66-naphthol-Ala in the presence of buffers of various pH show no difference. (Fig 4C),

As potential tools in super-resolution imaging for investigating dynamic processes in living cells, GFP and some of its mutants undergo efficient photoconversion to a red fluorescent state. Under anaerobic conditions, the redding state of GFP shows excitation–emission maxima at 525 and 600 nm, respectively. In the presence of potassium ferricyanide, EGFP shows red-shifted fluorescence spectra with excitation and emission peaks at 575 and 607 nm, respectively (Bogdanov et al. 2009). We thus investigated whether cpsfGFP-66-naphthol-Ala can facilitate the oxidation redding process in the presence of potassium ferricyanide. After incorporating naphthol-Ala into the 66th position of cpsfGFP and irradiating the mutant protein using a 450-nm laser (100 mW/cm^2) for 2 min in the presence of 1 mmol/L potassium ferricyanide, cpsfGFP-66-naphthol-Ala showed a remarkable red-shift with excitation and emission spectral maxima at 620 and 695 nm, respectively (Fig. 3A). These results indicate that oxidation redding can also occur in the presence of cpsfGFP-66-naphthol-Ala. Although the mechanism of GFP redding remains under investigation, it likely involves a two-electron oxidation process, where the amino acids near the chromophore, such as Glu222 (Saha et al. 2013), may act as an electron donor, and potassium ferricyanide acts as an electron acceptor. Upon electron transfer between these groups, a Ds-red-like chromophore is formed.

Crystallography characterization of cpsfGFP-66-naphthol-Ala

To elucidate structural changes causing the red-shifted absorption and emission spectra of cpsfGFP-66-naphthol-Ala, we determined the structure of cpsfGFP-66-naphthol-Ala at 1.3 Å resolution. Figure 5 shows that substitution of Tyr66 by naphthol-Ala causes substantial crowding around the chromophore. The protein backbone of cpsfGFP-66-naphthol-Ala is highly similar

Table 3 Sequence of naphthol-Ala-specific aaRSs

Position	32	65	70	108	109	158	162	164	200
Tyr RS	Y	L	H	F	Q	D	L	Y	D
Naphthol-Ala RS(8)	R	Y	G	N	C	N	S		

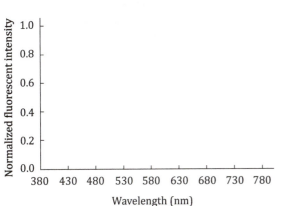
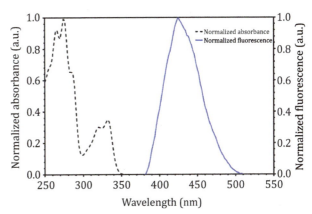

Fig. 3 **A** Absorption and emission spectra of cpsfGFP-66-naphthol-Ala in 20 mmol/L MOPS-citrate buffer at pH 5 before and after 450-nm laser irradiation for 2 min in the presence of 1 mmol/L potassium ferricyanide. **B** Normalized absorption and emission spectra of 10 mmol/L naphthol-Ala in 60 mmol/L Tris–HCl buffer at pH 7, Ex,max = 331 nm; Em,max = 424 nm

Table 4 Summary of the key characteristics of fluorescent proteins

	cpsfGFP-66-naphthol-Ala	cpsfGFP-66-naphthol-Ala after 405-nm irradation		eqFP670
Excitation peak (nm)	450	545	620 nm	605
Emission peak (nm)	637	640	695 nm	670
Fluorescence QY	0.15	0.02	–	0.06
Molar extinction coefficient (l/(mol·cm)) at excitation maximum (pH 7.4)	82,600	12,700	–	70,000
Brightness[a] (a.u.)	12,390	254	–	4200
Photostability, confocal[b] (s)	–	7.5[c]	–	75
pKa	7.7	7.7	–	4.5
Reference	This work	This work	This work	Shcherbo et al. (2010)

[a]Calculated as the product of the molar extinction coefficient and quantum yield
[b]Time to bleach 50% of the brightness of the fluorescence signal
[c]Excitation wavelength is 561 nm

to that of sfGFP. The 66th naphthol-Ala group presents two conformations due to the higher accessibility of the chromophore to solvents compared with sfGFP. The naphthol and imidazolinone rings are co-planar, thereby indicating that, similar to GFP, the chromophore is fully matured after nucleophilic attack of Gly67-Nα on the carbonyl group of Ser65, followed by dehydration and oxidation of the α–β bond in the naphthol-Ala group (Chudakov et al. 2010).

Confocal and fluorescence microscopy of the oxidative redding state of cpsfGFP-66-naphthol-Ala

We tested whether the oxidative redding phenomenon of cpsfGFP-66-naphthol-Ala could be visualized through confocal and fluorescence microscopy *in vivo*. Before irradiation, a thin (~ 10 μm) layer of *E. coli* cells expressed with cpsfGFP-66-naphthol-Ala in 20 mmol/L

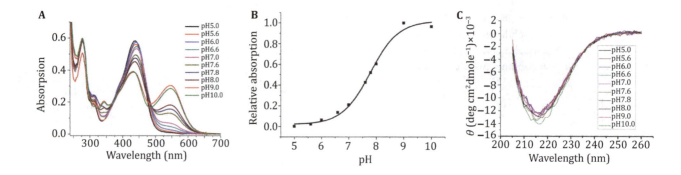

Fig. 4 **A** UV–Vis spectra of cpsfGFP-66-naphthol-Ala in different pH value buffer. **B** Anionic chromophore fraction of cpsfGFP-66-naphthol-Ala at varying pH. **C** CD spectra of cpsfGFP-66-naphthol-Ala in the presence of buffers of various pH

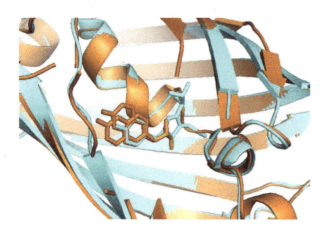

Fig. 5 Alignment of sfGFP and cpsfGFP-66-naphthol-Ala. The chromophores of cpsfGFP-66-naphthol-Ala (*orange*) and sfGFP (*blue*) show that the naphthol or phenol and imidazolinone rings are clearly coplanar

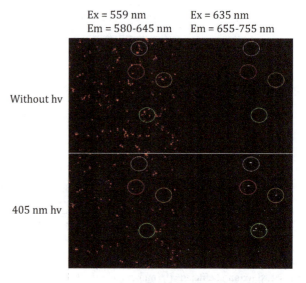

Fig. 6 Pre- and post-photoconversion images of a thin (~ 10 μm) layer of *E. coli* cells expressing cpsfGFP-66-naphthol-Ala in 20 mmol/L MOPS-citrate buffer at pH 5 under 405-nm excitation delivered through a 1.2-NA objective to the region within the circle in bleaching mode (Red channel: 580–645 nm; Red channel: 655–755 nm). Photoconversion was performed at 37% laser power and 30 iterations for 111 ms

MOPS-citrate buffer at pH 5 was prepared. The cells showed fluorescence at the detection window of 580–645 nm but were only slightly detectable at 655–755 nm. After irradiation with a 405-nm excitation laser for 2 min in the presence of 1 mmol/L potassium ferricyanide (highlighted by circles in Fig. 6), the fluorescence in the detection range of 655–755 nm increased. Overall, the oxidation redding phenomenon of cpsfGFP-66-naphthol-Ala could be observed *in vivo* through confocal and fluorescence microscopy.

CONCLUSION

In conclusion, we demonstrated that a red-shifted FP could be designed via a rational computational design and genetic code expansion. Our work establishes a platform for further optimization of FPs with superior properties for deep-tissue and super-resolution imaging. The method presented in this work will benefit further studies on FP engineering.

EXPERIMENTAL SECTION

Materials and reagents

6-Hydroxy-2-naphthoic acid was purchased from J&K chemical. All other chemicals used in this work were purchased from Sigma-Aldrich and applied without further purification. Silica gel chromatography purification was carried out using Silica Gel 60 (230–400 mesh). The PCR reagents, T4 DNA ligase, and restriction endonucleases were purchased from

Fermentas. The Ni-resin (NTA) affinity purification reagents and column were purchased from Qiagen. Genes and primers were synthesized by Sangon Biotech.

Instrument

All ^{1}H-NMR spectra are reported in parts per million (ppm) and were measured relative to the signals of DMSO (2.5 ppm). ^{13}C-NMR spectra are reported in ppm relative to those of residual DMSO (40 ppm). The mass spectra of the chemicals were obtained using a Waters LC–MS instrument equipped with a single-quadrupole mass detector and an electrospray ionization source (Waters ACQUITY QDa), while those of the proteins were obtained using Agilent 6100 equipment with a series of triple-quadrupole mass spectrometers (Agilent Technologies, CA, USA). Protein purification was performed using AKTA UPC 900 FPLC system (GE Healthcare), and absorption spectra were obtained at room temperature using a UV–visible spectrometer (Agilent 8453, Agilent Technologies). Fluorescence spectra were obtained using a microplate reader equipped with SkanIt 2.4.3 RE software for Varioskan Flash (Varioskan Flash, Thermo Fisher Scientific Inc.). Cell fluorescent images were recorded at room temperature by confocal and fluorescence microscopy (FV1000, OLYMPUS).

Theoretical calculation

The chromophores used for the calculations were taken from a structural model (PDB: 4JFG) and built with manual modifications. The excited-state geometry of each chromophore was calculated using the TDDFT and LRC methods. The TDDFT method greatly reduces computational costs compared with MCSCF techniques, while the LRC method can deal with long-range electron correlations, which are common in π-conjugated fluorescent chromophore transitions, with greater ease than conventional hybrid functionals, such as B3LYP. In Filippi's work, the LRC functionals CAM-B3LYP and LC-BLYP presented overall good agreement with the extrapolated results of solution experiments in vacuum conditions for wild-type GFP and wave function methods. All calculations were performed using the Gaussian 09 suite of programs (Frisch *et al.* 2016).

Synthesis of (6-methoxynaphthalen-2-yl)methanol (2)

6-Hydroxy-2-naphthoic acid (1) (3.764 g, 20 mmol) was dissolved in anhydrous DMF (30 ml), after which K_2CO_3 (6.91 g, 50 mmol, 2.5 equiv.) and MeI (2.80 ml, 45 mmol, 2.25 equiv.) were added to their mixture. The reaction mixture was stirred vigorously for 48 h at room temperature, diluted with ice-water (300 ml), and extracted with ethyl acetate (3 × 100 ml). The organic phase was washed with brine and dried with Na_2SO_4, and the solvent was evaporated under reduced pressure to yield an off-white solid precipitate. This solid was dissolved in anhydrous THF (50 ml), after which a solution of sodium bis(2-methoxyethoxy) aluminum hydride (Red-Al, 70 *wt*%) in toluene (14.6 ml, 40 mmol, 2 equiv.) was added dropwise to it under a nitrogen atmosphere. The mixture was refluxed at 60 °C for 4 h and then quenched with water (2.5 ml). The supernatant was obtained by centrifugation, and the solvent was removed under reduced pressure. Flash column chromatography (petrol ether/ethyl acetate 1:1) formed compound 2 (3.61 g, 19.2 mmol, 96%) as a white solid.

Synthesis of diethyl 2-acetamido-2-((6-methoxynaphthalen-2-yl)methyl) malonate (3)

Compound 2 (3.61 g, 19.2 mmol) was dissolved in anhydrous dichloromethane (30 ml) and then cooled in an ice-bath. Sulfoxide chloride (4.4 ml, 60 mmol, 3 equiv.) was added to the solution dropwise, and the mixture was stirred at 0 °C for 2 h. The solvent was removed, and the residue was co-evaporated twice with dichloromethane (20 ml) to yield a faint yellow solid.

A suspension of diethyl acetamidomalonate (6.26 g, 28.8 mmol, 1.5 equiv.) in anhydrous DMF (30 ml) was added to sodium hydride (1.38 g, 57.5 mmol, 3 equiv.) in small portions, and the mixture was stirred for 30 min under ice bath. Menaphthyl chloride dissolved in anhydrous DMF (30 ml) was then added dropwise to this solution. The mixture was stirred overnight at room temperature, diluted with ice-water (200 ml), and extracted with ethyl acetate (200 ml). The organic phase was washed with brine and dried with Na_2SO_4, and the solvent was removed under reduced pressure. The residue was purified by flash column chromatography (petrol ether/ethyl acetate 1:1) to yield compound 3 as a white solid (6.74 g, 17.4 mmol, 90.6%).

Synthesis of 2-amino-3-(6-hydroxynaphthalen-2-yl) propanoic acid hydrobromide (naphthol-Ala·HBr)

Compound 3 (6.74 g, 17.4 mmol) was dissolved in hydrogen bromide (40 ml, 48% *w/v*, aq.) and then refluxed at 110 °C overnight under a nitrogen atmosphere. The red solution was cooled to room temperature and then neutralized with 5 mol/L NaOH. The red precipitate was filtered, and the residue was co-evaporated thrice with menthol (3 × 10 ml) to afford the hydrobromide salt of naphthol-Ala (5.16 g, 16.5 mmol, 94.5%).

MS(ESI): mass calculated for $C_{13}H_{13}N_1O_3$ requires m/z: 231.09, found $[M + 1]^+$ m/z 232.19; $[M + Na]^+$ m/z 254.07.

^{1}H-NMR (500 MHz, D_2O) δ: 7.60 (d, J = 9.0 Hz, 1H), 7.54 (d, J = 8.5 Hz, 1H), 7.50 (s, 1H), 7.15 (dd, J = 8.5, 1.8 Hz, 1H), 7.03 (d, J = 2.4 Hz, 1H), 6.96 (dd, J = 8.9, 2.5 Hz, 1H), 4.20 (dd, J = 7.6, 5.7 Hz, 1H), 3.14 (dd, J = 14.7, 5.7 Hz, 1H), 3.15–3.08 (m, 1H).

^{13}C-NMR (126 MHz, D_2O) δ: 170.35, 152.66, 132.87, 128.97, 127.85, 127.60, 126.80, 126.55, 117.51, 108.35, 53.37, 34.76.

Plasmids and cell lines used

The plasmid pBK-lib-jw1 encodes a library of *Methanococcus jannaschii* tyrosyl tRNA synthetase (TyrRS) mutants randomized at residues Tyr32, Leu65, Phe108, Gln109, Asp158, and Leu162. Any one of Ile63, Ala67, His70, Try114, Ile159, and Val164 is also either mutated to Gly or kept unchanged. Plasmid pREP(2)/YC encodes $MjtRNA^{Tyr}_{CUA}$, the CAT gene with a TAG codon at residue 112, the GFP gene under the control of the T7 promoter, and a Tetr marker. Plasmid pLWJ17B3 encodes $MjtRNA^{Tyr}_{CUA}$ under the control of the *lpp* promoter and *rrnC* terminator, the barnase gene (with three amber codons at residues 2, 44, and 65) under the control of the *ara* promoter, and an Ampr marker. Plasmid pBAD/JYAMB-4TAG encodes the myoglobin gene of mutant sperm whale with an arabinose promoter and *rrnB* terminator, $MjtRNA^{Tyr}_{CUA}$ with an *lpp* promoter and *rrnC* terminator, and a tetracycline resistance marker.

Genetic selection of the mutant synthetase specific for naphthol-Ala

pBK-lib-jw1 consisting of 2 × 10^9 TyrRS-independent clones was constructed using standard PCR methods. *E. coli* strain DH10B harboring the pREP(2)/YC plasmid was used as the host strain for positive selection. Cells were transformed with the pBK-lib5 library, recovered in SOC for 1 h, and washed twice with glycerol minimal media with leucine (GMML) before plating on GMML-agar plates supplemented with 50 μg/mL kanamycin, 60 μg/mL chloramphenicol, 15 μg/mL tetracycline, and 1 mmol/L naphthol-Ala. The plates were incubated at 37 °C for 60 h, the surviving cells were scraped, and the plasmid DNA was extracted and purified by gel electrophoresis. pBK-lib-jw1 DNA was then transformed into electro-competent cells harboring the negative selection plasmid pLWJ17B3, recovered for 1 h in SOC, and then plated on LB-agar plates containing 0.2% arabinose, 50 μg/mL ampicillin, and 50 μg/mL kanamycin. The plates were incubated at 37 °C for 8–12 h, followed by a similar extraction of pBK-lib5 DNA from the surviving clones. The library was then carried through a subsequent round of positive selection, followed by negative selection and a final round of positive selection (with 70 μg/mL chloramphenicol). At this stage, 96 individual clones were selected, suspended in 50 ml of GMML in a 96-well plate, and replica-spotted on two sets of GMML plates. A set of GMML-agar plates was supplemented with 15 μg/mL tetracycline; 50 μg/mL kanamycin; 60, 80, 100 or 120 μg/mL chloramphenicol; and 1 mmol/L naphthol-Ala. Another set of identical plates without naphthol-Ala was produced with 0, 20, 40, or 60 μg/mL chloramphenicol. After 60 h of incubation at 37 °C, one clone was found to survive at 100 μg/mL chloramphenicol in the presence of 0.5 mmol/L naphthol-Ala, and at 20 μg/mL chloramphenicol in the absence naphthol-Ala.

Generation of mutants and site-directed mutagenesis analysis

Plasmid pET22b containing cpsfGFP was used to generate mutant cpsfGFPY66amber. Mutagenesis was confirmed through DNA sequencing analysis.

Expression and purification of cpsfGFP-66-naphthol-Ala

To express mutant cpsfGFP, plasmid pET22b-cpsfGFP-66-TAG was co-transformed with pEOVL-naphthol-Ala synthetase (RS) into BL21(DE3) *E. coli* cells. Cells were amplified in 5 ml of LB media supplemented with 50 μg/mL ampicillin and 30 μg/mL chloramphenicol. A starter culture (1 ml) was used to inoculate 100 ml of liquid LB supplemented with appropriate antibiotics and 0.5 mmol/L naphthol-Ala. The cells were then grown at 37 °C to an OD_{600} of 1.1, and protein production was induced by the addition of 0.2% arabinose and 1 mmol/L IPTG. After 12 h, the cells were harvested by centrifugation and then sonicated. The supernatant was collected and incubated with Ni–NTA agarose beads for 1 h at 4 °C, filtered, and washed with wash buffer containing 50 mmol/L HEPES (pH 7.5), 500 mmol/L NaCl, and 20 mmol/L imidazole. The protein was eluted with the wash buffer containing 50 mmol/L HEPES, pH 7.5, 500 mmol/L NaCl, and 250 mmol/L imidazole. cpsfGFP-66-naphthol-Ala was loaded onto a MonoQ column and eluted with a NaCl gradient. cpsfGFP-66-naphthol-Ala was further purified by a Superdex 75 column (GE Healthcare) in buffer containing 20 mmol/L HEPES–NaOH (pH 7.5) and concentrated to 16 mg/mL. Protein concentration was quantified using the Bradford assay.

Structural determination of cpsfGFP-66-Naphthol-Ala

Crystallization of cpsfGFP-66-naphthol-Ala was achieved through vapor diffusion against 100 mmol/L Tris (pH 6.0) and 20% PEG3K at 4 °C. The crystals appeared within 3 d and were flash-frozen in liquid nitrogen. Diffraction data were collected at a wavelength of 0.979 Å at BL17U of the Shanghai Synchrotron Radiation Facility with a Q315CCD detector. Data processing and reduction were carried out by using the HKL2000 package. The initial structure of cpsfGFP-66-naphthol-Ala was determined by molecular replacement with Molrep from the CCP4 suite by using the atomic coordinates of wild-type sfGFP (PDB code: 2B3P) as the search model. Molecular replacement solutions were modified and refined with alternative cycles of manual refitting. Structural refinement was carried out using PHENIX. During refinement, the Coot tool in the CCP4 program suite was used for the model building, ligand and water location, and real-space refinement of side chains and zones. The final structure of cpsfGFP-66-naphthol-Ala was checked for geometrical correctness with PROCHECK. The collected data and structural refinement statistics are summarized in Table 5. Cartoons and other protein structure representations were generated using PyMOL (http://www.pymol.org), and the atomic coordinates and structure factors were deposited in Protein Data Bank.

UV spectra of cpsfGFP-66-Naphthol-Ala

7 µmol/L solution of purified mutant cpsfGFP-66-naphthol-Ala protein was added to different NaiP/citric acid buffers of various pH, and their UV–Vis spectra were recorded using a quartz cuvette (100 µl, 1 cm path) at room temperature with Agilent 8453 UV–visible spectrophotometer.

Excitation and emission spectra of cpsfGFP-66-naphthol-Ala

The fluorescence spectra of a 2 µmol/L solution of purified mutant cpsfGFP-66-Naphthol-Ala protein were obtained using a Thermo Varioskan Flash instrument equipped with SkanIt 2.4.3 RE software.

Table 5 Summary of data-collection and refinement statistics

Space group	$P2_1$
Unit-cell parameters (Å)	$a = 46.8, b = 49.3, c = 49.2$
Resolution range (Å)	39.1–1.25 (1.29–1.25)
Number of unique reflections	58,574
Data completeness (%)	99.7 (100)
$\langle I \rangle / \langle \sigma(I) \rangle$	17.7 (2.8)
R_{merge} (%)[a]	6.4 (42.5)
R-factor/R_{free} (%)[b]	19.4/21.9
r.m.s.d. bond length (Å)	0.007
r.m.s.d. bond angles (°)	1.276
Number of atoms modeled	2102
Number of water molecules	215
Mean B factor (Å^2)	16.2
Protein main-chain atoms	14.4
Protein side-chain atoms	16.1
Water molecules	25.3
Ramachandran plot statistics	
Residues in most favored region (%)	96.0
Residues in additional allowed region (%)	3.6
Residues in disallowed region (%)	0.4

[a] $R_{merge} = \sum_{hkl} \sum_i |I(hkl) - \langle I(hkl) \rangle| / \sum_{hkl} \sum_i I(hkl)$, where $\langle I(hkl) \rangle$ is the main value of $I(hkl)$

[b] R-factor $= \sum ||F_{obs}| - |F_{calc}|| / \sum |F_{obs}|$, where F_{obs} and F_{calc} are the observed and calculated structure factors, respectively

The free R factor was calculated using 5% of the reflections omitted from the refinement. Numbers in parentheses represent the value of the highest resolution shell

Circular dichroism (CD) experiments

10 μmol/L solution of purified mutant cpsfGFP-66-naphthol-Ala in 10 mmol/L NaiP/citric acid buffer at various pH was transferred into a quartz cuvette (200 μl, 1 cm path), and spectra were obtained at room temperature by using a CD spectrometer (Chirascan Plus). Scanning was done thrice at a scan speed of 120 nm/min and 1-nm intervals. Data smoothing was not performed.

Confocal and fluorescence microscopy

Pre- and post-photoconversion images were obtained from a thin (~10 μm) layer of *E. coli* cells expressing the target protein in MOPS-citrate buffer (pH 5) using 405-nm excitation delivered using a 1.2-NA objective to the region within the circle in bleaching mode (Red channel: 580–645 nm; Red channel: 655–755 nm; scale bar: 50 μm). Photoconversion was performed at 37% laser power and 30 iterations at 111 ms. All observations were obtained by confocal and fluorescence microscopy (FV1000, Olympus).

For photostability determination, samples were obtained by overnight induction of cpsfGFP-66-naphthol-Ala and eqFP650 expressed in the BL21 strain. Bacterial solution (1 ml) was centrifuged for 4 min at 4000 r/min, and the pellet obtained was washed thrice with PBS buffer and suspended in 1 ml of PBS buffer. Thereafter, 100 μl of this solution was placed on a glass that had been preprocessed with polylysine. After 30 min of incubation, the glass was washed with 300 μl of PBS buffer for microscopic observation. The samples were photobleached by 30 W/cm^2 of 561 nm light, and the time was recorded. All observations were obtained using an Olympus 71 microscopy.

Acknowledgements This work was financially supported in part by Grants from the National Natural Science Foundation of China (21750003, 91527302, U1632133, 31628004, 21473237, and 31628004); the National Key Research and Development Program of China (2017YFA0503704 and 2016YFA0501502); and the Key Research Program of Frontier Sciences, CAS (QYZDB-SSW-SMC032). We thank S.S. Zang and X.H. Liu for their help with the NMR spectral analysis; Z. Xie for helping with the protein mass spectral analysis; J.H. Li for helping with the CD experiments; A.Y. Liu for helping with the optical microscopy experiments; and Y. Teng for helping with the confocal and fluorescence microscopy experiments.

Compliance with Ethical Standards

Conflict of interest Li Wang, Xian Chen, Xuzhen Guo, Jiasong Li, Qi Liu, Fuying Kang, Xudong Wang, Cheng Hu, Haiping Liu, Weimin Gong, Wei Zhuang, Xiaohong Liu, and Jiangyun Wang declare that they have no conflicts of interest.

Human and animal rights and informed consent This article does not contain any studies with human or animal subjects performed by the any of the authors.

References

Bogdanov AM, Mishin AS, Yampolsky IV, Belousov VV, Chudakov DM, Subach FV, Verkhusha VV, Lukyanov S, Lukyanov KA (2009) Green fluorescent proteins are light-induced electron donors. Nat Chem Biol 5:459–461

Chang H, Zhang MS, Ji W, Chen JJ, Zhang YD, Liu B, Lu JZ, Zhang JL, Xu PY, Xu T (2012) A unique series of reversibly switchable fluorescent proteins with beneficial properties for various applications. Proc Natl Acad Sci USA 109:4455–4460

Chen S, Tsao ML (2013) Genetic incorporation of a 2-naphthol group into proteins for site-specific azo coupling. Bioconjugate Chem 24:1645–1649

Chica RA, Moore MM, Allen BD, Mayo SL (2010) Generation of longer emission wavelength red fluorescent proteins using computationally designed libraries. Proc Natl Acad Sci USA 107:20257–20262

Chudakov DM, Matz MV, Lukyanov S, Lukyanov KA (2010) Fluorescent proteins and their applications in imaging living cells and tissues. Physiol Rev 90:1103–1163

Elowitz MB, Surette MG, Wolf PE, Stock J, Leibler S (1997) Photoactivation turns green fluorescent protein red. Curr Biol 7:809–812

Enterina JR, Wu LS, Campbell RE (2015) Emerging fluorescent protein technologies. Curr Opin Chem Biol 27:10–17

Gaussian 16, Revision B.01, Frisch MJ, Trucks GW, Schlegel HB, Scuseria GE, Robb MA, Cheeseman JR, Scalmani G, Barone V, Petersson GA, Nakatsuji H, Li X, Caricato M, Marenich AV, Bloino J, Janesko BG, Gomperts R, Mennucci B, Hratchian HP, Ortiz JV, Izmaylov AF, Sonnenberg JL, Williams-Young D, Ding F, Lipparini F, Egidi F, Goings J, Peng B, Petrone A, Henderson T, Ranasinghe D, Zakrzewski VG, Gao J, Rega N, Zheng G, Liang W, Hada M, Ehara M, Toyota K, Fukuda R, Hasegawa J, Ishida M, Nakajima T, Honda Y, Kitao O, Nakai H, Vreven T, Throssell K, Montgomery JA Jr, Peralta JE, Ogliaro F, Bearpark MJ, Heyd JJ, Brothers EN, Kudin KN, Staroverov VN, Keith TA, Kobayashi R, Normand J, Raghavachari K, Rendell AP, Burant JC, Iyengar SS, Tomasi J, Cossi M, Millam JM, Klene M, Adamo C, Cammi R, Ochterski JW, Martin RL, Morokuma K, Farkas O, Foresman JB, Fox DJ (2016) Gaussian, Inc., Wallingford CT

Grimm JB, English BP, Chen JJ, Slaughter JP, Zhang ZJ, Revyakin A, Patel R, Macklin JJ, Normanno D, Singer RH, Lionnet T, Lavis LD (2015) A general method to improve fluorophores for live-cell and single-molecule microscopy. Nat Methods 12:244

Laurent AD, Jacquemin D (2013) TD-DFT benchmarks: a review. Int J Quantum Chem 113:2019–2039

Liu XH, Li JS, Hu C, Zhou Q, Zhang W, Hu MR, Zhou JZ, Wang JY (2013) Significant expansion of the fluorescent protein chromophore through the genetic incorporation of a metal-chelating unnatural amino acid. Angew Chem Int Edit 52:4805–4809

Mou Y, Huang PS, Hsu FC, Huang SJ, Mayo SL (2015) Computational design and experimental verification of a symmetric protein homodimer. Proc Natl Acad Sci USA 112:10714–10719

Newman RH, Fosbrink MD, Zhang J (2011) Genetically encodable fluorescent biosensors for tracking signaling dynamics in living cells. Chem Rev 111:3614–3666

Nienhaus K, Nienhaus GU (2014) Fluorescent proteins for live-cell imaging with super-resolution. Chem Soc Rev 43:1088–1106

Niu W, Guo JT (2013) Expanding the chemistry of fluorescent

protein biosensors through genetic incorporation of unnatural amino acids. Mol BioSyst 9:2961–2970

Saha R, Verma PK, Rakshit S, Saha S, Mayor S, Pal SK (2013) Light driven ultrafast electron transfer in oxidative redding of Green Fluorescent Proteins. Sci Rep 3:1580

Sengupta P, van Engelenburg SB, Lippincott-Schwartz J (2014) Superresolution imaging of biological systems using photoactivated localization microscopy. Chem Rev 114:3189–3202

Shaner NC, Patterson GH, Davidson MW (2011) Advances in fluorescent protein technology (vol 120, pg 4247, 2007). J Cell Sci 124:2321

Shcherbo D, Shemiakina II, Ryabova AV, Luker KE, Schmidt BT, Souslova EA, Gorodnicheva TV, Strukova L, Shidlovskiy KM, Britanova OV, Zaraisky AG, Lukyanov KA, Loschenov VB, Luker GD, Chudakov DM (2010) Near-infrared fluorescent proteins. Nat Methods 7:827-U1520

Shu XK, Royant A, Lin MZ, Aguilera TA, Lev-Ram V, Steinbach PA, Tsien RY (2009) Mammalian expression of infrared fluorescent proteins engineered from a bacterial phytochrome. Science 324:804–807

Subach FV, Verkhusha VV (2012) Chromophore transformations in red fluorescent proteins. Chem Rev 112:4308–4327

Subach FV, Piatkevich KD, Verkhusha VV (2011) Directed molecular evolution to design advanced red fluorescent proteins. Nat Methods 8:1019–1026

Docking-based inverse virtual screening: methods, applications, and challenges

Xianjin Xu[1,2,3,4], Marshal Huang[1,3], Xiaoqin Zou[1,2,3,4]

[1] Dalton Cardiovascular Research Center, University of Missouri, Columbia, MO 65211, USA
[2] Department of Physics and Astronomy, University of Missouri, Columbia, MO 65211, USA
[3] Informatics Institute, University of Missouri, Columbia, MO 65211, USA
[4] Department of Biochemistry, University of Missouri, Columbia, MO 65211, USA

Abstract Identifying potential protein targets for a small-compound ligand query is crucial to the process of drug development. However, there are tens of thousands of proteins in human alone, and it is almost impossible to scan all the existing proteins for a query ligand using current experimental methods. Recently, a computational technology called docking-based inverse virtual screening (IVS) has attracted much attention. In docking-based IVS, a panel of proteins is screened by a molecular docking program to identify potential targets for a query ligand. Ever since the first paper describing a docking-based IVS program was published about a decade ago, the approach has been gradually improved and utilized for a variety of purposes in the field of drug discovery. In this article, the methods employed in docking-based IVS are reviewed in detail, including target databases, docking engines, and scoring function methodologies. Several web servers developed for non-expert users are also reviewed. Then, a number of applications are presented according to different research purposes, such as target identification, side effects/toxicity, drug repositioning, drug–target network development, and receptor design. The review concludes by discussing the challenges that docking-based IVS needs to overcome to become a robust tool for pharmaceutical engineering.

Keywords Inverse virtual screening, Target fishing, Polypharmacology, Side effects, Drug repositioning, Molecular docking

INTRODUCTION

Identifying protein targets for a query ligand is a crucial aspect of drug discovery. Historically, natural products derived from plants, animals, micro-organisms, *etc.*, were used as medicines to cure many diseases. The accumulated experience and knowledge of their usages have become an abundant resource for modern drug discovery (Ji *et al.* 2009). Although purified compounds from these natural products present good therapeutic activities, molecular mechanisms of action including the identification of binding targets are often shrouded in mystery. The drug design process in modern times is highly dependent on Ehrlich's assumption (Kaufmann 2008), in which drugs work as "magic bullets" modulating one target of particular relevance to a disease. Great success has been achieved with this simple assumption, while disadvantages are also emerging in recent years. The most visible disadvantage is the high attrition rate (about 90%) of potential compounds at the late stage of clinical trials due to certain efficacy and clinical safety problems (Nwaka and Hudson 2006). A number of drugs have been withdrawn from the market because of serious side effects or life-threatening toxicities. Recent studies also suggest that each existing

✉ Correspondence: zoux@missouri.edu (X. Zou)

drug binds to, on average, about six target proteins instead of one (Azzaoui et al. 2007; Mestres et al. 2008). If all the targets of an interested ligand can be identified at the early stage of new drug design, the side effects and toxicities that appear in the later stages of clinical trials can be effectively avoided. Thus, a prescreening process can significantly increase the success rate and reduce the development cost for the overall drug pipeline. However, the lack of effective experimental tools in identifying all the potential targets for a small molecule on a proteome-wide scale remains a daunting challenge to overcome.

Recently, an inverse virtual screening (IVS) technology based on molecular docking methods has been developed and widely used for the process of target identification (Chen and Zhi 2001). A molecular docking method is defined as the prediction of both the binding mode and binding affinity of a query ligand (such as a small-molecule drug) against a receptor (such as a target protein) (Brooijmans and Kuntz 2003; Sousa et al. 2006; Grinter and Zou 2014a, b). In the IVS method, a molecular docking process is employed to screen a protein database for a query ligand, and then an enriched subset containing possible targets of the ligand is provided. Figure 1 shows a flowchart of the docking-based IVS procedure.

To run a docking-based IVS study, at least two components are required, a protein database and a molecular docking program. The target database is a collection of structures of proteins or active sites. With the rapidly increasing number of structures deposited in the Protein Data Bank (PDB) (Berman et al. 2000), a desirable target database can be constructed for docking-based IVS. The target database can also be extended through homology modeling techniques. Then, a potentially interesting small molecule is docked to each element of the target database by a docking program. Generally, a docking program consists of two main components—the sampling algorithm and the scoring function. The sampling component generates sufficient putative binding modes. The scoring function further ranks these modes based on binding energy evaluations. The ability of the existing scoring functions to accurately predict binding energies remains limited (Brooijmans and Kuntz 2003; Huang et al. 2010). Fortunately, the purpose of IVS studies (and of virtual screening of potent ligands against a query target) is in pursuit of an enriched subset of potential candidates (e.g., top 1% of the ranked proteins in the IVS case or top 1% of the ranked ligands in the virtual screening case), which is a relatively less challenging task than binding energy prediction for a scoring function.

In addition to docking-based IVS, there are several other computational methods that can be used for target identification, including ligand-based methods, binding site comparisons, protein–ligand interaction fingerprints, and so on (Rognan 2010; Koutsoukas et al. 2011; Xie et al. 2011; Ma et al. 2013). Ligand-based methods are based on the molecular similarity principle, which states that molecules with similar structures tend to have similar biological activities (Willett et al. 1998; Bender and Glen 2004). These methods heavily rely on the pre-existing knowledge about the molecules in the database, and require a database of small molecules with known binding targets. Although ligand-based methods are widely used for target identification and have achieved a great amount of success, they become utterly useless for the remaining "unknown space" (i.e., dissimilar ligands). Similarly, for the methods of binding site comparison and protein–ligand interaction fingerprinting, at least one protein–ligand complex structure of the query small molecule is required (Rognan 2010). All the aforementioned approaches are classified as "knowledge-based" IVS methods. By contrast, docking-based IVS is the only method that does not rely on such preliminary information, rendering it a more attractive option in the field of target identification.

Ever since the first docking-based IVS program was developed by Chen et al. (Chen and Zhi 2001), the method has been improved and utilized widely for various purposes in the field of drug discovery. Here, we review the method of docking-based IVS, including the target database, docking engine, and scoring function

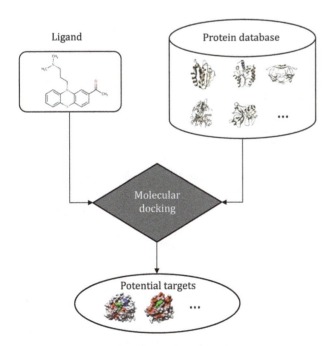

Fig. 1 A flowchart of the docking-based inverse virtual screening

components of this method. We also review the web servers that integrate the complex process of IVS for non-expert users. Then, we present published studies in which docking-based IVS played an important role. These application studies are classified into target identification, side effect/toxicity assessments, drug repositioning, multi-target therapy/drug–target network, and receptor design. Finally, we discuss about current challenges that docking-based IVS needs to overcome in order to become a robust tool for far-reaching applications.

DOCKING-BASED IVS

In docking-based IVS, a given small molecule is docked to the binding site of each protein in a target database through a docking engine. Then, target proteins are ranked according to the binding scores estimated by a scoring function. This complex process has been integrated and presented as online web servers for non-expert utilization. These components are explained in detail as follows.

Target databases

A database consisting of three-dimensional protein structures is required for the implementation of docking-based IVS. Owing to the development of technologies in structural biology, such as X-ray crystallography and NMR spectroscopy, an increasing number of protein crystal structures have been resolved and deposited in a publicly accessible database, the PDB (Berman *et al.* 2000). Up to the present (16th March 2017), the number of protein entries in the PDB has reached up to 118,663, which provides an abundant resource for constructing a sub-database for IVS.

For example, screening-PDB (sc-PDB) (Kellenberger *et al.* 2006) is a sub-database extracted from the PDB for the purpose of virtual screening. sc-PDB collects all the high-resolution crystal structures of protein–ligand complexes in which ligands are nucleotides (<4-mer), peptides (<9-mer), cofactors, and organic compounds. In the latest version v.2013, sc-PDB contains 9283 entries corresponding to 3678 different proteins and 5608 different ligands. The known protein–ligand complex structures in the database embed the information about the binding sites (*i.e.*, the pocket where the ligand binds), which would significantly reduce the sampling space for docking. The authors' indiscriminate collections enrich the sc-PDB database, but also complicate the subsequent analysis of the screening results. To address this issue, several databases that focus on specific topics have been constructed, and are introduced as follows.

Therapeutic target database (TTD) (Chen *et al.* 2002) focuses on known and potential therapeutic targets, which are proteins and nucleic acids collected from literature. Important information, such as targeted diseases, pathway information, and corresponding drugs/ligands, is provided in the database. After the latest update in 2015 (Yang *et al.* 2016), TTD contains 2589 targets, including 397 successful, 723 clinical trial, and 1469 research targets. However, the TTD database does not provide 3D structures of the targets, which need to be downloaded from the PDB database by users.

Potential drug–target database (PDTD) (Gao *et al.* 2008) is another database focusing on therapeutic targets. Different to TTD, PDTD contains only protein targets. Impressively, cleaned 3D structures for both protein and active sites are provided, minimizing the complexity of docking preparation for users. After the latest update in 2008, PDTD contains 1207 entries, covering 841 known and potential drug targets. Targets in the PDTD database were further categorized into several subsets according to two criteria: therapeutic areas and biochemical criteria. These subsets could be very effective for studies on a special topic. The database was implemented in an online web server TarFisDock (Li *et al.* 2006), which will be introduced later in this review.

Drug adverse reaction database (DART) (Ji *et al.* 2003) focuses on known and potential targets corresponding to the adverse effects of drugs. Information such as physiological function, binding affinity of known ligands, and corresponding adverse effects is provided. Currently, the DART database contains entries for 147 ADR targets and 89 potential targets. The structures of the targets and the active sites in the database need to be prepared by users.

Recently, our group presented a small molecule-transcription factor (SM-TF) database containing all the targetable TFs with known 3D structures (Xu *et al.* 2016). SM-TF contains 934 entries, covering 176 TFs from a variety of species. Besides the protein structures, the co-bound ligands are also provided in the SM-TF database. Therefore, the database is suitable for both docking-based IVS and ligand-based IVS.

In addition to the aforementioned freely accessible databases, researchers often construct highly specialized datasets. For example, a dataset containing enzymes was constructed by Macchiarulo *et al.* to study the selectivity and competition of metabolites between enzymes (Macchiarulo *et al.* 2004). Zahler *et al.* collected a dataset of protein kinase structures for identifying the targets of kinase inhibitors (Zahler *et al.* 2007).

Lauro et al. (2011) collected a dataset of proteins involved in cancer and tumor development for antitumor target identification of natural bioactive compounds. These individualized datasets can be either directly derived from a protein–ligand complex structure database like sc-PDB, or constructed by collecting information from publically accessible drug–target databases such as SuperTarget (Günther et al. 2008), BindingDB (Liu et al. 2007), and DrugBank (Wishart et al. 2006), as listed in Table 1. It should be noted that information in the later databases is redundant. The 3D structures of proteins need to be downloaded from the PDB database by users, and further preparations are necessary to fit the input file format of docking methods.

Docking engines

Prediction of protein–ligand complex structures plays an essential role in docking-based IVS. The credibility of predicted binding patterns of a ligand against each protein target is crucial to the final success. Fortunately, plenty of programs have been developed for the purpose of structure prediction of protein–ligand complexes (Brooijmans and Kuntz 2003; Sousa et al. 2006). Here, we focus on the issues closely related to IVS. Interested readers are referred to other recent reviews on molecular docking methods for more information (Brooijmans and Kuntz 2003; Sousa et al. 2006; Huang and Zou 2010; Grinter and Zou 2014a, b).

Briefly, a molecular docking program is designed to predict a complex structure based on the known 3D structures of its components. In other words, a docking method is a problem of searching for the ligand location on a given protein target (referred to as binding site prediction) and then for the ligand conformations and orientations in the binding site. Although methods of global blind docking are provided by most docking programs, they suffer from time-consuming execution and a low success rate compared to dockings into a known binding site. Considering the large number of proteins in the target database, protein structures with known active sites are preferred in the preparation of a target database.

In the early stages of the development of the docking methods, both the ligand and the receptor were treated rigidly. A shape matching method was employed to place a ligand in the binding site of a receptor. Only six degrees of freedom (three translational and three rotational) of a ligand conformation are considered, which is computationally efficient. However, binding of a ligand to a receptor is a mutual fitting progress, with conformational changes in both components. Thus, conformational search is necessary for both the ligand and the receptor during docking.

According to the searching method, ligand flexibility algorithms can be divided into three types: systematic, stochastic, and deterministic search. Systematic search generates all possible ligand binding conformations by exploring the whole conformational space. Despite the completeness of sampling, the number of evaluations increases rapidly as the number of degrees of freedom are increased (i.e., the number of rotatable bonds in a ligand). Examples of systematic search include exhaustive search implemented in Glide (Friesner et al. 2004), and a fragmentation method named incremental construction algorithm implemented in LUDI (Bohm 1992) and DOCK (DesJarlais et al. 1986). Stochastic algorithms sample the ligand conformational space by making random changes, which will be accepted or rejected according to a probabilistic criterion. This type of methods significantly reduces computational efforts for large systems; however, the uncertainty of convergence is a major concern. Examples of stochastic algorithms are Monte Carlo (MC) methods implemented in MCDOCK (Liu and Wang 1999), and evolutionary algorithms implemented in GOLD (Jones et al. 1997) and AutoDock (Morris et al. 1998). For deterministic search, the final state of the system depends on the initial state. Examples are energy minimization methods and molecular dynamics (MD) simulations. Systems are thus guided to states with lower energies. However, it is difficult to cross energy barriers, and systems are often trapped in local minima with these methods.

The flexibility of the receptor remains a big challenge for docking, because of the huge number of degrees of freedom in the system. Some methods for ligand flexibility are also applicable for receptor flexibility, such as the aforementioned evolutionary algorithms, MC, and MD methods. In addition, several approaches accounted for partial flexibility within the receptor, such as soft docking and conformer libraries. Soft docking allows an overlap between the ligand and the receptor by softening the interatomic van der Waals (vdW) interactions (Jiang and Kim 1991). The methods based on conformer libraries can be further divided into two different types. The first type describes the side-chain conformations by a rotamer library and keeps the backbones fixed (Leach 1994). The second type is referred to docking with multiple receptor structures, using pre-generated receptor conformers (Knegtel et al. 1997). Other methods, such as induced fit docking (IFD), change both protein and ligand conformations to fit each other during the docking process (Sherman et al. 2006). Theoretically, these methods can account for receptor flexibility in terms of either the side chains or the

Table 1 Publicly available databases containing the information about targetable proteins

Database	Description	URL
PDB	A pool of 3D structures of macromolecules, including proteins, nucleic acids, and complex assemblies. The total number of structures deposited in the database is more than 12,000	http://www.rcsb.org
sc-PDB	A subset of PDB with the collection of protein–ligand complexes. In the latest version v.2013, the database contains 9283 entries corresponding to 3678 different proteins and 5608 different ligands	http://bioinfo-pharma.u-strasbg.fr/scPDB
TTD	Therapeutic target database (TTD) contains 2360 targets, including 2589 targets, including 397 successful, 723 clinical trial, and 1469 research targets	http://bidd.nus.edu.sg/group/ttd
PDTD	Potential drug–target database (PDTD) contains 1207 entries covering 841 known and potential drug targets, which can be further categorized into subsets according to two criteria: therapeutic areas and biochemical criteria. Structures for both protein and active site are available	http://www.dddc.ac.cn/pdtd
DART	Drug adverse reaction database (DART) contains 147 ADR targets and 89 potential targets	http://bidd.nus.edu.sg/group/drt
SM-TF	A database of 3D structures of small molecule-transcription factor complexes. The database contains 934 entries, covering 176 TFs from a variety of species	http://zoulab.dalton.missouri.edu/SM-TF
SuperTarget	A database contains the information about drug–target relations. The database contains >6000 target proteins, 196,000 compounds, 282 drug–target-related pathways, and >6000 drug–target-related ontologies	http://bioinformatics.charite.de/supertarget
BindingDB	A database of measured binding affinities for drug–targets with small, drug-like molecules. Until now, the database contains more than 1,000,000 binding data, for about 7997 protein targets and 453,657 small molecules	http://www.bindingdb.org/bind
DrugBank	In the latest version (5.0), the database contains 8261 drug entries including 2021 FDA-approved small-molecule drugs, 233 FDA-approved biotech (protein/peptide) drugs, 94 nutraceuticals, and over 6000 experimental drugs. 4338 non-redundant protein sequences are linked to these drug entries	http://www.drugbank.ca

Some of them can be directly used for docking-based IVS studies. Others are abundant resources for constructing an individualized target dataset

backbones, or both. However, the rapidly growing degrees of freedom make even a single docking event very time-consuming, and make the hopes of implementing IVS a mirage.

According to a recent review that exhaustively presented the programs available for protein–ligand docking, the number of available docking programs was more than 50 and kept increasing (Sousa et al. 2013). It is difficult to say which docking program is better than the others, because the performance of most docking programs is highly dependent on the system of study, e.g., the characteristics of both the receptor and the ligand (Sousa et al. 2013). In the published literature related to docking-based IVS, the choice of a docking engine is quite arbitrary.

Scoring functions

The scoring function is another important component of protein–ligand docking protocols. It is for evaluation and ranking of the binding conformations generated by the searching algorithms described in the last section. In fact, scoring functions are usually implemented in docking programs. Here, we artificially separate scoring functions from docking engines, not only because scoring functions play an essential role in every docking protocol, but also because they are employed to pick potential targets out of a database in IVS.

Scoring functions for molecular docking can be grouped into three major classes according to how they are derived: force field-based, empirical, and knowledge-based. Parameters in force field-based scoring functions are derived from molecular mechanical force fields used in MD simulations, including contributions from vdW interactions, electrostatic interactions, and bond stretching/bending/torsional potentials. The desolvation effects can be considered by using implicit solvent models like the Poisson–Boltzmann/surface area (PB/SA) model (Baker et al. 2001; Grant et al. 2001; Rocchia et al. 2002) and the generalized-Born/surface area (GB/SA) model (Still et al. 1990; Hawkins et al. 1995; Qiu et al. 1997). However, the solvent models would significantly slow down the computational speed, which must be considered in screening studies. In addition, the absence of entropic terms is also a weakness of this type of scoring functions. For example, force-based scoring functions are used in docking programs such as DOCK (Meng et al. 1992) and GOLD (Jones et al.

1997). The second kind of scoring functions are empirical scoring functions, which are a sum of different energy terms such as vdW, electrostatics, hydrogen bond, desolvation, entropy, hydrophobicity, and so on. The weight of each energy term is generated based on a training set of experimental affinity data. The empirical scoring functions are easy to calculate and take much less computational time than force-filed-based scoring functions. However, the accuracy of an empirical scoring function heavily relies on the training set of experimental affinity data. Examples can be found in docking programs such as FlexX (Rarey et al. 1996), Glide (Friesner et al. 2004), ICM (Abagyan et al. 1994), and LUDI (Bohm 1994, 1998). The third kind of scoring functions are knowledge-based, which are also known as statistical potential-based scoring functions. They are developed by statistical analysis of the atom pair occurrence frequencies in a training set of experimentally determined protein–ligand complex structures. Briefly summarized, the frequency of structural features (such as atom pairs) that appear in a training dataset is used to derive the scoring functions. The relationship between the frequency of the structural features and the interaction energies assigned to those features relies on the inverse-Boltzmann equation (Thomas and Dill 1996). Compared to the previous two types of scoring functions, knowledge-based scoring functions hold a good balance between accuracy and speed. However, a weakness of knowledge-based scoring functions is that it is still training set-dependent. Examples of knowledge-based scoring functions are potential of mean force (PMF) (Muegge and Martin 1999; Muegge 2006) and ITScore (Huang and Zou 2006a, b; Grinter et al. 2013; Grinter and Zou 2014a, b; Yan et al. 2016). The interested reader is recommended to read recent reviews on scoring functions for protein–ligand docking (Huang et al. 2010; Grinter and Zou 2014a, b).

Generally, the best (i.e., the lowest) docking score from each protein–ligand docking is used for ranking the proteins in the database. Proteins with low docking scores are potential targets for the ligand. Then, proteins among the top 1% (or 5%) of the ranking list can be used for further analysis. However, this arbitrary cutoff results in enormous false positive targets, significantly increasing the degree of difficulty. Meanwhile, some real targets beyond the cutoff will be ignored. Although false positives and false negatives remain an open question in IVS, several efforts have been made to reduce false positive and false negative targets in the final predicted list.

In a pioneer work of docking-based IVS by Chen et al. (Chen and Zhi 2001), an energy threshold was introduced to filter the proteins in the ranking list. The method was based on an analysis of the known protein–ligand complexes in the PDB, which showed that the computed protein–ligand interaction energy was generally less than $\Delta E_{\text{Threshold}} = -\alpha N$ kcal/mol. Here, N is the number of ligand atoms, and α is a constant (~ 1.0) which can be determined by fitting the equation for a large set of PDB structures. Proteins with calculated binding energies less than $\Delta E_{\text{Threshold}}$ were predicted as potential targets. Furthermore, to consider competitive binding against natural ligands in vivo, another energy threshold, $\Delta E_{\text{Competitor}}$, was introduced. $\Delta E_{\text{Competitor}}$ is the binding energy of a competitive natural ligand interacting with each protein for a query ligand. The calculation of $\Delta E_{\text{Competitor}}$ was based on the experimental complex structure of the protein and the natural ligand. The calculated binding energy of the query ligand was required to be lower than $\beta \Delta E_{\text{Competitor}}$ for each protein, where $\beta \leq 1$. A value of 0.8 for β was recommended by the authors for both weak and strong binders.

In addition to the use of a threshold for binding scores obtained from the known protein–ligand complexes, Li et al. (2011) introduced consensus scoring to an IVS study. Consensus scoring is a combination of multiple scoring functions. Since every scoring function has its advantages and limitations, consensus scoring provides a way to combine the advantages from different scoring functions. In the work by Li et al. two different scoring functions, an empirical scoring function (ICM) and a knowledge-based scoring function (PMF), were employed for consensus scoring, leading to a clear enhancement in hit-rates.

In the web server SePreSA developed by Yang et al. (2009), a 2-directional Z-transformation (2DIZ) algorithm was used to process a docking-score matrix. Briefly, 79 proteins with co-crystallized ligands in the target database were selected to dock with 86 ligands, generating a docking-score matrix of 79×86 elements. Then, the Z-score was calculated by $Z_{ij} = (X_{ij} - \overline{X_j})/\text{SD}_{X_j}$, where X_{ij} is the docking score of ligand j to protein i, and $\overline{X_j}$ is the average docking score of ligand j against 79 proteins. SD_{X_j} is the standard deviation of docking scores for ligand j with those proteins. The Z-score matrix could be further normalized to a Z'-score matrix, in which the vector for each protein is normalized to a mean of zero and a standard deviation of one. According to results presented in the work, the 2DIZ algorithm significantly improved the prediction accuracy, compared to simply using docking score functions.

Another approach of the normalization of binding energies introduced by Lauro et al. (2011) was studying docking of multiple ligands against multiple proteins.

The normalization was based on the equation $V = V_0/[(M_L + M_R)/2]$, where V_0 is the binding energy calculated by the scoring function for each protein–ligand complex, M_L is the average binding energy of each ligand with different proteins, and M_R is the average binding energy of each protein with different ligands. Then, V was a normalized value associated with each ligand. The approach effectively avoided the selection of false positive results.

In a recent work by Santiago et al. (2012), a selected ligand dataset, the National Cancer Institute (NCI) Diversity Set I containing 1990 drug-like molecules, was used to calibrate binding scores of a query ligand against the proteins in a database. Specifically, the molecules in the NCI Diversity Set I were docked to each protein in the protein database. Then, the top-200, top-20, and Boltzmann-weighted averages of the binding scores were calculated, which served as the references for each protein. If the calculated binding score of the query ligand against a protein was lower than the reference score, the protein was considered as a hit. According to the work, the reference using the top-20 average performed better than the other two averages.

Web servers

To run an IVS, in addition to the time-consuming and labor-intensive process for the construction of a target database, programming skills and experiences are required to handle hundreds of dockings and to conduct post analysis, which could be tough for researchers focusing on experimental methods. Therefore, several web servers were developed for public use. The only thing that a user would need to do is to provide a small molecule of interest. Then the server automatically runs the IVS and outputs a list of potential targets. Available web servers of docking-based IVS are reported in Table 2.

Target fishing dock (TarFisDock) (Li et al. 2006) is the earliest freely accessible web server using the docking-based IVS technique. In this web server, PDTD is used as the target database, which contains 841 known and potential drug targets. DOCK4.0 (Ewing et al. 2001) is employed as the docking engine, and a force field-based scoring function implemented in DOCK is used for binding energy calculation. During docking, ligand flexibility is taken into account, whereas the protein under consideration is treated as rigid. Top 2%, 5%, or 10% of the ranking list can be output for users. Two multi-target ligands, vitamin E (14 known targets) and 4H-tamoxifen (ten known targets), were tested in the study. Top 2% of the ranking list covered 30% of known targets for the two cases. Moreover, 50% of the known targets of vitamin E and 4H-tamoxifen were covered by 10% and 5% of the ranking list, respectively. The TarFisDock server provides a convenient and rapid way to identify potential targets for a given small molecule. Because many of the proteins in PDTD are involved in different therapeutic areas, TarFisDock is a desirable tool for drug repositioning.

SePreSA (Yang et al. 2009) is the first docking-based web server focusing on targets related to severe adverse drug reactions (SADRs). The database contains 91 SADR proteins consisting of major phase I and II drug-metabolite enzymes, several human MHC I proteins, and pharmacodynamic proteins. DOCK4.0 is employed as the docking engine. Besides the scoring function implemented in DOCK, the 2DIZ algorithm is applied to generate a Z-score matrix or Z'-score matrix, which calculates the relative ligand–protein interaction strength. In a test of prediction for true and unidentified binding compounds, the value of the area under the curve (AUC) increases from 0.62 (using only the docking-score matrix) to 0.82 (using the 2DIZ algorithm). Therefore, SePreSA is a desirable tool to predict possible side effects of an interesting molecule in the early stage of drug design.

Drug repositioning potential and ADR via chemical–protein interaction (DRAR-CPI) (Luo et al. 2011) is another web server provided by the same group who developed SePreSA. The server was designed for drug repositioning by taking ADR into account. The target database contains 353 targetable human proteins with 385 binding sites. Also collected were the information of 254 forms of 166 small molecules with known ADR. Similar to SePreSA, DOCK6.0 (Lang et al. 2009) is employed as the docking engine of DRAR–CPI, and the 2DIZ algorithm is applied to generate a Z-score matrix or Z'-score matrix based on docking scores. Furthermore, the server uses an approach to evaluate the drug–drug associations based on gene-expression profiles, searching for similar or opposite drugs from the database for a query ligand. Because the drug–drug association method is beyond this review, the interested reader is recommended to read the original paper (Luo et al. 2011).

Recently, Wang et al. (2012a) released another docking-based IVS web server named idTarget. The docking engine is maximum-entropy based docking (MEDock) (Chang et al. 2005), which was also published as a web server by the same group. AutoDock4[RAP] (Wang et al. 2011), an improved version of the scoring function AutoDock4 (Huey et al. 2007), is used for the evaluation of potential targets. The Z-score of a ligand against a protein pocket is calculated based on an affinity profile of the binding pocket (Wang et al.

Table 2 Available web servers of the docking-based IVS

Web server	Description	URL
TarFisDock	Using DOCK4.0 as the docking engine and PDTD as the target database. Scores calculated by a force-based scoring function implemented in DOCK4.0 are used for the ranking of targets. Top 2%, 5%, or 10% of the ranking list can be output	http://www.dddc.ac.cn/tarfisdock
SePreSA	Focusing on targets related to SADRs. DOCK4.0 is employed as the docking engine and the database contains 91 SADR proteins. In addition to the scoring function implemented in DOCK, Z-scores are also calculated for the selection of potential targets	http://sepresa.bio-x.cn
DRAR-CPI	Provided by the same groups of SePreSA. The server was designed for drug repositioning by taking ADR into account. DOCK6.0 is employed as the docking engine and the target database contains 353 targetable human proteins. Similar strategy of scoring as in SePreSA is used for the selection of potential targets	http://cpi.bio-x.cn/drar
idTarget	Using MEDock as docking engine and AutoDock4^{RAP} as scoring function. Z-scores calculated based on affinity profiles of binding pockets are used for the selection of potential targets. A "contraction-and-expansion" strategy is used to extend the searching space	http://idtarget.rcas.sinica.edu.tw
DockoMatic	DockoMatic is a local program with GUI. AutoDock and AutoDock Vina can be selected as docking engine. BLAST and MODELER programs are implemented, allowing the user to easily extend the target database based on homology modeling	https://sourceforge.net/projects/dockomatic

2012a). Then, the ranking of the potential targets for a query ligand is based on their Z values. To screen a large protein structure database, such as the whole PDB database, the authors introduced a "contraction-and-expansion" strategy. In the contraction stage, the target database contains 2091 targets, which were constructed based on sc-PDB. Briefly, 3046 mean points of sc-PDB were clustered with a cutoff of 40% protein sequence identity. In sc-PDB, a mean point is a representative of a cluster containing entries of a protein bound with different ligands. The query ligand is firstly docked to the contracted database, and half of the targets with lower docking energies will be used for the next expansion stage. In the expansion stage, proteins that are homologous or contain similar binding pockets collected from both sc-PDB and PDB are also selected for screening.

In addition to the web servers described above, Bullock et al. provided a free and open source program DockoMatic2.0 (Bullock et al. 2013), with which the user is able to perform docking-based IVS through a graphical user interface (GUI). AutoDock (Morris et al. 1998) or AutoDock Vina (Trott and Olson 2010) can be selected as the docking engine, and the target database is provided by the user. Although the program DockoMatic2.0 is less convenient to use than web servers which only require a user to upload a query ligand, DockoMatic2.0 can be applied to a user-customized target database which is usually not allowed by web servers. It is worthy to note that the basic local alignment search tool (BLAST) (Altschul et al. 1997) and MODELER program (Sali and Blundell 1993) are also implemented in DockoMatic2.0. Thus, a user can extend the target database based on homology modeling.

APPLICATIONS

Target identification

Natural products have become an abundant resource for new drug discovery, due to the accumulation of ancient medical knowledge for thousands of years (Ji et al. 2009). Identification of the targets for these natural products can not only demystify traditional medicines, but also provide meaningful targets for modern drug design. There are a number of successful stories that utilize docking-based IVS to assist in identifying targets for natural ligands. Do et al. used an in-house developed strategy named Selnergy (Do and Bernard 2004), which is based on using the FlexX docking program (Rarey et al. 1996) to identify targets for two natural products, ε-viniferin (Do et al. 2005) and meranzin (Do et al. 2007). From a manually collected database containing 400 targets, cyclic nucleotide phosphodiesterase 4 (PDE4) was identified as a target of ε-viniferin, and three targets, COX1, COX2, and PPARγ, were identified as the targets of meranzin. Lauro et al. applied the IVS method to a set of ten phenolic natural compounds (Lauro et al. 2012). The target database consists of 163 proteins that are involved in the cancer process. The AutoDock Vina program was employed as the docking engine and the binding energies were normalized to rank the targets. Protein kinases PDK1 and PKC were confirmed as the targets of xanthohumol and isoxanthohumol through *in vitro* biological tests. Recently, the method became popular in the studies of traditional Chinese medicine (TCM) (Yue et al. 2008; Feng et al. 2011; Chen and Ren 2014). In the study by Chen and Ren (2014), the idTarget server (Wang et al. 2012a) along with a ligand-based IVS server PharmMapper (Liu et al. 2010b) was employed to identify the potential anticancer targets of Danshensu, an active compound from a widely used TCM Danshen (*Salvia miltiorrhiza*). The screening proposed GTPase HRas as a potential target of Danshensu for further study.

Toledo-Sherman et al. (Slon-Usakiewicz et al. 2004; Toledo-Sherman et al. 2004) developed a chemical proteomics approach, combining (experimental) ultrasensitive mass spectrometry with (computational) docking-based IVS. This proteomics approach was applied to the exploration of the action mechanism of methotrexate (MTX), an important drug used in cancer, immunosuppression, rheumatoid arthritis, and other highly proliferative diseases. Besides the three main known targets dihydrofolate reductase, thymidylate synthetase, and glycinamide ribonucleotide transformylase, at least eight other proteins were identified as the potential targets of MTX. By using a frontal affinity chromatography with mass spectrometry detection, the authors further confirmed one of these predicted targets, hypoxanthine–guanine amidophosphoribosyltransferase (HGPRT), as a real binder of MTX with a K_d of 4.2 μmol/L.

In another early application, Muller et al. applied IVS to searching for protein targets for a novel chemotype that uses five representative molecules from a combinatorial library that share a 1,3,5-triazepan-2,6-dione scaffold (Muller et al. 2006). A collection of 2148 binding sites (Release 1.0 of the sc-PDB (Kellenberger et al. 2006)) extracted from the PDB database was screened by the GOLD 2.1 docking program (Jones et al. 1997). Five proteins were selected from the top 2% scoring targets by some customized criteria for further experimental evaluation. Two secreted phospholipase A2 isoforms were successfully identified as the real targets of 1,3,5-triazepan-2,6-diones.

Moreover, high throughput screening (HTS) can quickly screen for potential drug candidates; however, the action mechanisms of the resulting candidates are elusive and further improvement of the potency is therefore difficult. IVS can be used to identify the potential targets of these compounds. An example is PRIMA-1 (p53 reactivation and induction of massive apoptosis). PRIMA-1 has the ability to restore the tumor suppressor function of mutant p53, leading to apoptosis in several types of cancer cells. Our group (Grinter et al. 2011) used MDock (Huang and Zou 2007a; Yan and Zou 2016) as the docking engine and ITScore (Huang and Zou 2006a, b) as the scoring function to screen the PDTD target database (Gao et al. 2008). The highest ranked human protein oxidosqualene cyclase (OSC) was suggested to be the primary binding target of PRIMA-1 and a novel anticancer therapeutic target.

Besides the wide applications in the drug design pipeline, IVS is applied to other fields such as environmental engineering and biosafety of nanomaterials. For example, Xu et al. has applied IVS to identifying the potential targets of persistent organic pollutants (POPs) such as dichlorodiphenyldichloroethylene (4,4'-DDE) and polychlorinated biphenyls (PCBs) (Xu et al. 2013). The toxicity mechanism of these POPs could be further illustrated. Calvaresi and Zerbetto have also used IVS to identify the protein targets of nanoparticle fullerene C_{60} (Calvaresi and Zerbetto 2010).

Side effects and toxicity

Side effects and toxicity are mainly responsible for the failure of the compounds in clinical trials, and also for the restricted use or withdrawal of approved drugs. Therefore, taking side effects into account in the initial step of new drug design could significantly increase the final success rate of drug development and drug safety.

Chen et al. first tested their in-house, docking-based IVS program named INVDOCK (Chen and Zhi 2001), on the side effects and toxicity of eight clinical agents, aspirin, gentamicin, ibuprofen, indinavir, neomycin, penicillin G, 4H-tamoxifen, and vitamin C (Chen and Ung 2001). It was found that 83% of the experimentally known side effects and toxicity targets could be predicted. Lately, the authors applied the approach to 11 marketed anti-HIV drugs, including protease, nucleoside reverse transcriptase, and non-nucleoside reverse transcriptase inhibitors (Ji et al. 2006). The results showed that over 86% of the adverse drug reactions predicted by INVDOCK were consistent with the adverse reactions reported in literature. The agreement between the predicted results and the experimental data was also achieved in the work of Rockey and Elcock's (Rockey and Elcock 2002), in which three clinically relevant inhibitors (Gleevec, purvalanol A, and hymenialdisine) were analyzed against a set of protein kinase targets (76 GDP receptors and 113 ADP receptors) by the AutoDock program (Morris et al. 1998). The success of these pioneering studies brings confidence to the use of a docking-based IVS approach in practice.

Recently, Ma et al. (2011) used INVDOCK to investigate potential toxicity mechanisms of melamine, which was found in infant formula and is responsible for the outbreak of nephrolithiasis among children in China. Four target proteins (glutathione peroxidase 1, beta-hexosaminidase subunit beta, l-lactate dehydrogenase, and lysozyme C) were suggested to be related to nephrotoxicity induced by melamine and its metabolite cyanuric acid. In addition, the authors also found three target proteins (superoxide dismutase, glucose-6-phosphate 1-dehydrogenase, glutathione reductase) that were related to lung toxicity. Furthermore, a biological signal cascade network was constructed based on these predicted target proteins. However, the results need to be verified experimentally.

The IVS approach has also been applied to clozapine, one of the most effective medications for the treatment of schizophrenia. The usage of clozapine is limited by its life-threatening adverse drug reaction (ADR), mainly agranulocytosis. Yang et al. (2011) used an IVS approach via the DRAR-CPI server to investigate the ADR across a panel of human proteins (381 unique human proteins with 410 binding pockets) for clozapine. As a reference, olanzapine, an analog of clozapine which has a much lower incidence of agranulocytosis, was also analyzed. With the hypothesis that targets related to agranulocytosis tend to bind clozapine but not olanzapine, HSPA1A (the gene of Hsp70) was identified as the off-target of clozapine. The result was confirmed by the comparison of mRNA expression studies on HSPA1A-related genes inside a leukemia cell line with and without the clozapine treatment.

Drug repositioning

As aforementioned, even officially approved drugs sometimes bind to off-targets and cause side effects. If the off-target of an approved drug happens to be the therapeutic target for another disease, the drug has a chance for a new use, namely drug repositioning. There are a number of repositioned drugs in the market. For example, sildenafil was primarily developed for angina but later approved for erectile dysfunction. Thalidomide was initially marketed for morning sickness but was later approved for leprosy and also for multiple myeloma. More examples can be found in a review by Ashburn and Thor (2004). Although docking-based IVS seems to be a tailor-made tool for drug repositioning, there have been few successful stories until now.

Recently, Li et al. (2011) performed a large-scale molecular docking of small-molecule drugs against protein drug targets, in order to find novel targets for the existing drugs. The drugs and targets in the study were based on the data deposited in the DrugBank 2.5 database (Wishart et al. 2006). Overall, 252 human protein drug targets and 4621 approved and experimental small-molecule drugs were collected. The ICM program (Abagyan et al. 1994) was employed as the docking engine. The large-scale cross dockings (4621 ligands against 252 receptors) were run on a powerful computer cluster with 1000 processors. A consensus score, consisting of an empirical scoring function ICM (Abagyan et al. 1994) and a knowledge-based scoring function PMF (Muegge and Martin 1999; Muegge 2006), was used to evaluate the docking poses. The consensus score performed much better than either the ICM score or the PMF score alone, with the percentage of the known interactions in the prediction set improved from 1.1% (ICM score) or 2.0% (PMF score) to 10.3%. Furthermore, by combining with the ranks of the proteins and drugs, the percentage value for the consensus score reached up to 48.8%, giving the confidence that the other 51.2% proteins were indeed novel targets. Successfully, the cancer drug nilotinib was further confirmed as a potent inhibitor of MAPK14 ($IC_{50} = 40$ nmol/L) by biological tests. MAPK14, also known as p38 alpha, is a target in inflammation, suggesting that nilotinib has a chance for being repurposed for the treatment of rheumatoid arthritis.

Multi-target therapy/drug–target network

In novel drug design, compounds are usually engineered to bind to a specific target, with the assumption that one drug binds to one target to treat one condition. However, this assumption is now in question, with the high failure rate during the late stage of clinical trials due to efficacy and clinical safety problems (Xie et al. 2011) being the main source of the scrutiny. Recent studies suggest that each existing drug binds to, on average, about six target proteins (Azzaoui et al. 2007; Mestres et al. 2008) instead of one. This phenomenon can be easily understood in a biological network, in which each node represents a protein and a link between two proteins means a direct interaction. Considering the robustness of biological systems, acting on multiple nodes should, in theory, be more effective in affecting the system overall than when only considering one node. Therefore, a multi-target therapy is expected to be able to break the bottleneck of current single-target drug design paradigms. However, the development of multi-target drugs proceeds slowly, partially due to the lack of experimental tools to identify targets on a proteome-wide scale (Xie et al. 2011). Thus, computational approaches, such as IVS described in this review, were developed to narrow down the targets of interest for further experimental validation.

An example of docking-based IVS for multi-target identification can be found in a recent work by Zhao et al. (2012). The INVDOCK program (Chen and Zhi 2001) was employed to search potential protein targets for astragaloside-IV (AGS-IV). The AGS-IV is one of the main active ingredients of *Astragalus membranaceus* Bunge, a traditional Chinese medicine for cardiovascular diseases (CVD). The protein targets of approved small-molecule drugs for CVD deposited in the DrugBank database (Wishart *et al.* 2006) were collected as the target database, consisting of 188 proteins. Among the 39 predicted targets, three proteins (calcineurin, angiotensin-converting enzyme, and c-Jun N-terminal kinase) were experimentally validated at a molecular level. By mapping the 39 proteins onto the protein–protein interaction network of the human genome, 34 of them can be linked into a sub-network, which can be further divided into six topologically compact modules. The effects of AGS-IV on CVD were supposed to act through binding to multiple targets, for example, by directly binding to the hubs of six modules. The results were further confirmed by the comparison with the drug–target networks of the approved CVD drugs that share common targets with AGS-IV.

Receptor design

In addition, the docking-based IVS method could be used for receptor design. Steffen *et al.* (2007) successfully improved the property of a synthetic receptor for a binding ligand. In this study, camptothecin (CPT) was chosen as the investigated ligand. Although CPT presents remarkable anticancer activity in preliminary clinical trials, its therapeutic potential is hampered by its low solubility and stability. Thus, hosts or so-called receptors were designed for the solubilization of the ligand. In particular, a set of β-cyclodextrin (β-CD) derivatives (a total of 1846 entities) was generated from the β-CD core and thiol building blocks as the receptor candidates (from the target database). CPT was docked to each β-CD derivative in the target database by two different docking programs, AutoDock 3.05 (Morris *et al.* 1998) and GlamDock 1.0 (Tietze and Apostolakis 2007). Nine receptors from the top 10% candidates were selected for experimental validation. Successfully, five of them significantly improved the solubility of CPT, and their ability to do so was significantly better than any other known CD derivative.

CHALLENGES

In summary, during the last decade, the entire field of docking-based IVS, including the construction of target databases, scoring functions, and post analysis, has been significantly improved by researchers from all over the world. A number of successful applications as described in this review have proved that docking-based IVS is a powerful technique for drug discovery. However, several challenges remain to be solved for docking-based IVS to become a robust tool.

The first challenge is the incompleteness of available target databases. Using the data in DrugPort (http://www.ebi.ac.uk/thornton-srv/databases/drugport/) as an example, there are a total of 1664 known druggable protein targets in the database, but only about half of them have 3D structures in the PDB. If unknown targets are considered, this rate could be much lower. Furthermore, these targets with known-structures are not evenly distributed among different superfamilies, due to experimental limitations. For example, the superfamily of membrane proteins, the G-protein-coupled receptors (GPCRs), is one of the most important targets in drug design, given the fact that they account for over a quarter of the known drug targets (Overington *et al.* 2006), and about half of the drugs on the market target GPCRs specifically (Klabunde and Hessler 2002). However, only a fraction of the GPCRs have experimental structures (Venkatakrishnan *et al.* 2013), because the structural resolution of membrane proteins like GPCRs is much more complicated and difficult to elucidate than global proteins such as enzymes. Fortunately, the current databases can be significantly improved through homology modeling techniques, and the incompleteness problem can be gradually solved with time as more and more complete structures are determined by experimental methods.

Another challenge is from the vantage point of protein flexibility. As aforementioned, protein–ligand binding is a mutual fitting process. The existing docking programs are able to account for the flexibility of small molecules very well, but the overall flexibility of the entire protein remains a great challenge. Efforts have been made to partially consider protein flexibility during docking. For example, the side chains of the residues in the active site can be treated to be flexible with the induced-fit docking strategies (Sherman *et al.* 2006). In another example, an ensemble of protein structures are used for docking in MDOCK (Huang and Zou 2007a, b). However, flexible docking using the induced-fit strategy is time-consuming. For the ensemble docking using MDOCK, an ensemble of experimentally determined protein structures are not always available. These methods are usually difficult to be directly applied to IVS studies which involve hundreds of different proteins. To the best of our knowledge, the proteins were all treated as rigid bodies in the published docking-based IVS studies. Thus, it would be useful to develop efficient protein flexibility algorithms for IVS studies.

At this stage, IVS and the more traditional VS work as an enrichment method rather than an accurate prediction tool, mainly due to the inaccuracy of the scoring functions. Simply selecting the top targets in the ranking list could result in many false positive candidates. As reviewed in the subsection on scoring functions, efforts have been made to improve the success rate, including setting a threshold for each target, using consensus scoring functions, or normalizing binding scores. However, all these methods can be regarded as post analysis, which are highly dependent on the scoring values calculated by the existing inaccurate scoring functions. In fact, the scoring function could be the biggest challenge for molecular docking. A detailed review about scoring functions for protein–ligand docking can be found in a recent review (Huang et al. 2010). Recently, Wang et al. (2012b) evaluated the performance of Glide scoring functions in IVS based on the Astex diverse set. Interestingly, "interprotein noises" were found in the Glide scores, suggesting that scoring functions that are developed for conformational (the same complex) ranking could result in over- or underestimated scores when they are directly used for the ranking of different protein–ligand complexes. By introducing a correction term based on a given protein characteristic, the ratio of the relative hydrophobic and hydrophilic character of the binding site, the accuracy of target prediction was improved by 27% (i.e., from 57% to 72%). The study could be used as a reference in the optimization of the existing scoring functions for IVS studies.

An efficient way to address the above challenges (i.e., protein flexibility and scoring function) could be the use of more accurate yet more time-consuming sampling/scoring strategies for the enriched subset (e.g., top 5% of the targets). Regarding the sampling aspect, protein flexibility could be partially considered by using ensemble docking or induced-fit docking strategies. Regarding the scoring aspect, contributions from the solvent effect and from the conformational entropic effect could be considered. Well-studied strategies are molecular dynamics (MD)-based binding free energy calculation methods, such as MM/PBSA and MM/GBSA (Srinivasan et al. 1998; Kollman et al. 2000; Wang et al. 2001). In addition, recent studies show that polarization effects are important for both binding mode and binding affinity predictions (Cho et al. 2005; Xu and Lill 2013). To efficiently consider polarization effects in the docking process, quantum mechanics (QM) or hybrid quantum mechanics/molecular mechanics (QM/MM) methods need to be employed. A QM-polarized ligand docking method has been implemented in a commercial software package, Schrödinger Suites (https://www.schrodinger.com).

There are many docking programs and scoring functions that can be used for an IVS study. As reviewed in this paper, some of them have already been used by different groups for different purposes with varying degrees of success. It would be interesting to find which programs are more effective for IVS studies than others. Such an attempt has been tried by Liu et al. (2010a). In their work, five schemes, GOLD (Jones et al. 1997) and FlexX (Rarey et al. 1996) implemented in Sybyl, TarFisDock (Li et al. 2006) which is based on DOCK4.0 (Ewing et al. 2001), and two in-house docking strategies, TarSearch-X and TarSearch-M (DOCK5.1 (Moustakas et al. 2006)) combined with two in-house scoring functions X-Score (Wang et al. 2002) and M-score (Yang et al. 2006), were tested for eight multi-target compounds extracted from DrugBank (Wishart et al. 2006). The target database was collected from the PDB, and contained 1714 entries from 1594 known drug targets. According to the order of the known targets in the rank list, their results show that TarSearch-X is the most efficient and GOLD is acceptable. However, the study has some limitations. Seven of the eight selected multi-target compounds have only two known targets. Another compound has three known targets. More convincing validation would be to use compounds that have many known targets, such as vitamin E with 14 known targets and 4H-tamoxifen with ten known targets which were used in the test for TarFisDock (Li et al. 2006). In addition, a number of other powerful docking programs and scoring functions are awaited to be assessed for IVS studies.

To effectively evaluate a method of docking-based IVS, a database is desired to contain both positive and negative results. However, negative data are difficult to collect because literature prefer to present successful cases rather than failed cases, i.e., in which a molecule does not interact with a protein. Fortunately, Schomburg and Rarey (2014) recently provided an example of such a database. Because of the limited data available for negative results, the authors constructed a small set with both positive and negative results. This small set, referred to as the selectivity dataset, consists of a total of eight proteins belonging to three target classes and 17 small molecules with defined selectivity in the respective target class. The selectivity dataset is suggested to be used for proof-of-concept studies. A large dataset containing 7992 protein structures and 72 drug-like ligands was also provided. The dataset, called Drugs/sc-PDB dataset, was constructed based on the data in DrugBank (Wishart et al. 2006) and sc-PDB (Kellenberger et al. 2006). The 72 drug-like ligands were selected based on the assumption that the selectivity and targets of the approved drugs have been well studied. The selectivity dataset and the Drugs/sc-PDB dataset form a benchmark for target identification methods.

The last challenge could potentially be the post-analysis problem. The output of IVS is an enriched subset, which contains at least tens of potential targets (including false positive targets). How to connect these predicted multiple targets to the mechanisms of the ligand remains an open question. Usually, the predicted targets need to be validated by biological experiments. Only then can biological functions of the true targets be connected to the phenotypic effects of the ligand. Recently, the biological network idea was employed for the analysis of IVS results. In the work by Zhao et al. (2012), predicted targets were mapped onto the protein–protein interaction network of the human genome. A sub-network was identified that could effectively explain a connection to the actual mechanisms of the ligand in question.

Acknowledgements This work was supported by the NSF CAREER Award (DBI-0953839), NIH (R01GM109980), and American Heart Association (Midwest Affiliate) (13GRNT16990076) to XZ. MH is supported by NIH T32LM012410 (PI: Chi-Ren Shyu).

Compliance with Ethical Standards

Conflict of interest Xianjin Xu, Marshal Huang, and Xiaoqin Zou declare that they have no conflict of interest.

Human and animal rights and informed consent This article does not contain any studies with human or animal subjects performed by any of the authors.

References

Abagyan R, Totrov M, Kuznetsov D (1994) ICM-A new method for protein modeling and design: applications to docking and structure prediction from the distorted native conformation. J Comput Chem 15:488–506

Altschul SF, Madden TL, Schaffer AA, Zhang J, Zhang Z, Miller W, Lipman DJ (1997) Gapped BLAST and PSI-BLAST: a new generation of protein database search programs. Nucleic Acids Res 25:3389–3402

Ashburn TT, Thor KB (2004) Drug repositioning: identifying and developing new uses for existing drugs. Nat Rev Drug Discov 3:673–683. https://doi.org/10.1038/nrd1468

Azzaoui K, Hamon J, Faller B, Whitebread S, Jacoby E, Bender A, Jenkins JL, Urban L (2007) Modeling promiscuity based on in vitro safety pharmacology profiling data. ChemMedChem 2:874–880. https://doi.org/10.1002/cmdc.200700036

Baker NA, Sept D, Joseph S, Holst MJ, McCammon JA (2001) Electrostatics of nanosystems: application to microtubules and the ribosome. Proc Natl Acad Sci USA 98:10037–10041. https://doi.org/10.1073/pnas.181342398

Bender A, Glen RC (2004) Molecular similarity: a key technique in molecular informatics. Org Biomol Chem 2:3204–3218. https://doi.org/10.1039/B409813G

Berman HM, Westbrook J, Feng Z, Gilliland G, Bhat TN, Weissig H, Shindyalov IN, Bourne PE (2000) The protein data bank. Nucleic Acids Res 28:235–242

Bohm HJ (1992) The computer program LUDI: a new method for the de novo design of enzyme inhibitors. J Comput Aided Mol Des 6:61–78

Bohm HJ (1994) The development of a simple empirical scoring function to estimate the binding constant for a protein–ligand complex of known three-dimensional structure. J Comput Aided Mol Des 8:243–256

Bohm HJ (1998) Prediction of binding constants of protein ligands: a fast method for the prioritization of hits obtained from de novo design or 3D database search programs. J Comput Aided Mol Des 12:309–323

Brooijmans N, Kuntz ID (2003) Molecular recognition and docking algorithms. Annu Rev Biophys Biomol Struct 32:335–373. https://doi.org/10.1146/annurev.biophys.32.110601.142532

Bullock C, Cornia N, Jacob R, Remm A, Peavey T, Weekes K, Mallory C, Oxford JT, McDougal OM, Andersen TL (2013) DockoMatic 2.0: high throughput inverse virtual screening and homology modeling. J Chem Inf Model 53:2161–2170. https://doi.org/10.1021/ci400047w

Calvaresi M, Zerbetto F (2010) Baiting proteins with C60. ACS Nano 4:2283–2299. https://doi.org/10.1021/nn901809b

Chang DT, Oyang YJ, Lin JH (2005) MEDock: a web server for efficient prediction of ligand binding sites based on a novel optimization algorithm. Nucleic Acids Res 33:W233–W238

Chen SJ, Ren JL (2014) Identification of a potential anticancer target of danshensu by inverse docking. Asian Pac J Cancer Prev 15:111–116

Chen YZ, Ung CY (2001) Prediction of potential toxicity and side effect protein targets of a small molecule by a ligand–protein inverse docking approach. J Mol Graph Model 20:199–218

Chen YZ, Zhi DG (2001) Ligand–protein inverse docking and its potential use in the computer search of protein targets of a small molecule. Proteins 43:217–226

Chen X, Ji ZL, Chen YZ (2002) TTD: therapeutic target database. Nucleic Acids Res 30:412–415

Cho AE, Guallar V, Berne BJ, Friesner R (2005) Importance of accurate charges in molecular docking: quantum mechanical/molecular mechanical (QM/MM) approach. J Comput Chem 26:915–931

DesJarlais RL, Sheridan RP, Dixon JS, Kuntz ID, Venkataraghavan R (1986) Docking flexible ligands to macromolecular receptors by molecular shape. J Med Chem 29:2149–2153

Do QT, Bernard P (2004) Pharmacognosy and reverse pharmacognosy: a new concept for accelerating natural drug discovery. IDrugs 7:1017–1027

Do QT, Renimel I, Andre P, Lugnier C, Muller CD, Bernard P (2005) Reverse pharmacognosy: application of selnergy, a new tool for lead discovery. The example of epsilon-viniferin. Curr Drug Discov Technol 2:161–167

Do QT, Lamy C, Renimel I, Sauvan N, André P, Himbert F, Morin-Allory L, Bernard P (2007) Reverse pharmacognosy: identifying biological properties for plants by means of their molecule constituents: application to meranzin. Planta Med 73:1235–1240. https://doi.org/10.1055/s-2007-990216

Ewing TJ, Makino S, Skillman AG, Kuntz ID (2001) DOCK 4.0: search strategies for automated molecular docking of flexible molecule databases. J Comput Aided Mol Des 15:411–428

Feng LX, Jing CJ, Tang KL, Tao L, Cao ZW, Wu WY, Guan SH, Jiang BH, Yang M, Liu X, Guo DA (2011) Clarifying the signal network of salvianolic acid B using proteomic assay and bioinformatic analysis. Proteomics 11:1473–1485. https://doi.org/10.1002/pmic.201000482

Friesner RA, Banks JL, Murphy RB, Halgren TA, Klicic JJ, Mainz DT, Repasky MP, Knoll EH, Shelley M, Perry JK, Shaw DE, Francis P, Shenkin PS (2004) Glide: a new approach for rapid, accurate docking and scoring. 1. Method and assessment of docking accuracy. J Med Chem 47:1739–1749. https://doi.org/10.1021/jm0306430

Gao Z, Li H, Zhang H, Liu X, Kang L, Luo X, Zhu W, Chen K, Wang X, Jiang H (2008) PDTD: a web-accessible protein database for drug target identification. BMC Bioinform 9:104. https://doi.org/10.1186/1471-2105-9-104

Grant JA, Pickup BT, Nicholls A (2001) A smooth permittivity function for Poisson-Boltzmann solvation methods. J Comput Chem 22:608–640

Grinter SZ, Zou X (2014a) A Bayesian statistical approach of improving knowledge-based scoring functions for protein-ligand interactions. J Comput Chem 35:932–943

Grinter SZ, Zou X (2014b) Challenges, applications, and recent advances of protein–ligand docking in structure-based drug design. Molecules 19:10150–10176. https://doi.org/10.3390/molecules190710150

Grinter SZ, Liang Y, Huang SY, Hyder SM, Zou X (2011) An inverse docking approach for identifying new potential anti-cancer targets. J Mol Graph Model 29:795–799. https://doi.org/10.1016/j.jmgm.2011.01.002

Grinter SZ, Yan C, Huang SY, Jiang L, Zou X (2013) Automated large-scale file preparation, docking, and scoring: evaluation of ITScore and STScore using the 2012 Community Structure-Activity Resource Benchmark. J Chem Inf Model 53:1905–1914

Günther S, Kuhn M, Dunkel M, Campillos M, Senger C, Petsalaki E, Ahmed J, Urdiales EG, Gewiess A, Jensen LJ, Schneider R, Skoblo R, Russell RB, Bourne PE, Bork P, Preissner R (2008) SuperTarget and Matador: resources for exploring drug-target relationships. Nucleic Acids Res 36:D919–D922. https://doi.org/10.1093/nar/gkm862

Hawkins GD, Cramer CJ, Truhlar DG (1995) Pairwise solute descreening of solute charges from a dielectric medium. Chem Phys Lett 246:122–129

Huang SY, Zou X (2006a) An iterative knowledge-based scoring function to predict protein–ligand interactions: I. Derivation of interaction potentials. J Comput Chem 27:1866–1875. https://doi.org/10.1002/jcc.20504

Huang SY, Zou X (2006b) An iterative knowledge-based scoring function to predict protein–ligand interactions: II. Validation of the scoring function. J Comput Chem 27:1876–1882. https://doi.org/10.1002/jcc.20505

Huang SY, Zou X (2007a) Ensemble docking of multiple protein structures: considering protein structural variations in molecular docking. Proteins 66:399–421. https://doi.org/10.1002/prot.21214

Huang SY, Zou X (2007b) Efficient molecular docking of NMR structures: application to HIV-1 protease. Protein Sci 16:43–51. https://doi.org/10.1110/ps.062501507

Huang SY, Zou X (2010) Advances and challenges in protein-ligand docking. Int J Mol Sci 11:3016–3034. https://doi.org/10.3390/ijms11083016

Huang SY, Grinter SZ, Zou X (2010) Scoring functions and their evaluation methods for protein–ligand docking: recent advances and future directions. Phys Chem Chem Phys 12:12899–12908. https://doi.org/10.1039/c0cp00151a

Huey R, Morris GM, Olson AJ, Goodsell DS (2007) A semiempirical free energy force field with charge-based desolvation. J Comput Chem 28:1145–1152. https://doi.org/10.1002/jcc.20634

Ji ZL, Han LY, Yap CW, Sun LZ, Chen X, Chen YZ (2003) Drug Adverse Reaction Target Database (DART): proteins related to adverse drug reactions. Drug Saf 26:685–690

Ji ZL, Wang Y, Yu L, Han LY, Zheng CJ, Chen YZ (2006) In silico search of putative adverse drug reaction related proteins as a potential tool for facilitating drug adverse effect prediction. Toxicol Lett 164:104–112. https://doi.org/10.1016/j.toxlet.2005.11.017

Ji HF, Li XJ, Zhang HY (2009) Natural products and drug discovery. Can thousands of years of ancient medical knowledge lead us to new and powerful drug combinations in the fight against cancer and dementia? EMBO Rep 10:194–200. https://doi.org/10.1038/embor.2009.12

Jiang F, Kim SH (1991) "Soft docking": matching of molecular surface cubes. J Mol Biol 219:79–102

Jones G, Willett P, Glen RC, Leach AR, Taylor R (1997) Development and validation of a genetic algorithm for flexible docking. J Mol Biol 267:727–748. https://doi.org/10.1006/jmbi.1996.0897

Kaufmann SH (2008) Paul Ehrlich: founder of chemotherapy. Nat Rev Drug Discov 7:373. https://doi.org/10.1038/nrd2582

Kellenberger E, Muller P, Schalon C, Bret G, Foata N, Rognan D (2006) sc-PDB: an annotated database of druggable binding sites from the Protein Data Bank. J Chem Inf Model 46:717–727. https://doi.org/10.1021/ci050372x

Klabunde T, Hessler G (2002) Drug design strategies for targeting G-protein-coupled receptors. ChemBioChem 3:928–944

Knegtel RM, Kuntz ID, Oshiro CM (1997) Molecular docking to ensembles of protein structures. J Mol Biol 266:424–440. https://doi.org/10.1006/jmbi.1996.0776

Kollman PA, Massova I, Reyes C, Kuhn B, Huo S, Chong L, Lee M, Lee T, Duan Y, Wang W, Donini O, Cieplak P, Srinivasan J, Case DA, Cheatham TE III (2000) Calculating structures and free energies of complex molecules: combining molecular mechanics and continuum models. Acc Chem Res 33:889–897

Koutsoukas A, Simms B, Kirchmair J, Bond PJ, Whitmore AV, Zimmer S, Young MP, Jenkins JL, Glick M, Glen RC, Bender A (2011) From in silico target prediction to multi-target drug design: current databases, methods and applications. J Proteomics 74:2554–2574. https://doi.org/10.1016/j.jprot.2011.05.011

Lang PT, Brozell SR, Mukherjee S, Pettersen EF, Meng EC, Thomas V, Rizzo RC, Case DA, James TL, Kuntz ID (2009) DOCK 6: combining techniques to model RNA-small molecule complexes. RNA 15:1219–1230. https://doi.org/10.1261/rna.1563609

Lauro G, Romano A, Riccio R, Bifulco G (2011) Inverse virtual screening of antitumor targets: pilot study on a small database of natural bioactive compounds. J Nat Prod 74:1401–1407. https://doi.org/10.1021/np100935s

Lauro G, Masullo M, Piacente S, Riccio R, Bifulco G (2012) Inverse virtual screening allows the discovery of the biological activity of natural compounds. Bioorg Med Chem 20:3596–3602. https://doi.org/10.1016/j.bmc.2012.03.072

Leach AR (1994) Ligand docking to proteins with discrete side-chain flexibility. J Mol Biol 235:345–356

Li H, Gao Z, Kang L, Zhang H, Yang K, Yu K, Luo X, Zhu W, Chen K, Shen J, Wang X, Jiang H (2006) TarFisDock: a web server for identifying drug targets with docking approach. Nucleic Acids Res 34:W219–W224. https://doi.org/10.1093/nar/gkl114

Li YY, An J, Jones SJ (2011) A computational approach to finding novel targets for existing drugs. PLoS Comput Biol 7:e1002139. https://doi.org/10.1371/journal.pcbi.1002139

Liu M, Wang S (1999) MCDOCK: a Monte Carlo simulation approach to the molecular docking problem. J Comput Aided Mol Des 13:435–451

Liu T, Lin Y, Wen X, Jorissen RN, Gilson MK (2007) BindingDB: a web-accessible database of experimentally determined protein–ligand binding affinities. Nucleic Acids Res 35:D198–D201. https://doi.org/10.1093/nar/gkl999

Liu H, Qing S, Zhang J, Fu W (2010a) Evaluation of various inverse docking schemes in multiple targets identification. J Mol Graph Model 29:326–330. https://doi.org/10.1016/j.jmgm.2010.09.004

Liu X, Ouyang S, Yu B, Liu Y, Huang K, Gong J, Zheng S, Li Z, Li H, Jiang H (2010b) PharmMapper server: a web server for potential drug target identification using pharmacophore mapping approach. Nucleic Acids Res 38:W609–W614. https://doi.org/10.1093/nar/gkq300

Luo H, Chen J, Shi L, Mikailov M, Zhu H, Wang K, He L, Yang L (2011) DRAR-CPI: a server for identifying drug repositioning potential and adverse drug reactions via the chemical-protein interactome. Nucleic Acids Res 39:W492–W498. https://doi.org/10.1093/nar/gkr299

Ma C, Kang H, Liu Q, Zhu R, Cao Z (2011) Insight into potential toxicity mechanisms of melamine: an in silico study. Toxicology 283:96–100. https://doi.org/10.1016/j.tox.2011.02.009

Ma DL, Chan DS, Leung CH (2013) Drug repositioning by structure-based virtual screening. Chem Soc Rev 42:2130–2141. https://doi.org/10.1039/c2cs35357a

Macchiarulo A, Nobeli I, Thornton JM (2004) Ligand selectivity and competition between enzymes in silico. Nat Biotechnol 22:1039–1045. https://doi.org/10.1038/nbt999

Meng EC, Shoichet BK, Kuntz ID (1992) Automated docking with grid-based energy evaluation. J Comput Chem 13:505–524

Mestres J, Gregori-Puigjane E, Valverde S, Sole RV (2008) Data completeness—the Achilles heel of drug–target networks. Nat Biotechnol 26:983–984. https://doi.org/10.1038/nbt0908-983

Morris GM, Goodsell DS, Halliday RS, Huey R, Hart WE, Belew RK, Olson AJ (1998) Automated docking using a Lamarckian genetic algorithm and an empirical binding free energy function. J Comput Chem 19:1639–1662

Moustakas DT, Lang PT, Pegg S, Pettersen E, Kuntz ID, Brooijmans N, Rizzo RC (2006) Development and validation of a modular, extensible docking program: DOCK 5. J Comput Aided Mol Des 20:601–619. https://doi.org/10.1007/s10822-006-9060-4

Muegge I (2006) PMF scoring revisited. J Med Chem 49:5895–5902. https://doi.org/10.1021/jm050038s

Muegge I, Martin YC (1999) A general and fast scoring function for protein–ligand interactions: a simplified potential approach. J Med Chem 42:791–804. https://doi.org/10.1021/jm980536j

Muller P, Lena G, Boilard E, Bezzine S, Lambeau G, Guichard G, Rognan D (2006) In silico-guided target identification of a scaffold-focused library: 1,3,5-triazepan-2,6-diones as novel phospholipase A2 inhibitors. J Med Chem 49:6768–6778. https://doi.org/10.1021/jm0606589

Nwaka S, Hudson A (2006) Innovative lead discovery strategies for tropical diseases. Nat Rev Drug Discov 5:941–955. https://doi.org/10.1038/nrd2144

Overington JP, Al-Lazikani B, Hopkins AL (2006) How many drug targets are there? Nat Rev Drug Discov 5:993–996. https://doi.org/10.1038/nrd2199

Qiu D, Shenkin PS, Hollinger FP, Still WC (1997) The GB/SA continuum model for solvation. a fast analytical method for the calculation of approximate born radii. J Phys Chem A 101:3005–3014

Rarey M, Kramer B, Lengauer T, Klebe G (1996) A fast flexible docking method using an incremental construction algorithm. J Mol Biol 261:470–489. https://doi.org/10.1006/jmbi.1996.0477

Rocchia W, Sridharan S, Nicholls A, Alexov E, Chiabrera A, Honig B (2002) Rapid grid-based construction of the molecular surface and the use of induced surface charge to calculate reaction field energies: applications to the molecular systems and geometric objects. J Comput Chem 23:128–137. https://doi.org/10.1002/jcc.1161

Rockey WM, Elcock AH (2002) Progress toward virtual screening for drug side effects. Proteins 48:664–671. https://doi.org/10.1002/prot.10186

Rognan D (2010) Structure-based approaches to target fishing and ligand profiling. Mol Inform 29:176–187

Sali A, Blundell TL (1993) Comparative protein modelling by satisfaction of spatial restraints. J Mol Biol 234:779–815. https://doi.org/10.1006/jmbi.1993.1626

Santiago DN, Pevzner Y, Durand AA, Tran M, Scheerer RR, Daniel K, Sung SS, Woodcock HL, Guida WC, Brooks WH (2012) Virtual target screening: validation using kinase inhibitors. J Chem Inf Model 52:2192–2203. https://doi.org/10.1021/ci300073m

Schomburg KT, Rarey M (2014) Benchmark data sets for structure-based computational target prediction. J Chem Inf Model 54:2261–2274. https://doi.org/10.1021/ci500131x

Sherman W, Day T, Jacobson MP, Friesner RA, Farid R (2006) Novel procedure for modeling ligand/receptor induced fit effects. J Med Chem 49:534–553

Slon-Usakiewicz JJ, Pasternak A, Reid N, Toledo-Sherman LM (2004) New targets for an old drug: II. Hypoxanthine-guanine amidophosphoribosyltransferase as a new pharmacodynamic target of methotrexate. Clin Proteom 1:227–234

Sousa SF, Fernandes PA, Ramos MJ (2006) Protein–ligand docking: current status and future challenges. Proteins 65:15–26. https://doi.org/10.1002/prot.21082

Sousa SF, Ribeiro AJ, Coimbra JT, Neves RP, Martins SA, Moorthy NS, Fernandes PA, Ramos MJ (2013) Protein–ligand docking in the new millennium—a retrospective of 10 years in the field. Curr Med Chem 20:2296–2314

Srinivasan J, Cheatham TE, Cieplak P, Kollman PA, Case DA (1998) Continuum solvent studies of the stability of DNA, RNA, and phosphoramidate–DNA helices. J Am Chem Soc 120:9401–9409

Steffen A, Thiele C, Tietze S, Strassnig C, Kämper A, Lengauer T, Wenz G, Apostolakis J (2007) Improved cyclodextrin-based receptors for camptothecin by inverse virtual screening. Chem Eur J 13:6801–6809. https://doi.org/10.1002/chem.200700661

Still WC, Tempczyk A, Hawley RC, Hendrickson T (1990) Semi-analytical treatment of solvation for molecular mechanics and dynamics. J Am Chem Soc 112:6127–6129

Thomas PD, Dill KA (1996) An iterative method for extracting energy-like quantities from protein structures. Proc Natl Acad Sci USA 93:11628–11633

Tietze S, Apostolakis J (2007) GlamDock: development and validation of a new docking tool on several thousand protein–ligand complexes. J Chem Inf Model 47:1657–1672. https://doi.org/10.1021/ci7001236

Toledo-Sherman LM, Desouza L, Hosfield CM, Liao L, Boutillier K, Taylor P, Climie S, McBroom-Cerajewski L, Moran MF (2004) New targets for an old drug: a chemical proteomics approach to unraveling the molecular mechanism of action of methotrexate. Clin Proteom 1:45–67

Trott O, Olson AJ (2010) AutoDock Vina: improving the speed and accuracy of docking with a new scoring function, efficient optimization, and multithreading. J Comput Chem 31:455–461. https://doi.org/10.1002/jcc.21334

Venkatakrishnan AJ, Deupi X, Lebon G, Tate CG, Schertler GF, Babu MM (2013) Molecular signatures of G-protein-coupled receptors. Nature 494:185–194. https://doi.org/10.1038/nature11896

Wang W, Donini O, Reyes CM, Kollman PA (2001) Biomolecular simulations: recent developments in force fields, simulations of enzyme catalysis, protein–ligand, protein–protein, and protein–nucleic acid noncovalent interactions. Annu Rev Biophys Biomol Struct 30:211–243

Wang R, Lai L, Wang S (2002) Further development and validation of empirical scoring functions for structure-based binding affinity prediction. J Comput Aided Mol Des 16:11–26

Wang JC, Lin JH, Chen CM, Perryman AL, Olson AJ (2011) Robust scoring functions for protein–ligand interactions with quantum chemical charge models. J Chem Inf Model 51:2528–2537. https://doi.org/10.1021/ci200220v

Wang JC, Chu PY, Chen CM, Lin JH (2012a) idTarget: a web server for identifying protein targets of small chemical molecules with robust scoring functions and a divide-and-conquer docking approach. Nucleic Acids Res 40:W393–W399. https://doi.org/10.1093/nar/gks496

Wang W, Zhou X, He W, Fan Y, Chen Y, Chen X (2012b) The interprotein scoring noises in glide docking scores. Proteins 80:169–183. https://doi.org/10.1002/prot.23173

Willett P, Barnard JM, Downs GM (1998) Chemical similarity searching. J Chem Inf Comput Sci 38:983–996

Wishart DS, Knox C, Guo AC, Shrivastava S, Hassanali M, Stothard P, Chang Z, Woolsey J (2006) DrugBank: a comprehensive resource for in silico drug discovery and exploration. Nucleic Acids Res 34:D668–D672. https://doi.org/10.1093/nar/gkj067

Xie L, Xie L, Bourne PE (2011) Structure-based systems biology for analyzing off-target binding. Curr Opin Struct Biol 21:189–199. https://doi.org/10.1016/j.sbi.2011.01.004

Xu M, Lill MA (2013) Induced fit docking, and the use of QM/MM methods in docking. Drug Discov Today Technol 10:e411–e418

Xu X-J, Su J-G, Liu B, Li C-H, Tan J-J, Zhang X-Y, Chen W-Z, Wang C-X (2013) Reverse virtual screening on persistent organic pollutants 4,4′-DDE and CB-153. Acta Phys Chim Sin 29:2276–2285

Xu X, Ma Z, Sun H, Zou X (2016) SM-TF: a structural database of small molecule–transcription factor complexes. J Comput Chem 37:1559–1564. https://doi.org/10.1002/jcc.24370

Yan C, Zou X (2016) An ensemble docking suite for molecular docking, scoring and in silico screening. In: Zhang W (ed) Methods in pharmacology and toxicology. Springer, New York, pp 153–166

Yan C, Grinter SZ, Merideth BR, Ma Z, Zou X (2016) Iterative knowledge-based scoring functions derived from rigid and flexible decoy structures: evaluation with the 2013 and 2014 CSAR benchmarks. J Chem Inf Model 56:1013–1021

Yang CY, Wang R, Wang S (2006) M-score: a knowledge-based potential scoring function accounting for protein atom mobility. J Med Chem 49:5903–5911. https://doi.org/10.1021/jm050043w

Yang L, Luo H, Chen J, Xing Q, He L (2009) SePreSA: a server for the prediction of populations susceptible to serious adverse drug reactions implementing the methodology of a chemical–protein interactome. Nucleic Acids Res 37:W406–W412. https://doi.org/10.1093/nar/gkp312

Yang L, Wang K, Chen J, Jegga AG, Luo H, Shi L, Wan C, Guo X, Qin S, He G, Feng G, He L (2011) Exploring off-targets and off-systems for adverse drug reactions via chemical–protein interactome-clozapine-induced agranulocytosis as a case study. PLoS Comput Biol 7:e1002016. https://doi.org/10.1371/journal.pcbi.1002016

Yang H, Qin C, Li YH, Tao L, Zhou J, Yu CY, Xu F, Chen Z, Zhu F, Chen Y (2016) Therapeutic target database update 2016: enriched resource for bench to clinical drug target and targeted pathway information. Nucleic Acids Res 44:D1069–D1074. https://doi.org/10.1093/nar/gkv1230

Yue QX, Cao ZW, Guan SH, Liu XH, Tao L, Wu WY, Li YX, Yang PY, Liu X, Guo DA (2008) Proteomics characterization of the cytotoxicity mechanism of ganoderic acid D and computer-automated estimation of the possible drug target network. Mol Cell Proteom 7:949–961. https://doi.org/10.1074/mcp.M700259-MCP200

Zahler S, Tietze S, Totzke F, Kubbutat M, Meijer L, Vollmar AM, Apostolakis J (2007) Inverse in silico screening for identification of kinase inhibitor targets. Chem Biol 14:1207–1214. https://doi.org/10.1016/j.chembiol.2007.10.010

Zhao J, Yang P, Li F, Tao L, Ding H, Rui Y, Cao Z, Zhang W (2012) Therapeutic effects of astragaloside IV on myocardial injuries: multi-target identification and network analysis. PLoS One 7:e44938. https://doi.org/10.1371/journal.pone.0044938

Evaluation of RNA secondary structure prediction for both base-pairing and topology

Yunjie Zhao[1,2], Jun Wang[1], Chen Zeng[3], Yi Xiao[1]

[1] Institute of Biophysics, School of Physics and Key Laboratory of Molecular Biophysics of the Ministry of Education, Huazhong University of Science and Technology, Wuhan 430074, China
[2] Institute of Biophysics and Department of Physics, Central China Normal University, Wuhan 430079, China
[3] Department of Physics, The George Washington University, Washington, DC 20052, USA

Graphical abstract

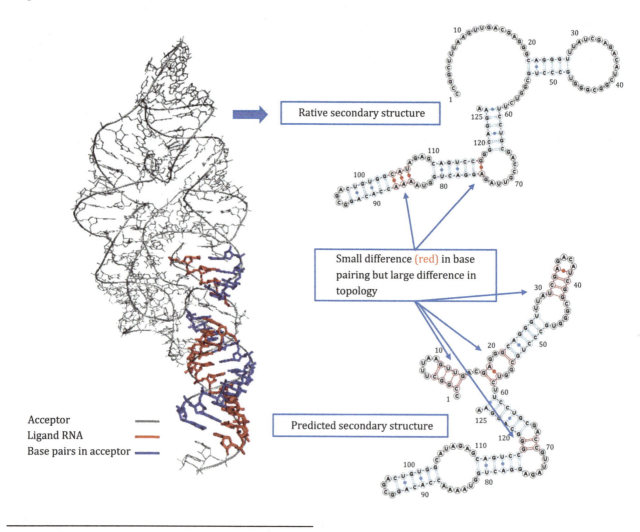

Correspondence: yxiao@hust.edu.cn (Y. Xiao)

Abstract Secondary structures of RNAs are crucial to the understanding of their tertiary structures and functions. At present, many theoretical methods are widely used to predict RNA secondary structures. The performance of these methods has been evaluated but only for their ability of base-pairing prediction. However, the topology of a RNA secondary structure is more important for understanding its tertiary structure and function, especially for long RNAs. In this paper, we constructed a new non-redundant RNA database containing 73 RNA with lengths of 50–300 nucleotides and benchmarked four popular algorithms for both base pairing and topology. The results show that the prediction accuracy of secondary structure topology is only 38%, in contrast to 70% for that of base pairing. Furthermore, the topological consistency is not strongly correlated to the base-pairing consistency. Our results will be helpful to understand the limitations of RNA secondary structure prediction methods from a different point of view and also to their improvements in future.

Keywords RNA secondary structure prediction, Base pairs, Structural topology, Prediction accuracy

INTRODUCTION

Now it is recognized that RNA plays more important roles in the life process than expected (Li *et al.* 2017; Zhao *et al.* 2016). Besides the messenger RNA, transfer RNA, and ribosomal RNA in the genetic central dogma, many new non-coding RNAs have been discovered to have important roles in various biological processes. Among them, there are large RNAs, like ribonuclease, ribozyme, signal recognition particle RNA (SRP RNA), and riboswitches. The tertiary structures of these RNA molecules are very important for their biological functions (Zhao *et al.* 2013). For example, the riboswitches are capable of binding the metabolites to regulate gene expressions using a variety of secondary and tertiary structures (Gong *et al.* 2014; Weinberg *et al.* 2017). Due to the technology limitations, it is still a challenge to determine RNA tertiary structures experimentally. So, many groups developed computational strategies to predict RNA tertiary structures. Since RNA folding is considered as a hierarchical process (Tinoco and Bustamante 1999), most successful approaches of predicting RNA tertiary structures are based on secondary structures (Cao and Chen 2011; Popenda *et al.* 2012; Wang *et al.* 2015, 2017; Xu and Chen 2015; Zhao *et al.* 2012) and this can improve the accuracy of RNA tertiary structure predictions significantly.

REVIEW

The methods of RNA secondary structure prediction have been well established (Mathews *et al.* 2016). The most accurate secondary structure prediction method is to use the multiple sequence analysis or shape-directed to find the conserved motifs (Tan *et al.* 2017). However, this is not always practical due to some unique sequences and limited information at present. Therefore, the physics-based free energy minimization approach of RNA secondary structure prediction is still widely used by biologists. The performances of this type of secondary structure predictions have been evaluated previously and were found to be about 70% when comparing predicted and native base pairs on an RNA database without riboswitches and ribozymes (Mathews *et al.* 2004; Xu *et al.* 2012). Under such an accuracy of base-pair prediction, if using the predicted secondary structure information in the tertiary structure prediction, the topology of the former is critical to the performance of the latter. Therefore, it is also important to know how accurately the current algorithms predict the topologies of RNA secondary structures besides base pairing. Here, we introduce a topology consistency metrics to evaluate the consistency of the topology of a predicted secondary structure with that of the corresponding experimental one. We shall use a database consisting of 73 RNAs with length of 50–300 nt extracted from the RCSB Protein Data Bank (Westbrook *et al.* 2003) and including all types of RNAs known by now, and predict their secondary structures using four popular methods based on free energy minimization: Mfold (Zuker 2003), RNAfold (Gruber *et al.* 2015), RNAstructure (Bellaousov *et al.* 2013), and RNAshapes (Janssen and Giegerich 2015). We also analyze some possible reasons for the difficulties of the prediction of correct topologies of RNA secondary structures.

RESULTS AND DISCUSSION

Features of native secondary structures

There are several types of RNAs in the non-redundant RNA database, including riboswitch, tRNA (transfer RNA), HCV-IRES (hepatitis C viral-internal ribosome entry site) RNA, Ribonuclease, Ribozyme, Ribosomal RNA, SRP (signal recognition particle) RNAs, Group introns, and others (listed in Table 1). RNA secondary structure can be divided into helical stems and various kinds of loops, such as internal loops, bulge loops, and multi-way junction loops. The helical stems are formed by complementary canonical Watson–Crick and non-canonical Watson–Crick base pairs. The internal loops or bulge loops are unpaired nucleotides between helical stems. The multi-way junction loops are the connections between three or more helical stems. Table 2 provides the features of the secondary structures of RNAs in the database. It shows that about 58% of the total 6936 nucleotides are involved in forming base pairs, and, therefore, about half of the nucleotides are involved in base pairs and other half in different types of loops. The correct formations of both helices and loops are crucial to the overall topology of the secondary structures.

Current algorithms prefer to consider the canonical base pairs (A–U, C–G, and G–U base pairs) in RNA secondary structure predictions. However, there are nearly 10% of the 2028 native base pairs that are non-canonical base pairs (A–G, A–C, A–A, U–U, C–U, C–C, and G–G base pairs). This implies that the prediction accuracy of base pairs cannot be higher than 90% by using current secondary structure prediction methods.

Base-pairing consistency of native and predicted secondary structures

Table 3 shows the base-pairing consistency of native and predicted secondary structures by the four methods stated above. It is shown that the mean values of the sensitivity and positive predictive value (PPV) are similar for the four algorithms and both are around 0.70. This is in agreement with previous results (Mathews et al. 2004) where they used a RNA database without riboswitches and ribozymes. Among the nine types of RNA, the values of the sensitivity (0.56–0.60) and PPV (0.44–0.49) for ribosomal RNA are significantly lower than other types for all the four algorithms. One of the reasons for this may be due to the base-pairing or tertiary interactions with proteins or other RNAs because of compact assembly of ribosomal RNAs and proteins in ribosomes. For HCV-IRES RNA, RNAshapes, Mfold, and RNAfold have much lower sensitivity (0.54–0.57) and PPV (0.60) except RNAstructure. For other types of RNA, the performances of the four algorithms are similar.

To further analyze the base-pairing consistency of the native and predicted secondary structures, the data in Tables 2 and 3 are also visualized in different ways in Fig. 1. Figure 1A shows the number of base pairs versus the lengths for the native secondary structures in the

Table 1 PDB ID of RNAs in the database

PDB	Type	Length	PDB	Type	Length
Unbound RNA					
3E5C	Riboswitch	54	1KH6	HCV-IRES	53
1U8D	Riboswitch	68	1P5O	HCV-IRES	77
3IVN	Riboswitch	69	1U9S	Ribonuclease	161
1Y26	Riboswitch	71	2QUS	Ribozyme	69
2GIS	Riboswitch	94	2OIU	Ribozyme	71
3F2Q	Riboswitch	112	1C2X	Ribosomal	120
2QBZ	Riboswitch	161	2IL9	Ribosomal	142
3DIG	Riboswitch	174	1Z43	SRP RNA	101
1YFG	tRNA	75	1KXK	Group intron	70
1EHZ	tRNA	76	1GRZ	Group intron	247
2 K4C	tRNA	76	1S9S	Others	101
2TRA	tRNA	79	1FOQ	Others	109
3A3A	tRNA	90			
Bound RNA in RNA-protein and RNA-RNA complexes					
3EGZ	Riboswitch	65	3A2K	tRNA	77
3K0J	Riboswitch	87	2ZUE	tRNA	78
3IWN	Riboswitch	93	1H3E	tRNA	86
2DLC	tRNA	69	1WZ2	tRNA	88
3EPH	tRNA	69	3ADB	tRNA	92
2DU3	tRNA	71	1M5 K	Ribozyme	92
2ZNI	tRNA	72	2GCS	Ribozyme	125
1B23	tRNA	74	2NZ4	Ribozyme	141
1GIX	tRNA	74	1DK1	Ribosomal	60
2D6F	tRNA	74	1UN6	Ribosomal	61
3AKZ	tRNA	74	3BBO_c	Ribosomal	103
1GAX	tRNA	75	3JYX_3	Ribosomal	113
2AZX	tRNA	75	3BBO_b	Ribosomal	117
1EIY	tRNA	76	1NKW	Ribosomal	124
1F7U	tRNA	76	1S1I	Ribosomal	125
2DER	tRNA	76	3JYX_4	Ribosomal	157
2WRN	tRNA	76	3KTW	SRP RNA	96
3KFU	tRNA	76	1L9A	SRP RNA	128
1C0A	tRNA	77	1U6B	Group intron	197
1H4Q	tRNA	77	2RKJ	Group intron	246
1J1U	tRNA	77	3HJW	Others	58
1J2B	tRNA	77	3HAY	Others	71
2FMT	tRNA	77	2IHX	Others	75
2WWL	tRNA	77	3CUL	Others	92

Table 2 The summary of native secondary structures

RNA type	Sequence number	Nucleotides	Base pairs
Riboswitch	11	1048	327
tRNA	31	2386	679
HCV-IRES	2	130	46
Ribonuclease	1	161	48
Ribozyme	5	498	140
Ribosomal RNA	10	1122	270
SRP RNA	3	325	122
Group intron	4	760	218
Others	6	506	178
Total	73	6936	2028

database. As expected, the base-pair number increases as the sequence length and has a high linear correlation coefficient of 0.86. The slope of the fitting line describes the probability of the base pairs formation, which is about one base pair for every four nucleotides as mentioned above. Figure 1B shows the consistent base-pair number (the number of true positives) of predicted base pairs versus that of native base pairs in the database. The two numbers also have a high linear correlation with a correlation coefficient of 0.83. It also shows that the base-pairing consistency does not decrease quickly with the number of the native base pairs or length of RNA because many short RNAs also have low base-pairing consistency. This is shown more clearly in Fig. 1C and D, which gives the base-pair sensitivity versus RNA length and the number of native base pairs, respectively. They show that about 20% short RNAs (<100 nt) has sensitivity less than 0.5. These results indicate that even for short RNAs (<100 nt) the base pairs cannot be correctly predicted by using current algorithms completely. For long RNAs (>100 nt), since the number of long RNA chains in the database is small, more long RNA sequences are needed to give a reliable conclusion about the base-pairing consistency.

Topological consistency of native and predicted secondary structures

Table 3 also shows the topological consistency of the native and predicted secondary structures. It shows that on average only about 38% of the predicted secondary structures have identical topologies with those of the native ones (Fig. 2A). This consistency rises up to 60%–70% if we include those predicted secondary structures that have similar topologies with the native ones (Fig. 2B). The topological consistency levels of the predicted secondary structures of the group intron and ribosomal RNA are the lowest ones for all the four algorithms besides ribonuclease which has only one sequence in the database. The reason for this low topological consistency is that the PPV, i.e., the percentage of predicted base pairs occurs in the native secondary structure, is only 69%. In other words, in the predicted base pairs there are about 31% that are inconsistent with the native ones. This inconsistent base pairing may form different internal bulge and multiple loops from the native ones and lead to different overall topologies of the secondary structures. Figure 2 gives three examples of topological consistency of the native and predicted secondary structures by RNAshapes. Figure 2A is an example (the ribozyme RNA (PDB ID: 2QUS)) that the topologies of the native and predicted secondary structures are identical and there are only two missing non-canonical base pairs C12–A53 and A38–A51 in the predicted secondary structure in comparison with the native one (marked in red color). In this case, the sensitivity of the predicted base pairs is about 0.91. Figure 2B shows an example (ribosomal RNA (PDB ID: 1DK1)) that the topologies of the native and predicted secondary structures are similar but there are base-pairing shifts (the region in red color). In this case, the sensitivity of the predicted base pairs is about 0.72. There is a bulge C50 in the native secondary structure. However, in the predicted secondary structure, C50 forms a base pair with G22 and there is an internal loop (U23, G24, A48, and C49) formed with shifted base pairs G25–C47, G26–C46, and A27–U45. Figure 2C is an example (tRNA (PDB ID: 3BBV)) that the topologies of the native and predicted secondary structures are completely different. The native secondary structure has a four-way junction but the predicted result is a long helical stem only with internal loops and bulge loops. In this case, the sensitivity of the predicted base pairs is about 0.52.

Some discussions

The results above show that it is still a challenge to predict the topology of RNA secondary structures correctly and the identical topology consistency of predicted secondary structures with native ones is only about 38% on average. Therefore, the RNA tertiary structure prediction based on pure predicted secondary structures is greatly limited. Even the similar topology consistency of the predicted secondary structures with the native ones can reach 60%–70%, but in this case the predicted secondary structures usually introduce different or additional bulge and internal loops that may also decrease the accuracy of

Table 3 The base-pairing and topological consistencies by different RNA secondary structure prediction algorithms

RNA type	RNAshapes			Mfold		
	Sensitivity	PPV	Identical(similar) topology consistency	Sensitivity	PPV	Identical(similar) topology consistency
Riboswitch	0.73	0.76	0.45(0.73)	0.71	0.74	0.45(0.64)
tRNA	0.69	0.67	0.39(0.48)	0.68	0.66	0.39(0.55)
HCV-IRES	0.54	0.60	0.50(0.50)	0.54	0.60	0.50(0.50)
Ribonuclease	0.88	0.84	0.00(1.00)	0.88	0.84	0.00(1.00)
Ribozyme	0.81	0.76	0.40(0.60)	0.81	0.75	0.40(0.60)
Ribosomal RNA	0.58	0.47	0.10(0.70)	0.60	0.49	0.10(0.70)
SRP RNA	0.75	0.85	0.33(0.67)	0.76	0.89	0.33(0.67)
Group intron	0.69	0.68	0.00(0.25)	0.70	0.69	0.00(0.25)
Others	0.89	0.92	1.00(1.00)	0.89	0.92	1.00(1.00)
Total	0.71	0.69	0.38(0.60)	0.71	0.69	0.38(0.62)
RNA type	RNAfold			RNAstructure		
	Sensitivity	PPV	Identical(similar) topology consistency	Sensitivity	PPV	Identical(similar) topology consistency
Riboswitch	0.69	0.71	0.45(0.73)	0.69	0.73	0.36(0.73)
tRNA	0.69	0.65	0.35(0.48)	0.78	0.75	0.48(0.81)
HCV-IRES	0.57	0.60	0.50(0.50)	0.85	0.91	0.50(1.00)
Ribonuclease	0.88	0.84	0.00(1.00)	0.88	0.82	0.00(1.00)
Ribozyme	0.72	0.67	0.20(0.40)	0.80	0.74	0.20(0.60)
Ribosomal RNA	0.56	0.44	0.10(0.70)	0.57	0.44	0.00(0.50)
SRP RNA	0.70	0.81	0.00(0.67)	0.66	0.77	0.00(0.33)
Group intron	0.70	0.67	0.00(0.25)	0.69	0.66	0.00(0.25)
Others	0.83	0.86	0.83(0.83)	0.83	0.86	0.67(1.00)
Total	0.69	0.66	0.33(0.58)	0.73	0.70	0.34(0.71)

the RNA tertiary structure prediction based on them. Therefore, from a topological point of view, the success rate of RNA tertiary structure prediction based on pure predicted secondary structures can reach more than 38% or 60%–70% if wrong internal and bulge loops are ignored.

The results above (Fig. 2) also show that the topological consistency is unnecessarily always correlated to the base-pair consistency, as indicated in Fig. 3. For example (Fig. 4), the sensitivity and PPV of the predicted base pairs of a ribozyme (PDB ID: 2GCS) are about 0.86 and 0.65, respectively, but the small number of inconsistent base pairs makes the topology of the predicted secondary structure significantly different from the native one (Figs. 4B, C). Furthermore, we analyzed the statistical correlation of sensitivity/specificity and topological consistency (including both identical and similar topologies) for RNAshapes (0.64/0.54), Mfold (0.65/0.52), RNAfold (0.50/0.45), and RNAstructure (0.78/0.63), respectively (Fig. 3). The low correlation values indicate that the topological consistency is not strongly correlated to the base-pairing consistency and small number of wrong-paired bases can change the topologies significantly.

One of the reasons that affects the base-pairing and topological predictions may be due to the base-pairing or tertiary interactions with other molecules (Perederina et al. 2002). Figure 4A is the native complex structure of a ribozyme (PDB ID: 2GCS) and its amino RNA inhibitor. It shows that G3 to U13 and G14 form canonical and non-canonical base pairs with the amino RNA inhibitor in the native secondary structure and cannot form the hairpin structure in the predicted structure (Fig. 4C). Similarly, C16–A18 form tertiary interactions with U31–G33 and cannot form the internal loop G17–C58 and A18–U57 in the predicted structure. These indicate that accurate prediction of the topology of secondary structures of RNA molecules also needs to consider their interactions with other molecules, although the number of these interactions is

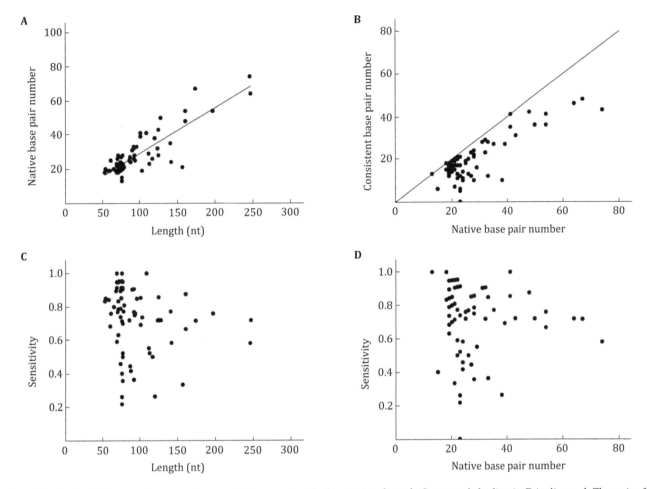

Fig. 1 Statistics of the secondary structure prediction results. The line in **A** is linearly fitting and the line in **B** is diagonal. The unit of length and base-pair number is nucleotide. Sensitivity describes what percentage of native base pairs occurs in the predicted secondary structure

small in comparison with that of the total native base pairs.

CONCLUSION

In this paper, we evaluate the physics-based free energy minimization secondary structure prediction methods by comparing the base-pairing as well as topological consistencies. We built a non-redundant RNA tertiary structure database consisting main types of RNAs to construct our statistical analysis. The benchmark tests show that the percentages of correct predictions of the base-pair predictions and topology are about 70% and 38% on average, respectively. Furthermore, the topological consistency is not strongly correlated to the base-pairing consistency under current accuracy of base-pair prediction. Relatively high accuracy of base-pair prediction does not mean correct topology of secondary structure. This suggests that experimental information about secondary structure is usually needed to build accurate tertiary structures of RNAs. Our results will be helpful to understand the limitations of RNA secondary structure prediction methods and their applications in RNA tertiary structure prediction.

MATERIALS AND METHODS

Database of experimental RNA tertiary structures

In order to evaluate the performance of different RNA secondary structure prediction methods fairly, we built a non-redundant RNA tertiary structure database from the experimental RNA tertiary structures in the RCSB Protein Data Bank (PDB) (Westbrook et al. 2003). The RNAs with size of 50–300 nt are practical for RNA tertiary structure prediction (Zhao et al. 2012). Therefore, we first collected all the RNA structures with lengths

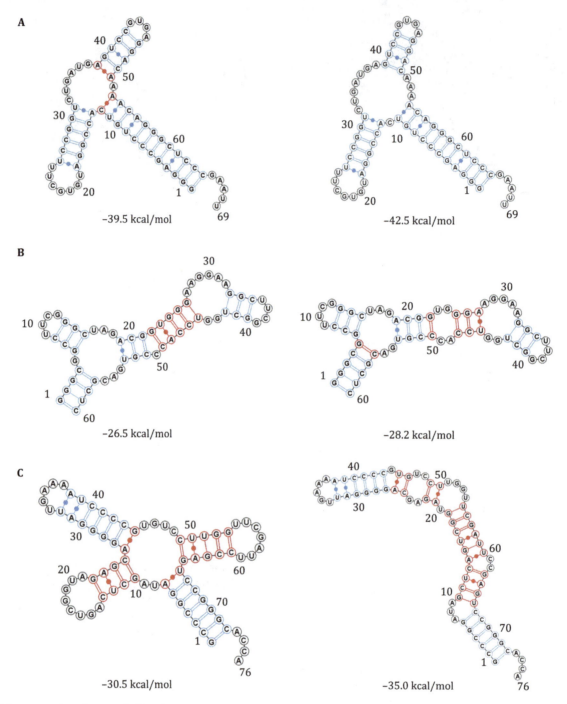

Fig. 2 Three examples of native (left) and predicted (right) secondary structures. Their overall topologies are identical (**A**), similar (**B**), and different (**C**). The consistent and inconsistent base pairs are colored in *blue* and *red*, respectively. Their free energies are also listed

between 50 and 300 nucleotides in the PDB database. The current non-redundant RNA tertiary structure databases of statistical potential used the sequence identity of 95% and 80% to reduce redundancy (Bernauer et al. 2011; Capriotti et al. 2011; Wang et al. 2015). Here, we used a lower sequence identity of 75% to remove possible homology structures in the selected RNAs. Although a lot of efforts, it is still a challenge to predict the pseudoknots. And so the RNAs with pseudoknots are not included in the non-redundant RNA tertiary structure database. The database totally contains 73 RNA tertiary structures (Table 1), including 25 in unbound state and 48 in RNA–protein and RNA–RNA complexes.

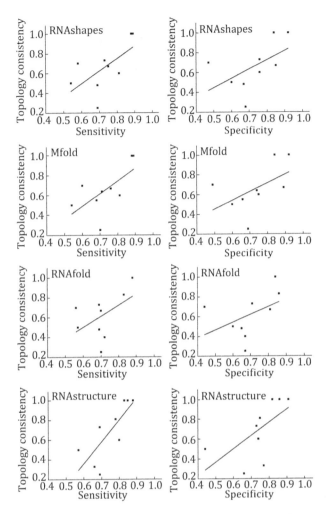

Fig. 3 Statistical linear fitting of sensitivity/specificity and topological consistency for RNAshapes, Mfold, RNAfold, and RNAstructure, respectively

Secondary structure prediction and analysis

The free energy minimization approach of RNA secondary structure prediction is the most popular method when RNA homologous information is limited (Mathews and Turner 2006), e.g., Mfold (Zuker 2003), RNAfold (Gruber et al. 2015), RNAstructure (Bellaousov et al. 2013), and RNAshapes (Janssen and Giegerich 2015). Mfold uses a dynamic programming algorithm to generate a set of candidates of secondary structure for an RNA, and then calculates their free energies by adding up those of independent subunits using the nearest neighborhood approximation. The free energies of the subunits were determined experimentally. RNAstructure is similar to Mfold but uses alternative set of thermodynamic parameters. RNAfold is also based on the minimum free energy model but it can compute the equilibrium partition functions and base-pairing probabilities. RNAshapes is different from the three algorithms above. It first clusters potential secondary structures of an RNA into different abstract shapes and finds a minimum free energy structure as the representative of each shape. Then, it can calculate the shape probability or use other biological information to identify the possible native secondary structure from the representatives. Here, we use these four algorithms to evaluate the performance of secondary structure prediction. The experimental secondary structures (called as "native secondary structures") were generated from the experimental tertiary structures in the RNA database using the sequence to structure (S2S) algorithm (Jossinet and Westhof 2005). S2S can display the RNA data, like sequences, secondary structures, and tertiary structures. In addition, we analyzed the Watson–Crick base pairs (Leontis and Westhof 2001) (A–U, C–G, and G–U base pairs) and non-Watson–Crick base pairs (Leontis et al. 2002) (A–G, A–C, A–A, U–U, C–U, C–C, and G–G base pairs). In order to observe the consequence of RNA–protein or RNA–RNA tertiary interactions on predicted RNA secondary structures, we also calculated the hydrogen bonds between RNA and protein or RNAs in the complexes by the hydrogen bond calculation algorithm HBPLUS (McDonald and Thornton 1994). The free energy of RNA secondary structure was calculated by using the RNAeval algorithm in Vienna RNA package (Gruber et al. 2015).

We analyzed base-pairing consistencies between native and predicted secondary structures of the RNA structures in the database. The base-pairing consistency was measured by both sensitivity (STY) and positive predictive value of precision (PPV) (Parisien et al. 2009).

$$STY(sensitivity) = \frac{TP}{TP + FN}$$

$$PPV(precision) = \frac{TP}{TP + FP}$$

The base pairs found in both native and prediction sets are true positives (TP). The base pairs found in native sets but not in prediction sets are false negatives (FN). The base pairs found in prediction sets but not in native sets are false positives (FP). Sensitivity (STY) describes what percentage of native base pairs occurs in the predicted secondary structure. Specificity (PPV) denotes what percentage of predicted base pairs occurs in the native secondary structure.

Topological consistency analysis

The topological consistency measures the topological similarity of the native and predicted secondary structures. Since it is difficult to find a simple index to clearly

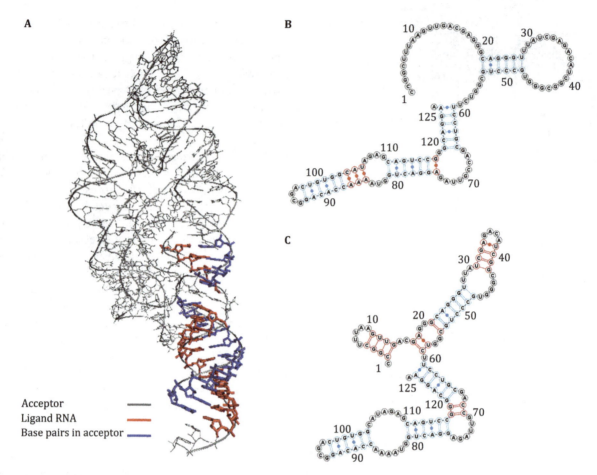

Fig. 4 **A** Tertiary interactions between a ribozyme (PDB ID: 2GCS) and ligand RNAs. **B** and **C** are the native and predicted secondary structures of 2GCS, respectively. The consistent and inconsistent base pairs between them are colored in *blue* and *red*, respectively

distinguish the difference of RNA secondary structures, we divide the topological consistency into three types: identical, similar, and different (Fig. 2). The topology of a predicted secondary structure is considered to be identical to that of the native one if the former has the same loops with the latter (Fig. 2A). In this case, the predicted secondary structures may have a few additional base pairs that do not occur in the native ones or a few of the native base pairs that do not form. The topology of a predicted secondary structure is considered to be similar to that of the native one if the former has the same multi-way junction loops with the latter but different internal and bulge loops (Fig. 2B). Finally, the topology of a predicted secondary structure is considered to be different from that of the native one if the former has different multi-way junction loops with the latter (Fig. 2C).

Acknowledgements This work is supported by the NSFC under Grant Nos. 31570722 and 11374113.

Compliance with Ethical Standards

Conflict of interest Yunjie Zhao, Jun Wang, Chen Zeng, and Yi Xiao declare that they have no conflict of interest.

Human and animal rights and informed consent This article does not contain any studies with human or animal subjects performed by any of the authors.

References

Bellaousov S, Reuter JS, Seetin MG, Mathews DH (2013) RNAstructure: web servers for RNA secondary structure prediction and analysis. Nucleic Acids Res 41:W471–W474

Bernauer J, Huang X, Sim AY, Levitt M (2011) Fully differentiable coarse-grained and all-atom knowledge-based potentials for RNA structure evaluation. RNA 17:1066–1075

Cao S, Chen SJ (2011) Physics-based *de novo* prediction of RNA 3D structures. J Phys Chem B 115:4216–4226

Capriotti E, Norambuena T, Marti-Renom MA, Melo F (2011) All-atom knowledge-based potential for RNA structure prediction and assessment. Bioinformatics 27:1086–1093

Gong Z, Zhao Y, Chen C, Duan Y, Xiao Y (2014) Insights into ligand binding to PreQ1 Riboswitch Aptamer from molecular dynamics simulations. PLoS ONE 9:e92247

Gruber AR, Bernhart SH, Lorenz R (2015) The ViennaRNA web services. Methods Mol Biol 1269:307–326

Janssen S, Giegerich R (2015) The RNA shapes studio. Bioinformatics 31:423–425

Jossinet F, Westhof E (2005) Sequence to structure (S2S): display, manipulate and interconnect RNA data from sequence to structure. Bioinformatics 21:3320–3321

Leontis NB, Westhof E (2001) Geometric nomenclature and classification of RNA base pairs. RNA 7:499–512

Leontis NB, Stombaugh J, Westhof E (2002) The non-Watson-Crick base pairs and their associated isostericity matrices. Nucleic Acids Res 30:3497–3531

Li X, Bu D, Sun L, Wu Y, Fang S, Li H, Luo H, Luo C, Fang W, Chen R, Zhao Y (2017) Using the NONCODE Database Resource. Curr Protoc Bioinform 58:12.16.1–12.16.19

Mathews DH, Turner DH (2006) Prediction of RNA secondary structure by free energy minimization. Curr Opin Struct Biol 16:270–278

Mathews DH, Disney MD, Childs JL, Schroeder SJ, Zuker M, Turner DH (2004) Incorporating chemical modification constraints into a dynamic programming algorithm for prediction of RNA secondary structure. Proc Natl Acad Sci USA 101:7287–7292

Mathews DH, Turner DH, Watson RM (2016) RNA secondary structure prediction. Curr Protoc Nucleic Acid Chem 67:1–19

McDonald IK, Thornton JM (1994) Satisfying hydrogen bonding potential in proteins. J Mol Biol 238:777–793

Parisien M, Cruz JA, Westhof E, Major F (2009) New metrics for comparing and assessing discrepancies between RNA 3D structures and models. RNA 15:1875–1885

Perederina A, Nevskaya N, Nikonov O, Nikulin A, Dumas P, Yao M, Tanaka I, Garber M, Gongadze G, Nikonov S (2002) Detailed analysis of RNA–protein interactions within the bacterial ribosomal protein L5/5S rRNA complex. RNA 8:1548–1557

Popenda M, Szachniuk M, Antczak M, Purzycka KJ, Lukasiak P, Bartol N, Blazewicz J, Adamiak RW (2012) Automated 3D structure composition for large RNAs. Nucleic Acids Res 40:e112

Tan Z, Fu Y, Sharma G, Mathews DH (2017) TurboFold II: RNA structural alignment and secondary structure prediction informed by multiple homologs. Nucleic Acids Res 45:11570–11581

Tinoco I Jr, Bustamante C (1999) How RNA folds. J Mol Biol 293:271–281

Wang J, Zhao Y, Zhu C, Xiao Y (2015) 3dRNAscore: a distance and torsion angle dependent evaluation function of 3D RNA structures. Nucleic Acids Res 43:e63

Wang J, Mao K, Zhao Y, Zeng C, Xiang J, Zhang Y, Xiao Y (2017) Optimization of RNA 3D structure prediction using evolutionary restraints of nucleotide-nucleotide interactions from direct coupling analysis. Nucleic Acids Res 45:6299–6309

Weinberg Z, Nelson JW, Lunse CE, Sherlock ME, Breaker RR (2017) Bioinformatic analysis of riboswitch structures uncovers variant classes with altered ligand specificity. Proc Natl Acad Sci USA 114:E2077–E2085

Westbrook J, Feng Z, Chen L, Yang H, Berman HM (2003) The protein data bank and structural genomics. Nucleic Acids Res 31:489–491

Xu X, Chen SJ (2015) Physics-based RNA structure prediction. Biophys Rep 1:2–13

Xu Z, Almudevar A, Mathews DH (2012) Statistical evaluation of improvement in RNA secondary structure prediction. Nucleic Acids Res 40:e26

Zhao Y, Huang Y, Gong Z, Wang Y, Man J, Xiao Y (2012) Automated and fast building of three-dimensional RNA structures. Sci Rep 2:734

Zhao Y, Wang J, Chen X, Luo H, Zhao Y, Xiao Y, Chen R (2013) Large-scale study of long non-coding RNA functions based on structure and expression features. Sci China Life Sci 56:953–959

Zhao Y, Li H, Fang S, Kang Y, Wu W, Hao Y, Li Z, Bu D, Sun N, Zhang MQ, Chen R (2016) NONCODE 2016: an informative and valuable data source of long non-coding RNAs. Nucleic Acids Res 44:D203–D208

Zuker M (2003) Mfold web server for nucleic acid folding and hybridization prediction. Nucleic Acids Res 31:3406–3415

Regulation of metabolism by the Mediator complex

Dou Yeon Youn[1,2], Alus M. Xiaoli[1,3], Jeffrey E. Pessin[1,2], Fajun Yang[1,3]

[1] Division of Endocrinology, Department of Medicine, Diabetes Research Center, Albert Einstein College of Medicine, Bronx, NY 10461, USA

[2] Department of Molecular Pharmacology, Albert Einstein College of Medicine, Bronx, NY 10461, USA

[3] Department of Developmental and Molecular Biology, Albert Einstein College of Medicine, Bronx, NY 10461, USA

Abstract The Mediator complex was originally discovered in yeast, but it is conserved in all eukaryotes. Its best-known function is to regulate RNA polymerase II-dependent gene transcription. Although the mechanisms by which the Mediator complex regulates transcription are often complicated by the context-dependent regulation, this transcription cofactor complex plays a pivotal role in numerous biological pathways. Biochemical, molecular, and physiological studies using cancer cell lines or model organisms have established the current paradigm of the Mediator functions. However, the physiological roles of the mammalian Mediator complex remain poorly defined, but have attracted a great interest in recent years. In this short review, we will summarize some of the reported functions of selective Mediator subunits in the regulation of metabolism. These intriguing findings suggest that the Mediator complex may be an important player in nutrient sensing and energy balance in mammals.

Keywords Transcription, Mediator, Cofactor, Metabolism, Insulin resistance, Obesity

INTRODUCTION

Gene transcription in eukaryotic cells is orchestrated through extremely complicated processes and multiple steps, including initiation, elongation, and termination, with the initiation being the most studied regulation step of gene expression. Disturbance of transcription initiation often results in serious diseases, such as cancer, in humans (Lee and Young 2013). Among the numerous different components in the initiation step of eukaryotic transcription, RNA polymerase II (Pol II) and the general transcription factors (TFIIB, TFIID, TFIIE, TFIIF, and TFIIH) constitute the basal transcription machinery, and specific sets of transcription factors are essential to determine the activation or repression of the target genes. However, it is generally believed that most of the transcription factors in eukaryotes are unable to directly interact with Pol II, which is responsible for the production of all mRNAs. Studies in the past several decades suggest that this crucial gap is often filled by various transcription cofactors, which critically regulate the activation or repression in gene expression.

There is no doubt that human beings are currently experiencing an epidemic of obesity. As a result, the prevalence of type 2 diabetes (T2D) has sharply increased in the past decades. Dysregulation of genes in the pathways controlling glucose and/or lipid metabolism is common in states of insulin resistance and T2D, especially when patients are also diagnosed with non-alcoholic fatty liver disease (NAFLD) or cardiovascular disease (Brown and Goldstein 2008; Oh et al. 2013). The DNA-binding transcription factors and their cofactors have been one focus of the studies in the hope of understanding the molecular mechanisms that cause metabolic diseases. A few notable examples of transcription factors that regulate glucose and/or lipid metabolism include cAMP-regulated enhancer-binding

Correspondence: fajun.yang@einstein.yu.edu (F. Yang)

protein (CREB), forkhead O box proteins (FoxOs), glucocorticoid receptor (GR), hepatic nuclear factors (HNFs), sterol response element binding protein-1c (SREBP-1c), carbohydrate response element binding protein (ChREBP), liver X receptors (LXRs), and peroxisome proliferator-activated receptor-γ (PPARγ) (Brown and Goldstein 2008; Oh et al. 2013; Lefterova et al. 2014). Among the transcription cofactors, a multi-subunit protein assembly called the Mediator complex has been linked to several of these transcription factors (Malik and Roeder 2010; Taatjes 2010; Conaway and Conaway 2011; Allen and Taatjes 2015). Here, we summarize the current understanding on the regulation of transcription factors by individual subunits of the Mediator complex, focusing mainly on the metabolic involvement of the mediator subunits in mammals.

The Mediator complex as a transcription cofactor

The Mediator complex is able to bind to various transcription factors and integrates the transcriptional signals to the basal transcription machinery (Malik and Roeder 2010; Taatjes 2010; Conaway and Conaway 2011; Allen and Taatjes 2015). Originally discovered in yeast as a transcription cofactor (Kelleher et al. 1990; Flanagan et al. 1991; Kim et al. 1994), the mammalian Mediator complex has been given several different names in the early literature, including TRAP (thyroid hormone receptor-associated protein) (Fondell et al. 1996), ARC (activator-recruited cofactor) (Naar et al. 1999), or DRIP (vitamin D receptor-interacting protein) (Rachez et al. 1998). In HeLa cells, the Mediator complex can be biochemically isolated in at least two forms: the small mediator, also known as the core mediator (with a mass of about 600 kDa), that comprises up to 26 subunits organized into the head, middle, and tail sub-modules, and the large mediator (with a mass of about 1.2 MDa), which additionally contains the kinase sub-module of four subunits—cyclin-dependent kinase 8 (CDK8), cyclin C (CycC), MED12, and MED13, but lacks the MED26 subunit (Taatjes et al. 2002). However, in other studies only the large Mediator complex was observed in the biochemical purifications (Wang et al. 2001; Sato et al. 2004). Nevertheless, it is generally believed that the Mediator complex is heterogeneous both structurally and functionally. In terms of functions, initial data suggested that the small mediator functions to activate transcription initiation in vitro, while the large mediator is inactive or acts as a transcription repressor (Taatjes et al. 2002). However, the large mediator has also been linked to transcription elongation (Donner et al. 2010). In addition, the mediator kinase module behaves in a context-specific manner to either repress or activate transcription, depending on the transcription factors and/or target gene promoters (Nemet et al. 2014).

Since the Mediator complex can directly bind to Pol II (Naar et al. 1999; Casamassimi and Napoli 2007; Taatjes 2010; Soutourina et al. 2011; Lariviere et al. 2012), collectively it may regulate many genes. However, studies on individual subunits of the Mediator complex revealed remarkable transcription factor and/or tissue-specific functions for certain subunits. One example is that the MED15 subunit binds to and regulates SREBP-dependent transcription, while the MED25 subunit primarily controls VP16-mediated transcription (Mittler et al. 2003; Yang et al. 2004, 2006). Although it is unclear how many transcription factors could physically interact with the Mediator complex, it is less likely that all transcription factors need this complex to regulate target gene expression. For instance, currently there is no evidence that CREB or Myb can directly bind to the Mediator complex, and depletion of some subunits such as MED15 or MED25 had no effect on Myb-dependent gene transcription in HEK293 cells (Yang et al. 2004, 2006). Moreover, the mediator subunit abundance and/or subcellular location may be different under different conditions. For example, feeding, obesity, NAFLD, and aging can cause a significant reduction of CDK8 and CycC proteins in the liver due to mTORC1 activation (Feng et al. 2015). Another example is that in response to stress, CycC undergoes translocation from the nucleus to the cytoplasm to regulate mitochondrial fission (Cooper et al. 2014; Wang et al. 2015). Although little is known about how the mediator subunit abundance is regulated, it may alter the Mediator complex composition and thus may have profound effects on the mediator-dependent functions.

Recent structure studies have demonstrated the presence of many different conformational assembly states of the Mediator complex (Tsai et al. 2014; Wang et al. 2014), indicating the intrinsic flexibility and heterogeneity. Moreover, the functions of the Mediator complex are now expanded to include the transcription elongation (Donner et al. 2010; Takahashi et al. 2011) and termination (Mukundan and Ansari 2011), and mRNA processing (Huang et al. 2012) and export (Schneider et al. 2015). Due to its massive size and complexity, however, the precise molecular mechanism(s) by which the Mediator complex regulates the gene-specific transcription remain poorly understood but are likely to play important functional roles in both normal physiological and pathophysiological states.

The head module

The head module includes MED6, MED8, MED11, MED17, MED18, MED20, MED22, MED27, MED28, MED29, and MED30. This sub-module maintains the overall structure of the Mediator complex. The importance of the head module has been demonstrated from studies showing that yeast loss of MED17 prevents nearly all mRNA synthesis (Holstege et al. 1998; Thompson and Young 1995). Although it is unclear whether MED17 plays a similarly essential role in mammalian cells, a recent study reported that MED17 in the liver regulates lipogenic gene expression and lipid metabolism through the LXR transcription factors (Kim et al. 2015), providing a mechanism for the mediator regulation of lipid metabolism. Moreover, MED17 also binds to VP16, p53, and the estrogen receptor (ER) (Burakov et al. 2000; Park et al. 2003; Mehta et al. 2009; Meyer et al. 2010; Kumafuji et al. 2014). Similarly, PPARγ interacts with both MED1 (Zhu et al. 1997; Yuan et al. 1998; Ge et al. 2008) and MED14 (Grontved et al. 2010) subunits of the middle and tail modules, respectively. Thus, it appears that a given transactivation domain sometimes recruits the Mediator complex by binding to more than one subunit.

It is the current understanding that the head and middle modules may bind to either Pol II or the kinase module in an exclusive manner (Knuesel et al. 2009a; Naar et al. 2002). The head module is thought to provide the greatest impact in controlling overall transcription primarily by shifting the transcription machinery from active to inactive state rather than serving as a binding site for other transcriptional regulators. The electron microscopy-derived structure and affinity pull-down experiments have identified MED17 as the main subunit that directly interacts with Pol II and promotes the transcription initiation of selected genes (Soutourina et al. 2011). Interestingly, the head module subunit MED30 is a metazoan-specific subunit, and a missense mutation of MED30 in mice resulted in a pleiotropic decrease in transcription of cardiac genes that are necessary for oxidative phosphorylation and mitochondrial integrity (Krebs et al. 2011). The mutation effect of MED30 can be partially protected by a ketogenic diet through increasing the expression of genes, such as *Pgc1a* and *Sod2* (Krebs et al. 2011). This study suggests a critical role of MED30 in metabolism, but the regulatory mechanism(s) shall be investigated in the future.

MED1 provides a docking surface for nuclear receptors

The middle module includes subunits of MED1, MED4, MED7, MED9, MED10, MED19, MED21 MED26, and MED31. The MED26 subunit is a specific subunit of the small mediator (Taatjes et al. 2002) and it is required for transcription elongation (Takahashi et al. 2011). The most studied subunit of the middle module is MED1, but metabolic functions of other subunits in the middle module remain to be investigated.

MED1 is able to interact with numerous transcription factors or cofactors, including PPARα, PPARγ, GR, C/EBPβ, and PGC1α, which are all implicated in metabolic regulation (Zhu et al. 1997; Yuan et al. 1998) (See Jia et al. 2014 for a recent review on MED1). Although deletion of MED1 results in embryonic lethality at E11.5, the MED1 conditional knockout or mutant mice play an important role in adipogenesis through PPARγ, fatty acid oxidation through PPARα, mammary gland development through ER, liver steatosis through GR and constitutive androstane receptor (CAR), thermogenesis regulation through uncoupling protein 1 (UCP1) up-regulation, skeletal muscle function, and insulin signaling (Ge et al. 2002; Jia et al. 2004, 2009; Jiang et al. 2010; Iida et al. 2015).

Similar to the steroid receptor co-activator (SRC) and PGC1 families of transcription cofactors, MED1 contains two LXXLL motifs [located within amino acid (aa) 589–593 and 630–634, respectively] that provide the binding surfaces for various nuclear receptors, and either one of the LXXLL motifs is sufficient for protein–protein interactions (Chang et al. 1999; Ge et al. 2008). A dominant negative form of MED1 with mutant LXXLL motifs reduces the transcription activity of nuclear receptors and suppresses PPARγ-induced adipogenesis (Ge et al. 2002). However, these LXXLL motifs are not required for MED1 regulation of PPARγ in cultured MEFs, suggesting MED1 regulation of gene transcription through alternative mechanisms in a context-dependent manner. Moreover, the conserved N-terminus (aa1–530) of MED1 is also important for PPARγ-target gene expression (Ge et al. 2008). Deletion of MED1 in mouse liver abrogates PPARα-activated peroxisomal proliferation (Jia et al. 2004) and acetaminophen-induced hepatotoxicity through CAR (Jia et al. 2005). Moreover, liver-specific knockout of MED1 protects mice from excessive fat accumulation under high-fat diet, whereas the wild-type mice exhibited fatty liver (Bai et al. 2011).

Recently, MED1 has been found to interact with PRDM16, a key inducer of brown adipose tissue (BAT)-selective genes (Harms et al. 2015; Iida et al. 2015). Directly interacting with the N-terminus of MED1, PRDM16 promotes *Ucp1* gene expression in brown adipocytes (Iida et al. 2015). ChIP-Seq and ChIP-qPCR analyses show that MED1 is recruited to the enhancer sites of BAT-selective genes such as *Ucp1*, *Cidea*, *Ppara*, and *Pgc1a* in the wild-type, but not in PRDM16 knockout brown adipocytes (Harms et al. 2015). Furthermore, skeletal muscle-specific knockout of MED1 increased gene expression of *Ucp1* and *Cidea*, and promoted mitochondrial density in white glycolytic skeletal muscles and respiratory uncoupling (Chen et al. 2010). The regulatory roles of MED1 in major metabolic organs, such as liver, adipose tissues, and skeletal muscle, suggest that MED1 is an important regulator for metabolic gene expression. However, it is difficult to know whether the function(s) of MED1 are all mediator dependent or not.

The tail module directly interacts with various transcription factors

The tail module includes subunits of MED14, MED15, MED16, MED23, MED24, and MED25. The metabolic roles of this module are well established, particularly for MED14, MED15, and MED23.

MED14 has been implicated in regulation of lipid homeostasis. It has been reported that MED14 binds to the N-terminal AF1 domain of GR (Hittelman et al. 1999) or PPARγ (Grontved et al. 2010) to activate target gene transcription. In vitro and in vivo assays show that MED14 directly interacts with GR, and this interaction increased GR-dependent transcription activation (Hittelman et al. 1999). In addition, MED14 directly interacts with PPARγ and promotes PPARγ-dependent transactivation as well as the recruitment of the Mediator complex (Grontved et al. 2010). Recently, it has been demonstrated that the Mediator complex is recruited to the PPARγ-target gene promoters without the LXXLL motifs of MED1, suggesting that MED14 may be more critically required for the transcription activation during adipogenesis (Grontved et al. 2010). Studies of MED14 knockdown revealed that the recruitment of PPARγ, MED6, MED8, and Pol II to the transcription start sites is dependent on MED14 in 3T3-L1 cells (Grontved et al. 2010). At present, it is unclear what the basis is for the apparent conflicting roles of MED1 and MED14 in the regulation of PPARγ-target genes. It remains to be established whether these differences reflect the different experimental interventions used, differences in cells examined, or differential interactions and physiologic outputs that occur during adipogenesis versus fully differentiated adipocytes. In addition, there may be undetermined cell-context effects resulting from the secondary transcription factor interactions. For example, MED14 was also reported to interact with the SREBP transcription factors (Toth et al. 2004).

The functions of MED15 and other mediator subunits in worms have been recently reviewed (Grants et al. 2015). In mammalian cells, MED15 acts as an important regulator of lipid biosynthesis through modulating the SREBPs (Yang et al. 2006). The interaction between MED15 and SREBPs is through the MED15-KIX domain, which is structurally similar to the KIX domains of CBP and p300 (Yang et al. 2006). Interestingly, the KIX domain of MED15 interacts only with SREBP-1a, but not with several other transcription factors examined, suggesting its binding specificity. The SREBP-target genes include the important lipogenic and cholesterogenic genes such as fatty acid synthase (*Fasn*) and HMG-CoA synthase (*Hmgcr*) (Amemiya-Kudo et al. 2002). By recruiting the Mediator complex upon binding to SREBPs, MED15 promotes SREBP-target gene expression (Naar et al. 1998, 1999). Interestingly, small-molecule inhibitors that block the interaction between the MED15-KIX domain and the transactivation domain of SREBP-1a protect mice from the metabolic dysregulation that occurs during diet-induced obesity (Zhao et al. 2014).

MED23 has been well studied in mammalian model systems. The importance of MED23 in viability has been shown with the embryonic lethality when MED23 was knocked out in mice (Balamotis et al. 2009). MED23 has been linked to insulin signaling in the adipogenesis transcription cascade (Wang et al. 2009). In 3T3-L1 cells, it has been shown that the interaction of MED23 with Elk1 is enhanced by the insulin-induced MAPK activation, resulting in further induction of Krox20, the initial transcription factor in the adipogenesis pathway (Wang et al. 2009). Interestingly, a recent study has shown that MED23 is also involved in regulating the differentiation of mesenchymal stem cells into smooth muscle cells or adipocytes (Yin et al. 2012). MED23 deficiency promoted mesenchymal stem cells into smooth muscle cells while preventing the differentiation into adipocytes (Yin et al. 2012). The mechanism by which MED23 controls the differentiation is via regulating the balance of serum response factor (SRF) downstream genes, *RhoA/MAL* or *Ras/ELK1*, by directly interacting with them in response to upstream signals (Stevens et al. 2002; Wang et al. 2009; Yin et al. 2012). MED23 favors the ELK1–SRF complex formation, which in turn facilitates the growth-related and adipogenic genes and suppresses the cytoskeleton and smooth muscle gene transcription (Yin et al. 2012). Furthermore, another recent report revealed that liver-specific knockdown or knockout of MED23 significantly improved glucose and lipid metabolism, especially for mice that were fed a high-fat diet (Chu et al. 2014). Reduced FoxO1-target gene expression was found in MED23-deficient primary hepatocytes, demonstrating that the regulation of MED23 on gluconeogenic gene expression is through FoxO1 (Chu et al. 2014).

Regulation of the balance between adipocyte and smooth muscle development as well as hepatic lipid/glucose metabolism by MED23 suggests the importance of this subunit in metabolism.

It is known that HNF4α regulates a various set of genes that are not only involved in early development but also in liver and pancreatic cell differentiation, glucose metabolism, and lipid homeostasis (Odom et al. 2004). In yeast two-hybrid assays, MED25 was found to bind to HNF4α, inducing HNF4α-dependent gene transcription that maintains the normal function of insulin secretion in pancreatic β-cells (Odom et al. 2004). As a subunit in the tail module, MED25 interaction with HNF4α recruits the Mediator complex and Pol II for transcriptional activation (Rana et al. 2011). Although the exact mechanism is still unclear, these functional studies also suggest a role of MED25 in metabolism and lipid homeostasis.

The kinase module functions in a context-dependent manner

Originally considered as a part of a transcriptional repressor in the large Mediator complex, the kinase module can repress or activate gene expression through kinase-dependent or kinase-independent mechanisms (Malik and Roeder 2010). Interestingly, the small mediator subunit MED26 is present in a mutually exclusive manner with CDK8 (Taatjes et al. 2002). Moreover, the kinase activity of CDK8 is not always required for its function in gene expression (Holstege et al. 1998). Besides the four conserved subunits in this module, paralogues have been identified for CDK8, MED12, and MED13, i.e., CDK19, MED12L, and MED13L (Daniels et al. 2013). Although their biological functions are less clear, these paralogues are subunits of the mammalian Mediator complex in a mutually exclusive manner with CDK8, MED12, and MED13 (Daniels et al. 2013). Among the four subunits, the evolutionarily conserved CDK8 and CycC have been studied for their functions in lipogenic gene expression (Zhao et al. 2012). The tissue-specific CDK19 is highly similar to CDK8 in amino acid sequence, but they may have both overlapping and distinct functions (Tsutsui et al. 2011). For instance, CDK8 but not CDK19 regulates HIF1-dependent gene expression (Galbraith et al. 2013). Recent studies also revealed the functional roles of MED12, MED13, and MED13L in cardiovascular and systemic metabolic regulation in both physiological and pathophysiological states.

The CDK8–CycC dimer negatively regulates de novo lipogenesis by reducing nuclear SREBP-1a or SREBP-1c protein stability (Zhao et al. 2012). SREBP-1a and SREBP-1c are key regulators of de novo lipogenesis, and post-translational modifications represent critical mechanisms that regulate their activity and/or abundance (Brown and Goldstein 2008). In addition to acetylation/deacetylation and ubiquitination, the phosphorylation of nuclear forms of SREBP-1 is critical for the proteasome-mediated degradation (Sundqvist et al. 2005). Therefore, the direct phosphorylation of threonine 426 or 402 in SREBP-1a and SREBP-1c isoforms, respectively, by CDK8 is important to understand the mechanism of SREBP-1 regulation of lipid metabolism (Zhao et al. 2012). The functions of CDK8–CycC dimer may be independent of the Mediator complex, as up to 30% of CDK8 exists as a free form from the Mediator complex although the presence of MED12 is required for maximal kinase activity of CDK8 (Knuesel et al. 2009b). Interestingly, insulin stimulation in vitro and in vivo can reduce the protein levels of CDK8 and CycC (Zhao et al. 2012). In mouse livers, CDK8 knockdown increased the blood lipid levels and NAFLD-like phenotypes (Zhao et al. 2012). Recent data have further suggested that mTORC1 activation upon feeding or in states of insulin resistance or NAFLD is responsible for the down-regulation of CDK8 and CycC in the liver (Feng et al. 2015).

In addition to CDK8 function in the negative regulation of de novo lipogenesis through its kinase activity, it has been shown that CDK8 has a positive role in response to serum, likely at the elongation step instead of the initiation step where CDK8 is a negative regulator (Donner et al. 2010) (see Nemet et al. 2014) for a recent review). The difficulty of identification of CDK8 substrates in vivo lies with embryonic lethality of CDK8-deficient mice, whereas there is no effect on cell viability in some cultured cells (Westerling et al. 2007). Nevertheless, using a selective inhibitor of CDK8 and CDK19 in tissue culture, a recent study has added more potential substrates of CDK8 and/or CDK19 to the increasing list (Poss et al. 2016), which includes Cyclin H (Akoulitchev et al. 2000), Notch (Fryer et al. 2004), histone H3 (Knuesel et al. 2009b), Smads (Alarcon et al. 2009), SREBP-1 (Zhao et al. 2012), E2F1 (Zhao et al. 2013), and STAT1 (Bancerek et al. 2013; Putz et al. 2013). However, further studies are necessary to understand the roles of CDK8 and CDK19 in metabolic regulation.

Through MED13, the MED12–MED13 dimer links the small mediator as well as the CDK8–CycC dimer (Tsai et al. 2013). Although the mechanism is unclear, MED12 can activate the Mediator complex-independent kinase activity of CDK8 in vitro (Knuesel et al. 2009b). Interestingly, MED13 in the heart regulates a subset of nuclear hormone receptor target genes that are major determinants of the metabolic rate and whole-body energy expenditure (Grueter et al. 2012). MED13 overexpression in the heart resulted in an increase in energy expenditure and resistance to diet-induced obesity (Grueter et al. 2012). MED13 is a target of miR-208a, one of the several microRNAs that are encoded by the cardiac-specific α-myosin heavy chain gene intron and are involved in heart disease and metabolic regulation (Montgomery et al. 2011; Grueter et al. 2012). Inhibition of miR-208a resulted in a similar metabolic phenotype that agrees with the finding that miR-208a negatively regulates MED13 (Grueter et al. 2012). Elevated protein levels of MED13 in the heart resulted in a significant increase in oxygen consumption, whereas the cardiac deletion of MED13 resulted in increased lipid accumulation without changes in food intake in mice (Grueter et al. 2012). MED13, but not MED13L, controls the whole-body metabolic homeostasis through altering metabolic profiles in white adipose tissue and liver when MED13 is altered in the heart (Baskin et al. 2014). Moreover, cardiac overexpression of MED13 in mice improves dysregulation of energy metabolism under high-fat diet likely through unknown circulating factors, as determined by heterotypic parabiosis experiments (Baskin et al. 2014). However, MED13 knockout in skeletal muscle resulted in resistance to hepatic steatosis in mice by activating a metabolic gene program that enhances muscle glucose uptake and storage as glycogen (Amoasii et al. 2016). Mechanistically, MED13 suppresses the expression of genes involved in glucose uptake and metabolism in skeletal muscle by inhibiting the nuclear receptor NURR1 and the MEF2 transcription factor (Amoasii et al. 2016). Although the roles of MED13 in other metabolic tissues have not been reported, the opposing metabolic regulation by MED13 in skeletal muscle and the heart further demonstrates the tissue-specific functions of the Mediator complex.

To date, not much information is available on the functions of MED12L and MED13L. Expressed in both heart and brain, MED13L is associated with early development of both heart and brain since its missense mutations and gene interruption are found in patients with congenital heart defect, learning disabilities, and facial anomalies (Muncke et al. 2003; Asadollahi et al. 2013; Davis et al. 2013). Both MED13 and MED13L are degraded by the SCF/Fbw7-dependent ubiquitination mechanism, and similar to MED13, MED13L may be also responsible for linking the kinase module to the small Mediator complex (Davis et al. 2013).

CONCLUSION

Since the discovery of the Mediator complex as a transcription cofactor of eukaryotes, its subunits have been associated with various biological processes and several diseases ranging from developmental defects to cancer in animal models and humans. The metabolic functions of the Mediator complex have become increasingly significant. In most cases, the Mediator complex functions as a bridge to connect and integrate specific transcription factors to the basal transcription machinery, resulting in expression changes of a selective set of genes. In addition to the presence of many subunits, tissue-specific expression and possible diverse assembly states at different physiological conditions further complicate the biological functions of the Mediator complex. Future studies on the role of the Mediator complex in metabolism may include identification of how the metabolic or nutrient signals may regulate the assembly states and subunit compositions and investigation of tissue-specific functions of each subunit in regulating nutrient and energy metabolism in vivo.

Abbreviations

Pol II	RNA polymerase II
T2D	Type 2 diabetes
NAFLD	Non-alcoholic fatty liver disease
CREB	cAMP-regulated enhancer-binding protein
FoxO	Forkhead O box protein
GR	Glucocorticoid receptor
HNF	Hepatic nuclear factor
SREBP	Sterol response element binding protein
LXR	Liver X receptor
PPAR	Peroxisome proliferator-activated receptor
CDK	Cyclin-dependent kinase
CycC	Cyclin C
ER	Estrogen receptor
CAR	Constitutive androstane receptor
UCP1	Uncoupling protein 1
SRC	Steroid receptor co-activator
BAT	Brown adipose tissue
SRF	Serum response factor

Acknowledgments This work was supported by Grants from the National Institutes of Health (DK020541, DK093623, DK098439, and DK110063). We apologize for being unable to cite all original studies due to the space limitation.

Compliance with Ethical Standards

Conflict of interest The authors declare that they have no conflict of interest.

Human and Animal Rights and Informed Consent This article does not contain any studies with human or animal subjects performed by any of the authors.

References

Akoulitchev S, Chuikov S, Reinberg D (2000) TFIIH is negatively regulated by cdk8-containing mediator complexes. Nature 407:102–106

Alarcon C, Zaromytidou AI, Xi Q, Gao S, Yu J, Fujisawa S, Barlas A, Miller AN, Manova-Todorova K, Macias MJ, Sapkota G, Pan D, Massagué J (2009) Nuclear CDKs drive Smad transcriptional activation and turnover in BMP and TGF-beta pathways. Cell 139:757–769

Allen BL, Taatjes DJ (2015) The Mediator complex: a central integrator of transcription. Nat Rev Mol Cell Biol 16:155–166

Amemiya-Kudo M, Shimano H, Hasty AH, Yahagi N, Yoshikawa T, Matsuzaka T, Okazaki H, Tamura Y, Iizuka Y, Ohashi K, Osuga J, Harada K, Gotoda T, Sato R, Kimura S, Ishibashi S, Yamada N (2002) Transcriptional activities of nuclear SREBP-1a, -1c, and -2 to different target promoters of lipogenic and cholesterogenic genes. J Lipid Res 43:1220–1235

Amoasii L, Holland W, Sanchez-Ortiz E, Baskin KK, Pearson M, Burgess SC, Nelson BR, Bassel-Duby R, Olson EN (2016) A MED13-dependent skeletal muscle gene program controls systemic glucose homeostasis and hepatic metabolism. Genes Dev 30:434–446

Asadollahi R, Oneda B, Sheth F, Azzarello-Burri S, Baldinger R, Joset P, Latal B, Knirsch W, Desai S, Baumer A, Houge G, Andrieux J, Rauch A (2013) Dosage changes of MED13L further delineate its role in congenital heart defects and intellectual disability. Eur J Hum Genet 21:1100–1104

Bai L, Jia Y, Viswakarma N, Huang J, Vluggens A, Wolins NE, Jafari N, Rao MS, Borensztajn J, Yang G, Reddy JK (2011) Transcription coactivator mediator subunit MED1 is required for the development of fatty liver in the mouse. Hepatology 53:1164–1174

Balamotis MA, Pennella MA, Stevens JL, Wasylyk B, Belmont AS, Berk AJ (2009) Complexity in transcription control at the activation domain-mediator interface. Sci Signal 2:ra20

Bancerek J, Poss ZC, Steinparzer I, Sedlyarov V, Pfaffenwimmer T, Mikulic I, Dolken L, Strobl B, Muller M, Taatjes DJ, Kovarik P (2013) CDK8 kinase phosphorylates transcription factor STAT1 to selectively regulate the interferon response. Immunity 38:250–262

Baskin KK, Grueter CE, Kusminski CM, Holland WL, Bookout AL, Satapati S, Kong YM, Burgess SC, Malloy CR, Scherer PE, Newgard CB, Bassel-Duby R, Olson EN (2014) MED13-dependent signaling from the heart confers leanness by enhancing metabolism in adipose tissue and liver. EMBO Mol Med 6:1610–1621

Brown MS, Goldstein JL (2008) Selective versus total insulin resistance: a pathogenic paradox. Cell Metab 7:95–96

Burakov D, Wong CW, Rachez C, Cheskis BJ, Freedman LP (2000) Functional interactions between the estrogen receptor and DRIP205, a subunit of the heteromeric DRIP coactivator complex. J Biol Chem 275:20928–20934

Casamassimi A, Napoli C (2007) Mediator complexes and eukaryotic transcription regulation: an overview. Biochimie 89:1439–1446

Chang C, Norris JD, Gron H, Paige LA, Hamilton PT, Kenan DJ, Fowlkes D, McDonnell DP (1999) Dissection of the LXXLL nuclear receptor-coactivator interaction motif using combinatorial peptide libraries: discovery of peptide antagonists of estrogen receptors alpha and beta. Mol Cell Biol 19:8226–8239

Chen W, Zhang X, Birsoy K, Roeder RG (2010) A muscle-specific knockout implicates nuclear receptor coactivator MED1 in the regulation of glucose and energy metabolism. Proc Natl Acad Sci USA 107:10196–10201

Chu Y, Gomez Rosso L, Huang P, Wang Z, Xu Y, Yao X, Bao M, Yan J, Song H, Wang G (2014) Liver Med23 ablation improves glucose and lipid metabolism through modulating FOXO1 activity. Cell Res 24:1250–1265

Conaway RC, Conaway JW (2011) Origins and activity of the Mediator complex. Semin Cell Dev Biol 22:729–734

Cooper KF, Khakhina S, Kim SK, Strich R (2014) Stress-induced nuclear-to-cytoplasmic translocation of cyclin C promotes mitochondrial fission in yeast. Dev Cell 28:161–173

Daniels DL, Ford M, Schwinn MK, Benink H, Galbraith MD, Amunugama R, Jones R, Allen D, Okazaki N, Yamakawa H, Miki F, Nagase T, Espinosa JM, Urh M (2013) Mutual exclusivity of MED12/MED12L, MED13/13L, and CDK8/19 paralogs revealed within the CDK-mediator kinase module. J Proteomics Bioinform. doi:10.4172/jpb.S4172-4004

Davis MA, Larimore EA, Fissel BM, Swanger J, Taatjes DJ, Clurman BE (2013) The SCF-Fbw7 ubiquitin ligase degrades MED13 and MED13L and regulates CDK8 module association with mediator. Genes Dev 27:151–156

Donner AJ, Ebmeier CC, Taatjes DJ, Espinosa JM (2010) CDK8 is a positive regulator of transcriptional elongation within the serum response network. Nat Struct Mol Biol 17:194–201

Feng D, Youn DY, Zhao X, Gao Y, Quinn WJ 3rd, Xiaoli AM, Sun Y, Birnbaum MJ, Pessin JE, Yang F (2015) mTORC1 down-regulates cyclin-dependent kinase 8 (CDK8) and cyclin C (CycC). PLoS One 10:e0126240

Flanagan PM, Kelleher RJ 3rd, Sayre MH, Tschochner H, Kornberg RD (1991) A mediator required for activation of RNA polymerase II transcription in vitro. Nature 350:436–438

Fondell JD, Ge H, Roeder RG (1996) Ligand induction of a transcriptionally active thyroid hormone receptor coactivator complex. Proc Natl Acad Sci USA 93:8329–8333

Fryer CJ, White JB, Jones KA (2004) Mastermind recruits CycC:CDK8 to phosphorylate the Notch ICD and coordinate activation with turnover. Mol Cell 16:509–520

Galbraith MD, Allen MA, Bensard CL, Wang X, Schwinn MK, Qin B, Long HW, Daniels DL, Hahn WC, Dowell RD, Espinosa JM (2013) HIF1A employs CDK8-mediator to stimulate RNAPII elongation in response to hypoxia. Cell 153:1327–1339

Ge K, Guermah M, Yuan CX, Ito M, Wallberg AE, Spiegelman BM, Roeder RG (2002) Transcription coactivator TRAP220 is required for PPAR gamma 2-stimulated adipogenesis. Nature 417:563–567

Ge K, Cho YW, Guo H, Hong TB, Guermah M, Ito M, Yu H, Kalkum M, Roeder RG (2008) Alternative mechanisms by which mediator subunit MED1/TRAP220 regulates peroxisome proliferator-activated receptor gamma-stimulated adipogenesis and target gene expression. Mol Cell Biol 28:1081–1091

Grants JM, Goh GY, Taubert S (2015) The Mediator complex of Caenorhabditis elegans: insights into the developmental and physiological roles of a conserved transcriptional coregulator. Nucleic Acids Res 43:2442–2453

Grontved L, Madsen MS, Boergesen M, Roeder RG, Mandrup S (2010) MED14 tethers mediator to the N-terminal domain of peroxisome proliferator-activated receptor gamma and is required for full transcriptional activity and adipogenesis. Mol Cell Biol 30:2155–2169

Grueter CE, van Rooij E, Johnson BA, DeLeon SM, Sutherland LB, Qi X, Gautron L, Elmquist JK, Bassel-Duby R, Olson EN (2012) A cardiac microRNA governs systemic energy homeostasis by regulation of MED13. Cell 149:671–683

Harms MJ, Lim HW, Ho Y, Shapira SN, Ishibashi J, Rajakumari S, Steger DJ, Lazar MA, Won KJ, Seale P (2015) PRDM16 binds MED1 and controls chromatin architecture to determine a brown fat transcriptional program. Genes Dev 29:298–307

Hittelman AB, Burakov D, Iniguez-Lluhi JA, Freedman LP, Garabedian MJ (1999) Differential regulation of glucocorticoid receptor transcriptional activation via AF-1-associated proteins. The EMBO journal 18:5380–5388

Holstege FC, Jennings EG, Wyrick JJ, Lee TI, Hengartner CJ, Green MR, Golub TR, Lander ES, Young RA (1998) Dissecting the regulatory circuitry of a eukaryotic genome. Cell 95:717–728

Huang Y, Li W, Yao X, Lin QJ, Yin JW, Liang Y, Heiner M, Tian B, Hui J, Wang G (2012) Mediator complex regulates alternative mRNA processing via the MED23 subunit. Mol Cell 45:459–469

Iida S, Chen W, Nakadai T, Ohkuma Y, Roeder RG (2015) PRDM16 enhances nuclear receptor-dependent transcription of the brown fat-specific Ucp1 gene through interactions with Mediator subunit MED1. Genes Dev 29:308–321

Jia Y, Qi C, Kashireddi P, Surapureddi S, Zhu YJ, Rao MS, Le Roith D, Chambon P, Gonzalez FJ, Reddy JK (2004) Transcription coactivator PBP, the peroxisome proliferator-activated receptor (PPAR)-binding protein, is required for PPARalpha-regulated gene expression in liver. J Biol Chem 279:24427–24434

Jia Y, Guo GL, Surapureddi S, Sarkar J, Qi C, Guo D, Xia J, Kashireddi P, Yu S, Cho YW, Rao MS, Kemper B, Ge K, Gonzalez FJ, Reddy JK (2005) Transcription coactivator peroxisome proliferator-activated receptor-binding protein/mediator 1 deficiency abrogates acetaminophen hepatotoxicity. Proc Natl Acad Sci USA 102:12531–12536

Jia Y, Viswakarma N, Fu T, Yu S, Rao MS, Borensztajn J, Reddy JK (2009) Conditional ablation of mediator subunit MED1 (MED1/PPARBP) gene in mouse liver attenuates glucocorticoid receptor agonist dexamethasone-induced hepatic steatosis. Gene Expr 14:291–306

Jia Y, Viswakarma N, Reddy JK (2014) Med1 subunit of the mediator complex in nuclear receptor-regulated energy metabolism, liver regeneration, and hepatocarcinogenesis. Gene Expr 16:63–75

Jiang P, Hu Q, Ito M, Meyer S, Waltz S, Khan S, Roeder RG, Zhang X (2010) Key roles for MED1 LxxLL motifs in pubertal mammary gland development and luminal-cell differentiation. Proc Natl Acad Sci USA 107:6765–6770

Kelleher RJ 3rd, Flanagan PM, Kornberg RD (1990) A novel mediator between activator proteins and the RNA polymerase II transcription apparatus. Cell 61:1209–1215

Kim YJ, Bjorklund S, Li Y, Sayre MH, Kornberg RD (1994) A multiprotein mediator of transcriptional activation and its interaction with the C-terminal repeat domain of RNA polymerase II. Cell 77:599–608

Kim GH, Oh GS, Yoon J, Lee GG, Lee KU, Kim SW (2015) Hepatic TRAP80 selectively regulates lipogenic activity of liver X receptor. J Clin Investig 125:183–193

Knuesel MT, Meyer KD, Bernecky C, Taatjes DJ (2009a) The human CDK8 subcomplex is a molecular switch that controls Mediator coactivator function. Genes Dev 23:439–451

Knuesel MT, Meyer KD, Donner AJ, Espinosa JM, Taatjes DJ (2009b) The human CDK8 subcomplex is a histone kinase that requires Med12 for activity and can function independently of mediator. Mol Cell Biol 29:650–661

Krebs P, Fan W, Chen YH, Tobita K, Downes MR, Wood MR, Sun L, Li X, Xia Y, Ding N, Spaeth JM, Moresco EM, Boyer TG, Lo CW, Yen J, Evans RM, Beutler B (2011) Lethal mitochondrial cardiomyopathy in a hypomorphic Med30 mouse mutant is ameliorated by ketogenic diet. Proc Natl Acad Sci USA 108:19678–19682

Kumafuji M, Umemura H, Furumoto T, Fukasawa R, Tanaka A, Ohkuma Y (2014) Mediator MED18 subunit plays a negative role in transcription via the CDK/cyclin module. Genes Cells 19:582–593

Lariviere L, Plaschka C, Seizl M, Wenzeck L, Kurth F, Cramer P (2012) Structure of the Mediator head module. Nature 492:448–451

Lee TI, Young RA (2013) Transcriptional regulation and its misregulation in disease. Cell 152:1237–1251

Lefterova MI, Haakonsson AK, Lazar MA, Mandrup S (2014) PPARgamma and the global map of adipogenesis and beyond. Trends Endocrinol Metab 25:293–302

Malik S, Roeder RG (2010) The metazoan Mediator co-activator complex as an integrative hub for transcriptional regulation. Nat Rev Genet 11:761–772

Mehta S, Miklos I, Sipiczki M, Sengupta S, Sharma N (2009) The Med8 mediator subunit interacts with the Rpb4 subunit of RNA polymerase II and Ace2 transcriptional activator in Schizosaccharomyces pombe. FEBS Lett 583:3115–3120

Meyer KD, Lin SC, Bernecky C, Gao Y, Taatjes DJ (2010) p53 activates transcription by directing structural shifts in Mediator. Nat Struct Mol Biol 17:753–760

Mittler G, Stuhler T, Santolin L, Uhlmann T, Kremmer E, Lottspeich F, Berti L, Meisterernst M (2003) A novel docking site on Mediator is critical for activation by VP16 in mammalian cells. EMBO J 22:6494–6504

Montgomery RL, Hullinger TG, Semus HM, Dickinson BA, Seto AG, Lynch JM, Stack C, Latimer PA, Olson EN, van Rooij E (2011) Therapeutic inhibition of miR-208a improves cardiac function and survival during heart failure. Circulation 124:1537–1547

Mukundan B, Ansari A (2011) Novel role for mediator complex subunit Srb5/Med18 in termination of transcription. J Biol Chem 286:37053–37057

Muncke N, Jung C, Rudiger H, Ulmer H, Roeth R, Hubert A, Goldmuntz E, Driscoll D, Goodship J, Schon K, Rappold G (2003) Missense mutations and gene interruption in PROSIT240, a novel TRAP240-like gene, in patients with congenital heart defect (transposition of the great arteries). Circulation 108:2843–2850

Naar AM, Beaurang PA, Robinson KM, Oliner JD, Avizonis D, Scheek S, Zwicker J, Kadonaga JT, Tjian R (1998) Chromatin, TAFs, and a novel multiprotein coactivator are required for synergistic activation by Sp1 and SREBP-1a in vitro. Genes Dev 12:3020–3031

Naar AM, Beaurang PA, Zhou S, Abraham S, Solomon W, Tjian R (1999) Composite co-activator ARC mediates chromatin-directed transcriptional activation. Nature 398:828–832

Naar AM, Taatjes DJ, Zhai W, Nogales E, Tjian R (2002) Human CRSP interacts with RNA polymerase II CTD and adopts a specific CTD-bound conformation. Genes Dev 16:1339–1344

Nemet J, Jelicic B, Rubelj I, Sopta M (2014) The two faces of Cdk8, a positive/negative regulator of transcription. Biochimie 97:22–27

Odom DT, Zizlsperger N, Gordon DB, Bell GW, Rinaldi NJ, Murray HL, Volkert TL, Schreiber J, Rolfe PA, Gifford DK, Fraenkel E, Bell GI, Young RA (2004) Control of pancreas and liver gene expression by HNF transcription factors. Science 303:1378–1381

Oh KJ, Han HS, Kim MJ, Koo SH (2013) Transcriptional regulators of hepatic gluconeogenesis. Arch Pharmacal Res 36:189–200

Park JM, Kim JM, Kim LK, Kim SN, Kim-Ha J, Kim JH, Kim YJ (2003) Signal-induced transcriptional activation by Dif requires the dTRAP80 mediator module. Mol Cell Biol 23:1358–1367

Poss ZC, Ebmeier CC, Odell AT, Tangpeerachaikul A, Lee T, Pelish HE, Shair MD, Dowell RD, Old WM, Taatjes DJ (2016) Identification of Mediator Kinase substrates in human cells using cortistatin A and quantitative phosphoproteomics. Cell reports 15:436–450

Putz EM, Gotthardt D, Hoermann G, Csiszar A, Wirth S, Berger A, Straka E, Rigler D, Wallner B, Jamieson AM, Pickl WF, Zebedin-Brandl EM, Müller M, Decker T, Sexl V (2013)

CDK8-mediated STAT1-S727 phosphorylation restrains NK cell cytotoxicity and tumor surveillance. Cell Rep 4:437–444

Rachez C, Suldan Z, Ward J, Chang CP, Burakov D, Erdjument-Bromage H, Tempst P, Freedman LP (1998) A novel protein complex that interacts with the vitamin D3 receptor in a ligand-dependent manner and enhances VDR transactivation in a cell-free system. Genes Dev 12:1787–1800

Rana R, Surapureddi S, Kam W, Ferguson S, Goldstein JA (2011) Med25 is required for RNA polymerase II recruitment to specific promoters, thus regulating xenobiotic and lipid metabolism in human liver. Mol Cell Biol 31:466–481

Sato S, Tomomori-Sato C, Parmely TJ, Florens L, Zybailov B, Swanson SK, Banks CA, Jin J, Cai Y, Washburn MP, Conaway JW, Conaway RC (2004) A set of consensus mammalian mediator subunits identified by multidimensional protein identification technology. Mol Cell 14:685–691

Schneider M, Hellerschmied D, Schubert T, Amlacher S, Vinayachandran V, Reja R, Pugh BF, Clausen T, Kohler A (2015) The nuclear pore-associated TREX-2 complex employs mediator to regulate gene expression. Cell 162:1016–1028

Soutourina J, Wydau S, Ambroise Y, Boschiero C, Werner M (2011) Direct interaction of RNA polymerase II and Mediator required for transcription in vivo. Science 331:1451–1454

Stevens JL, Cantin GT, Wang G, Shevchenko A, Shevchenko A, Berk AJ (2002) Transcription control by E1A and MAP kinase pathway via Sur2 mediator subunit. Science (New York, NY) 296:755–758

Sundqvist A, Bengoechea-Alonso MT, Ye X, Lukiyanchuk V, Jin J, Harper JW, Ericsson J (2005) Control of lipid metabolism by phosphorylation-dependent degradation of the SREBP family of transcription factors by SCF(Fbw7). Cell Metab 1:379–391

Taatjes DJ (2010) The human Mediator complex: a versatile, genome-wide regulator of transcription. Trends Biochem Sci 35:315–322

Taatjes DJ, Naar AM, Andel F 3rd, Nogales E, Tjian R (2002) Structure, function, and activator-induced conformations of the CRSP coactivator. Science 295:1058–1062

Takahashi H, Parmely TJ, Sato S, Tomomori-Sato C, Banks CA, Kong SE, Szutorisz H, Swanson SK, Martin-Brown S, Washburn MP, Florens L, Seidel CW, Lin C, Smith ER, Shilatifard A, Conaway RC, Conaway JW (2011) Human mediator subunit MED26 functions as a docking site for transcription elongation factors. Cell 146:92–104

Thompson CM, Young RA (1995) General requirement for RNA polymerase II holoenzymes in vivo. Proc Natl Acad Sci USA 92:4587–4590

Toth JI, Datta S, Athanikar JN, Freedman LP, Osborne TF (2004) Selective coactivator interactions in gene activation by SREBP-1a and -1c. Mol Cell Biol 24:8288–8300

Tsai KL, Sato S, Tomomori-Sato C, Conaway RC, Conaway JW, Asturias FJ (2013) A conserved Mediator-CDK8 kinase module association regulates Mediator-RNA polymerase II interaction. Nat Struct Mol Biol 20:611–619

Tsai KL, Tomomori-Sato C, Sato S, Conaway RC, Conaway JW, Asturias FJ (2014) Subunit architecture and functional modular rearrangements of the transcriptional mediator complex. Cell 157:1430–1444

Tsutsui T, Fukasawa R, Tanaka A, Hirose Y, Okhuma Y (2011) Identification of target genes for the CDK subunits of the Mediator complex. Genes Cells 16:1208–1218

Wang G, Cantin GT, Stevens JL, Berk AJ (2001) Characterization of mediator complexes from HeLa cell nuclear extract. Mol Cell Biol 21:4604–4613

Wang W, Huang L, Huang Y, Yin JW, Berk AJ, Friedman JM, Wang G (2009) Mediator MED23 links insulin signaling to the adipogenesis transcription cascade. Dev Cell 16:764–771

Wang X, Sun Q, Ding Z, Ji J, Wang J, Kong X, Yang J, Cai G (2014) Redefining the modular organization of the core Mediator complex. Cell Res 24:796–808

Wang K, Yan R, Cooper KF, Strich R (2015) Cyclin C mediates stress-induced mitochondrial fission and apoptosis. Mol Biol Cell 26:1030–1043

Westerling T, Kuuluvainen E, Makela TP (2007) Cdk8 is essential for preimplantation mouse development. Mol Cell Biol 27:6177–6182

Yang F, DeBeaumont R, Zhou S, Naar AM (2004) The activator-recruited cofactor/Mediator coactivator subunit ARC92 is a functionally important target of the VP16 transcriptional activator. Proc Natl Acad Sci USA 101:2339–2344

Yang F, Vought BW, Satterlee JS, Walker AK, Jim Sun ZY, Watts JL, DeBeaumont R, Saito RM, Hyberts SG, Yang S, Macol C, Iyer L, Tjian R, van den Heuvel S, Hart AC, Wagner G, Näär AM (2006) An ARC/Mediator subunit required for SREBP control of cholesterol and lipid homeostasis. Nature 442:700–704

Yin JW, Liang Y, Park JY, Chen D, Yao X, Xiao Q, Liu Z, Jiang B, Fu Y, Bao M, Huang Y, Liu Y, Yan J, Zhu MS, Yang Z, Gao P, Tian B, Li D, Wang G (2012) Mediator MED23 plays opposing roles in directing smooth muscle cell and adipocyte differentiation. Genes Dev 26:2192–2205

Yuan CX, Ito M, Fondell JD, Fu ZY, Roeder RG (1998) The TRAP220 component of a thyroid hormone receptor- associated protein (TRAP) coactivator complex interacts directly with nuclear receptors in a ligand-dependent fashion. Proc Natl Acad Sci USA 95:7939–7944

Zhao X, Feng D, Wang Q, Abdulla A, Xie XJ, Zhou J, Sun Y, Yang ES, Avinash M, Liu LP, Vaitheesvaran B, Bridges L, Kurland IJ, Strich R, Ni JQ, Wang C, Ericsson J, Pessin JE, Ji JY, Yang F (2012) Regulation of lipogenesis by cyclin-dependent kinase 8-mediated control of SREBP-1. J Clin Investig 122:2417–2427

Zhao J, Ramos R, Demma M (2013) CDK8 regulates E2F1 transcriptional activity through S375 phosphorylation. Oncogene 32:3520–3530

Zhao X, Xiaoli, Zong H, Abdulla A, Yang ES, Wang Q, Ji JY, Pessin JE, Das BC, Yang F (2014) Inhibition of SREBP transcriptional activity by a boron-containing compound improves lipid homeostasis in diet-induced obesity. Diabetes 63:2464–2473

Zhu Y, Qi C, Jain S, Rao MS, Reddy JK (1997) Isolation and characterization of PBP, a protein that interacts with peroxisome proliferator-activated receptor. J Biol Chem 272:25500–25506

MetaDP: a comprehensive web server for disease prediction of 16S rRNA metagenomic datasets

Xilin Xu[1,2], Aiping Wu[2], Xinlei Zhang[3], Mingming Su[4], Taijiao Jiang[2], Zhe-Ming Yuan[1]

[1] Hunan Provincial Key Laboratory for Biology and Control of Plant Diseases and Insect Pests, Hunan Agricultural University, Changsha 410128, China

[2] Center for Systems Medicine, Institute of Basic Medical Sciences, Chinese Academy of Medical Sciences & Peking Union Medical College, Beijing 100005; Suzhou Institute of Systems Medicine, Suzhou 215123, China

[3] Suzhou Geneworks Technology Company Limited, Suzhou 215123, China

[4] Institute of Basic Medical Sciences, Chinese Academy of Medical Sciences & Peking Union Medical College, Beijing 100005, China

Abstract High-throughput sequencing-based metagenomics has garnered considerable interest in recent years. Numerous methods and tools have been developed for the analysis of metagenomic data. However, it is still a daunting task to install a large number of tools and complete a complicated analysis, especially for researchers with minimal bioinformatics backgrounds. To address this problem, we constructed an automated software named MetaDP for 16S rRNA sequencing data analysis, including data quality control, operational taxonomic unit clustering, diversity analysis, and disease risk prediction modeling. Furthermore, a support vector machine-based prediction model for intestinal bowel syndrome (IBS) was built by applying MetaDP to microbial 16S sequencing data from 108 children. The success of the IBS prediction model suggests that the platform may also be applied to other diseases related to gut microbes, such as obesity, metabolic syndrome, or intestinal cancer, among others (http://metadp.cn:7001/).

Keywords Disease prediction, 16S rRNA, Metagenomics, Intestinal bowel syndrome

Xilin Xu, Aiping Wu have contributed equally to this work.

✉ Correspondence: taijiao@moon.ibp.ac.cn (T. Jiang), zhmyuan@sina.com (Z.-M. Yuan)

INTRODUCTION

A wide variety of microbes live in the human body. These microbes exist in oral, nasopharynx, skin, gut, and many other regions of the body and play an important role in human health (Human Microbiome Project 2012; Sankar et al. 2015). To date, there is still significant uncertainty about the relationships between resident microbes and human diseases.

Most microorganisms in the human body have remained uncultured. Therefore, traditional methods for the inspection and identification of the microbial species have significant limitations. In 1998, Handelsman et al. first put forward the concept of the "metagenome" (Handelsman et al. 1998), and defined it as the genes and genomes of all of the microorganisms in an environmental sample. With the rapid development of high-throughput sequencing technology and the establishment of numerous microbial databases, metagenomics has become an emerging topic of interest in biomedical research. Recently, multiple metagenomics studies have revealed that microbial communities are associated with human diseases. Turnbaugh et al. characterized the gut microbial communities of 154 individuals and found that obesity was associated with phylum-level change in the microbiota and reduction of bacterial diversity (Turnbaugh et al. 2009). Pushalkar et al. studied five saliva microbial samples and found fifteen unique phylotypes in three oral squamous cell carcinoma subjects (Pushalkar et al. 2011). The relationships between microorganisms and some other diseases have also been investigated, such as oral diseases (Belda-Ferre et al. 2012), neurological diseases (Hsiao et al. 2013), rheumatoid arthritis (Scher et al. 2013), and Crohn's disease (Gevers et al. 2014). Furthermore, some computational models have been constructed for disease classification and prediction based on metagenomic data. Qin et al. analyzed the differences between type 2 diabetes (T2D) patients and non-diabetic controls in 345 Chinese gut microbial samples. The researchers chose 50 gene markers to develop a T2D classifier model and used it for risk assessment and monitoring of T2D (Qin et al. 2012). Saulnier et al. compared the gut microbiomes of healthy children and pediatric patients with irritable bowel syndrome (IBS), and found some differences in the microbial communities in these two sample sets, which might suggest a novel technique for the diagnosis of pediatric patients with functional bowel disorders (Saulnier et al. 2011). Moreover, Qin et al. developed a support vector machine (SVM) model and indicated that microbiota-targeted biomarkers may serve as new tools for disease diagnoses (Qin et al. 2014). These prediction models indicate that metagenomics data can perhaps play an important role in the prevention and early diagnosis of disease.

Although numerous tools and methods have been developed to investigate the relationship between microbes and human diseases, there is still an absence of a general automated workflow from raw data to disease prediction. Some metagenomic data analysis tools, such as QIIME (Caporaso et al. 2010a, b), mother (Schloss et al. 2009), and RDP classifier (Wang et al. 2007), are readily amenable to running automated analyses, especially for biologists with minimal bioinformatics backgrounds. To address this problem, we developed a web-based platform called MetaDP, in which an automated analysis workflow was built for 16S rRNA sequences generated by both the 454 and Illumina platforms. The web server is constructed based on the open-source bioinformatics platform, Galaxy (Goecks et al. 2010) (https://galaxyproject.org/). In MetaDP, we integrated a number of metagenomics-associated tools and further built an automatic analysis pipeline. MetaDP also presents a user-friendly interface for one-stop automatic analysis and provides most of the output results in downloadable figure formats.

Previously reported 16S rRNA sequencing data from IBS disease were imported into the MetaDP platform. Based on microbial information from pediatric patients with IBS and healthy children, we constructed an IBS disease prediction model with a high degree of accuracy. This model is integrated into the MetaDP platform and may be helpful for IBS prevention and early diagnosis. The MetaDP web server is available publically (http://metadp.cn:7001/).

RESULTS

The MetaDP framework

The MetaDP webserver is freely available at (http://metadp.cn:7001/) (Fig. 1A, B). MetaDP provides pre-defined workflows and can be used without registration. It begins with a straightforward process whereby a user uploads sequencing data. The analysis mainly includes three parts: data pre-processing, traditional metagenomic data analysis, and disease prediction (Fig. 1C). Pre-processing includes filtering low-quality sequences, splitting libraries based on the barcodes, removing chimeric sequences, and assembling reads. Traditional metagenomic data analysis includes microbial composition taxonomic analysis, alpha diversity, and beta diversity. The disease prediction aspect classifies testing samples with our pre-defined disease prediction model. The essential purpose of the MetaDP web service is to provide a user-friendly automated analysis system, in which users simply upload their raw data generated from a high-throughput sequencing platform. Thereby, the MetaDP may be readily used. There is no need for installing, integrating, and designing individual tools. In addition, MetaDP provides some optional parameters for better analysis.

Metagenomic data analysis

Operational taxonomic unit (OTU) counting

For our dataset, after the pre-processing step, filtered sequences were clustered by the Uclust method (with a sequence similarity threshold of 97%). Then, the longest sequence from each cluster was chosen as its representative sequence. The OTU summary (http://metadp.cn:7001/metadp/F1/OTU_summary.txt) included 91,470 OTUs in a total of 2,448,155 sequence counts, in which the maximal OTU count among samples was 76,939. The microbial composition summary for each taxonomic level (from phylum to genus) is shown in Table 1.

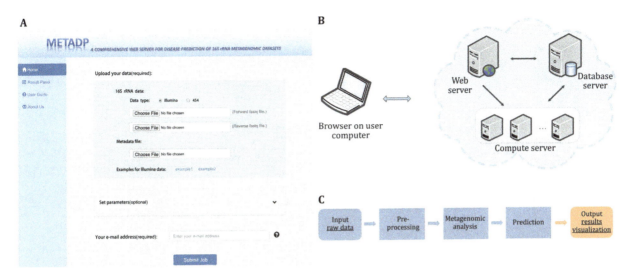

Fig. 1 The framework of MetaDP. **A** User interface of web server. **B** System architecture. **C** Main steps of analysis

Taxonomic abundance

Taxonomic binning of classified sequences was generated at five levels, from phylum to genus (http://metadp.cn:7001/metadp/F2/barchart_for_samples.html). Samples were grouped and averaged to plot the stacked bar charts (http://metadp.cn:7001/metadp/F3/barchart_for_groups.html). Another group of stacked bar plots were generated with the sorted taxonomic abundance data in samples (http://metadp.cn:7001/metadp/F4/OTU_sorted_barplot_for_samples.pdf) and in groups (http://metadp.cn:7001/metadp/F5/OTU_sorted_barplot_for_groups.pdf). Figure 2A shows the microbial stacked bar plot for the grouped sorted data of IBS versus noIBS samples at the order level. The analysis indicated that there is no obvious difference between the two groups, and the main microbes of these two groups are all *Bacteroidales* and *Clostridiales*, which is consistent with previously reported results (Riehle et al. 2012; Saulnier et al. 2011).

Alpha diversity

Alpha diversity analysis provides insight into differences in species abundance, richness, and evenness.

Table 1 The counts of microbial communities at different taxonomy levels

Level	Counts
Phylum	14
Class	25
Order	42
Family	82
Genus	169

Alpha diversity indices were analyzed with the default metrics, Chao1, ACE, Simpson, Shannon, Good's coverage, and PD whole tree (http://metadp.cn:7001/metadp/F6/alpha_index_table.txt). Plots were generated and exported for rank-abundance, rarefaction index, and species richness. The rank-abundance curve (http://metadp.cn:7001/metadp/F7/rank_abundance_plot.pdf) is a 2D chart with abundance rank on the X-axis and relative abundance on the Y-axis. The alpha rarefaction analysis was performed by computation with multiple metrics (defaults: chao1, Shannon, and observed species) (http://metadp.cn:7001/metadp/F8/alpha_rarefaction_plot.html). Figure 2B shows the rarefaction curve displayed by groups that were analyzed with the observed species metrics. This curve demonstrates that the number of species in the two groups increased gradually with increasing sample sequence number, eventually saturating. The curve also indicates that the species richness of the noIBS sample (the blue line) is higher than that of the IBS sample (the red line).

Beta diversity

Beta diversity analysis provides a measure of the distance between each sample. Both weighted and unweighted distance matrices were calculated and visualized with Principal coordinates analysis (PcoA) plots (http://metadp.cn:7001/metadp/F9/weighted_PCoA.html and http://metadp.cn:7001/metadp/F10/unweighted_PCoA.html). Figure 2C shows the weighted-distance distribution of samples in 3D space. In this figure, both IBS (red) and noIBS (blue) samples are mixed, indicating that it was difficult to classify the samples according to the distance matrix.

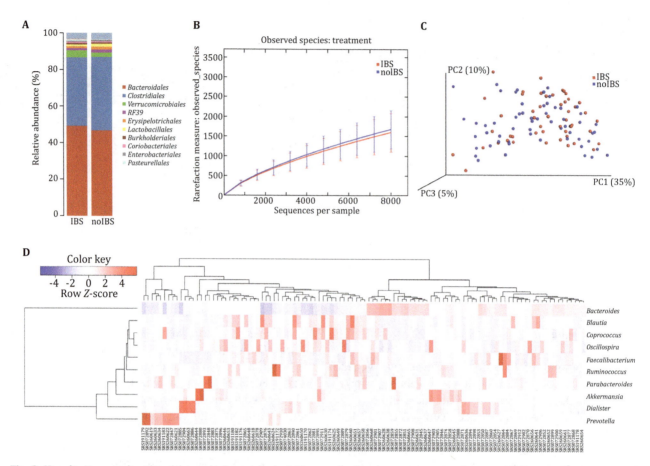

Fig. 2 Visualization results of metagenomic data analysis. **A** Taxonomic abundance comparison between children with IBS and healthy children. **B** Rarefaction curve by groups. **C** Weighted UniFrac PCoA. The IBS and healthy samples are colored as *red* and *blue*, respectively. **D** Heatmap analysis for the top ten microbes at the genus level. Microbes and samples are both clustered. *Each row* is scaled by *Z*-score

OTU heatmaps

A heatmap was used to visualize the relationships between the OTUs and samples (http://metadp.cn:7001/metadp/F11/raw_OTUs_heatmap.html). In the heatmap, raw OTU counts per sample are displayed. Figure 2D presents the heatmap for the top 10 microbes at the genus level (other classification levels are listed in http://metadp.cn:7001/metadp/F12/top10_heatmaps.pdf). Both samples in columns and OTUs in rows were clustered by relative abundance, and the rows were scaled by *Z*-score.

Prediction model

In total, 91,470 OTUs were obtained among 108 samples by OTU picking. After filtering for zero values (percentage >80% in all samples), 1726 OTUs were selected. Then, a *t* test was used to examine the discriminatory ability of each feature. Finally, 110 OTU feature sets were selected for the construction of the next model (http://metadp.cn:7001/metadp/F13/filtered_OTU_table.txt). The top 20 most significant features and their *p*-values are listed in Table 2. Within these features, *Bacteroides*, *Dorea*, and *Faecalibacterium* have been reported to be associated with IBS (Saulnier et al. 2011; Ghoshal et al. 2012; Rajilić-Stojanović et al. 2015).

Then, the quantified feature vector could be input into LIBSVM. The radial basis function (RBF) kernel was used in LIBSVM, and a grid search program (grid.py) was used to obtain the optimized parameter combination $C = 4.0$, $\gamma = 0.125$. Thereby, the IBS prediction model was constructed successfully. To test the performance of the IBS model, tenfold cross-validation was adopted. The results show that the accuracy and the AUC score were 0.93 and 0.95, respectively (Fig. 3).

DISCUSSION

The MetaDP platform is a one-stop 16S rRNA sequencing data analysis flowchart with a friendly user interface that aims to help researchers investigate the structure

Table 2 Information for the top 20 most significant features

OTU ID	p-value	Taxonomy
358944	0.001*	k_Bacteria;p_Firmicutes;c_Clostridia;o_Clostridiales;f_;g_;s_
New.CleanUp.ReferenceOTU327210	0.0014*	k_Bacteria;p_Bacteroidetes;c_Bacteroidia;o_Bacteroidales;f_Bacteroidaceae;g_Bacteroides;s_
189384	0.0034*	k_Bacteria;p_Bacteroidetes;c_Bacteroidia;o_Bacteroidales;f_Bacteroidaceae;g_Bacteroides;s_
199283	0.0034*	k_Bacteria;p_Firmicutes;c_Clostridia;o_Clostridiales;f_Ruminococcaceae;g_Faecalibacterium;s_prausnitzii
179665	0.0074*	k_Bacteria;p_Firmicutes;c_Clostridia;o_Clostridiales;f_Lachnospiraceae;g_Dorea;s_
New.ReferenceOTU288	0.0076*	k_Bacteria;p_Bacteroidetes;c_Bacteroidia;o_Bacteroidales;f_Rikenellaceae;g_;s_
New.CleanUp.ReferenceOTU389203	0.0077*	k_Bacteria;p_Firmicutes;c_Clostridia;o_Clostridiales;f_Lachnospiraceae;g_;s_
589277	0.0085*	k_Bacteria;p_Bacteroidetes;c_Bacteroidia;o_Bacteroidales;f_Bacteroidaceae;g_Bacteroides;s_
189855	0.0095*	k_Bacteria;p_Firmicutes;c_Clostridia;o_Clostridiales;f_Lachnospiraceae;g_;s_
187251	0.0103	k_Bacteria;p_Firmicutes;c_Clostridia;o_Clostridiales;f_Lachnospiraceae
New.ReferenceOTU7	0.0103	k_Bacteria;p_Bacteroidetes;c_Bacteroidia;o_Bacteroidales;f_Bacteroidaceae;g_Bacteroides;s_
2653002	0.0112	k_Bacteria;p_Bacteroidetes;c_Bacteroidia;o_Bacteroidales;f_Bacteroidaceae;g_Bacteroides;s_ovatus
New.CleanUp.ReferenceOTU262311	0.0114	k_Bacteria;p_Bacteroidetes;c_Bacteroidia;o_Bacteroidales;f_Bacteroidaceae;g_Bacteroides;s_
New.CleanUp.ReferenceOTU387907	0.013	k_Bacteria;p_Firmicutes;c_Clostridia;o_Clostridiales;f_Ruminococcaceae
New.ReferenceOTU412	0.0133	k_Bacteria;p_Bacteroidetes;c_Bacteroidia;o_Bacteroidales;f_Bacteroidaceae;g_Bacteroides;s_
New.ReferenceOTU156	0.0134	k_Bacteria;p_Bacteroidetes;c_Bacteroidia;o_Bacteroidales;f_Bacteroidaceae;g_Bacteroides;s_
196713	0.0136	k_Bacteria;p_Firmicutes;c_Clostridia;o_Clostridiales;f_Lachnospiraceae;g_;s_
1835779	0.0153	k_Bacteria;p_Firmicutes;c_Clostridia;o_Clostridiales;f_Lachnospiraceae
198990	0.0161	k_Bacteria;p_Firmicutes;c_Clostridia;o_Clostridiales;f_Lachnospiraceae;g_;s_
New.CleanUp.ReferenceOTU320393	0.0163	k_Bacteria;p_Bacteroidetes;c_Bacteroidia;o_Bacteroidales;f_Bacteroidaceae;g_Bacteroides;s_

* Represents extremely significant difference ($p < 0.01$)

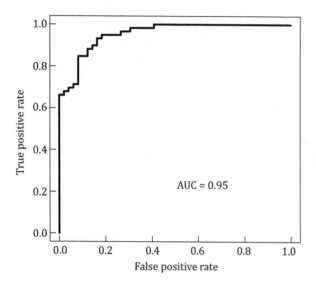

Fig. 3 ROC curve of the SVM model of IBS disease. X and Y axes represent the false positive rates (1−sensitivity) and the true positive rates (sensitivity), respectively. The AUC score is 0.95

and diversity of human microbial flora and provide deep insight into microorganisms associated with the disease. An automatic analysis workflow can be performed once users upload their raw sequencing data with barcodes. In this version, our platform provides a set of universal 16S rRNA data analysis tools to constitute a workflow for data from the 454 and Illumina platforms. The workflow outputs the bacterial distribution, alpha diversity, beta diversity, and disease risk assessment with a plug-in prediction model. To build the prediction model, we used IBS as an example with a total of 108 microbial samples. In the near future, we will increase the sample size of intestinal microbial diseases and improve the prediction model.

In future work, MetaDP will provide an open API interface, so that researchers can easily integrate other bioinformatics tools and data analysis workflows with our platform. We will also integrate more metagenomic data analysis tools, data analysis workflows, and machine learning models, making our platform useful for the analysis of more diseases. Users can also perform custom/personalized data analysis processes according to their own requirements. The MetaDP platform can be easily used for microorganism-associated diseases, such as diabetes, obesity, and colorectal cancer, among others. We will collect more intestinal microbial sequencing data to expand disease prediction models for better disease prevention and diagnosis.

MATERIALS AND METHODS

MetaDP provides pre-defined workflows for metagenomic data analysis and disease prediction modeling based on the Galaxy platform (Fig. 4). Users simply need to upload their raw 16S sequencing data generated by 454 pyrosequencing or by the Illumina platform and another metadata mapping file with detailed sample information, including sample names, barcodes, descriptions of the columns. The core analysis pipeline consists of demultiplexing, quality filtering, OTU picking, taxonomic assignment, phylogenetic reconstruction, diversity analysis, and visualization. In addition, a configured SVM-based prediction model has been constructed for intestinal bowel syndrome.

Data pre-processing

First, sample isolation and quality control must be performed from multiplexed Standard Flowgram Format (SFF) file or FASTQ files. The four main steps for raw data pre-processing are as follows. (1) Sample demultiplexing: the multiplexed reads are assigned to samples based on their unique nucleotide barcodes in the mapping file. (2) Primer removal: during demultiplexing, the primer sequences and barcodes have to be removed at the same time. (3) Quality filtering: short or low average quality score reads are removed using customized thresholds, and any sequence with the first nucleotide as "N" or "n" is cut. (4) Denoising and chimera removal: before sequence clustering, denoising and chimera removal are required for 454 and Illumina datasets. In this platform, chimera detection is based on the USEARCH 6.1 algorithm (Edgar 2010). The above steps are all run by calling QIIME (Caporaso et al. 2010a, b). Paired-end reads for the Illumina platform are trimmed using Trimmomatic (Bolger et al. 2014). Then, FLASH software (Magoc and Salzberg 2011) is used to assemble the trimmed paired-end reads, and the resulting contigs are compiled into an input file to use for the next sample demultiplexing step.

Metagenomic data analysis

OTU picking

OTUs are normally used for analyzing microbial composition and diversity. Pre-processing sequences are grouped into a cluster when their sequence similarities are greater than the threshold value, such as 97% at the species level. In this study, we chose Uclust (Edgar

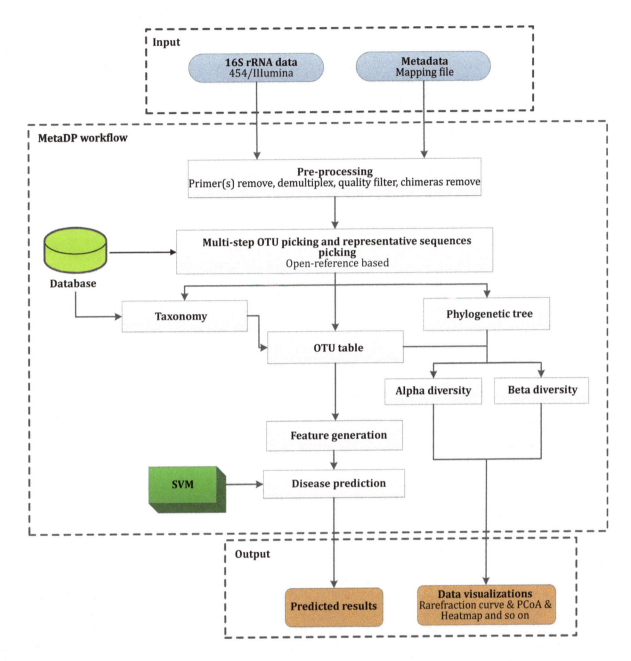

Fig. 4 Overview of the MetaDP workflow for 16S rRNA sequences analysis and disease prediction. The workflow supports the input of 16S rRNA sequencing data and sample metadata. The analysis includes sequence pre-processing, OTU picking, biodiversity analysis, and disease prediction with the configured SVM model. Predicted results and visualized data are returned

2010) as the default OTU clustering tool. The five steps for OTU picking are given as follows. (1) Pre-filtering: the sequences are searched against the GreenGenes reference database (DeSantis et al. 2006), filtered for at least a low percent identity (default: 0.60), and discarded if they fail to match. (2) Multi-step OTU picking: the pre-filtered sequences are aligned with an existing database, and added to the database as new reference sequences if the sequences are mismatched. (3) Representative sequence picking: the longest sequence is chosen as the representative sequence. (4) Taxonomic assignment: a taxonomic classification is assigned to each sequence of the representative set with the GreenGenes database and newly defined taxonomies from step 2. (5) OTU table generation: an OTU table is constructed in the Genomics Standards Consortium candidate standard Biological Observation Matrix (BIOM) format. The BIOM format file can be converted to other formats with a series of scripts available from the BIOM project (McDonald et al. 2012).

Phylogenetic analysis

Representative sequences are assigned to the core set of the GreenGenes database (DeSantis et al. 2006) with PyNAST (Caporaso et al. 2010a, b). Then, the sequence alignment is filtered by removing the gap regions from every sequence. The FastTree method (Price et al. 2009) is utilized to construct phylogenetic trees based on the filtered sequence files. The phylogenetic tree can be interactively displayed through an online tool named Interactive Tree of Life (iTOL, http://itol.embl.de/) (Ciccarelli et al. 2006).

Taxa summaries

A taxa summary summarizes the relative abundance of different taxonomic levels (from phylum to genus) among all samples based on an OTU table. Sequences are taxonomically binned based on the output of a local copy of the ribosomal database project (RDP) classifier. Normalized data are produced from the relative abundances of taxa present in each sample. Any unclear taxa are combined and named "other." The results from the taxonomic binning of classified sequences are displayed as bar charts, which make it easier to convey the main compositions of the samples.

Biodiversity

Two types of diversity measurements (alpha diversity and beta diversity) are usually used for assessing the relatedness of metadata attributes on OTU tables. Alpha diversity is mainly used to estimate the diversity of a microbial community within a group of samples, through a series of statistical indices such as Chao1, ACE, Shannon, Simpson, Good's coverage, and so on (Navas-Molina et al. 2013). Rarefaction curves are plotted by counting the OTU numbers from random reads of the samples based on these diversity metrics. Beta diversity is mainly used to compare the differences of microbial communities between samples. UniFrac (Lozupone et al. 2011) is always used for comparing biological communities. Both weighted and unweighted variants of UniFrac are widely used. The former accounts for the abundance of OTUs, while the latter only considers their presence or absence. The distance metrics are investigated through PCoA, and an interactive 3D plot is generated.

OTU heatmaps

For the composition analysis of OTUs among samples, two types of OTU heatmaps are provided. The first type of heatmap is an interactive plot and is directly colored to reflect the absolute abundance of raw OTUs. The other type of heatmap is a bi-directional map, in which both the samples and the taxa summary are clustered. Users can set the threshold for the microbes at different classification levels (the default is top ten microbes at the genus level).

Disease prediction model

Feature selection

Feature selection (Saeys et al. 2007), also known as variable selection or attribute selection in machine learning, is the selection of a subset of redundancy features for the construction of a prediction model. In this study, feature selection is used mainly for the simplification of models for better feature interpretation, and for the reduction of overfitting. In our training set, the values of the OTU tables generated in the metagenomics data analysis were used. For each feature, a value with zero is deleted first, then feature selection is performed based on the statistical comparison.

Support vector machine (SVM)

SVMs are important supervised learning algorithms for classification and regression analysis. In recent years, SVMs have been widely used in life sciences research, such as for studies on alternative splice site recognition, biomarker selection, remote homology detection, gene function prediction, and protein–protein interaction prediction, among others (Pavlidis et al. 2002; Liao and Noble 2003; Ben-Hur and Noble 2005; Ratsch et al. 2005; Sonnenburg et al. 2007). Some useful software packages have also been developed (Bottou 2007; Fan et al. 2008; Chang and Lin 2011). In this study, we use LIBSVM (Chang and Lin 2011) (http://www.csie.ntu.edu.tw/~cjlin/libsvm), which is an integrated software package for support vector classification, regression, and distribution estimation. An SVM can efficiently perform a non-linear classification through a so-called kernel function, thus implicitly mapping inputs into high-dimensional feature spaces. The RBF kernel was chosen for our study. The penalty parameter C and kernel parameters γ in the RBF kernel are optimized to result in the best prediction performance. To obtain the optimal C and γ values, we used the grid search method. The main steps for a grid search can be described as follows. First, M and N numbers of C and γ values are assigned, respectively. Then, different SVM models with $M \times N$ (C, γ) numbers of parameters combined are trained. Finally, the optimal pair of parameters is selected.

Evaluation

Cross-validation (tenfold) is used to estimate the performance of our prediction model. In this study, we use SVM-train with parameter $-v$ 10, as it will randomly split samples into ten subsamples; each subsample is used once as the validation data for testing the model, and the remaining nine subsamples are used as training data; finally, the average accuracy will be reported. A receiver-operating characteristic (ROC) curve is used to illustrate the performance of the classifier model. The ROC curve plots the true positive rate (TPR) against the false positive rate (FPR) at various threshold values. The TPR and FPR are given by $TPR = TP/(TP + FN)$ and $FPR = FP/(FP + TN)$, respectively. The area under the ROC curve (AUC) score is used to estimate the overall classifier performance. The ROCR package from CRAN (http://cran.r-project.org/) was used to calculate the TPR and FPR values and to draw ROC curves, the AUC scores were also provided to estimate this classifier model performance.

Implementation

MetaDP has been implemented in a local Galaxy instance running under a GNU/Linux operating system. Galaxy was obtained from http://wiki.galaxyproject.org/Admin/GetGalaxy and intentionally installed as a normal user ("galaxy") for easy migration and security. The advantage of using the Galaxy framework for MetaDP is that Galaxy provides a web-accessible platform to integrate different command-line tools and has a customized workflow configuration system. Additionally, Galaxy provides some useful functional dependencies, such as a web service (Nginx), database storage (MySQL), a job queuing system, and history management. In MetaDP, we integrated the metagenomic data analysis package QIIME, the SVM model, and NGS quality control tools. The applications of all tools were implemented with XML files, Python, Perl, and Shell wrappers. These tools consisted of the specific workflow for library splitting, OTU picking, taxonomy analysis, rarefaction analysis, and disease prediction. For ease of use, we simplified the operations of the web applications and designed a more interactive and user-friendly website. The user simply needs to upload input files and run the workflow through a web interface.

Datasets

In total, 108 samples (49 samples from pediatric patients with IBS and another 59 samples from healthy children) of 16S rRNA 454 sequencing data were downloaded from the NCBI database (http://www.ncbi.nlm.nih.gov/sra, SRP002457) (Saulnier et al. 2011).

Abbreviation

MetaDP   Disease prediction of metagenomic datasets

Acknowledgements This work was supported by the Science and Technology Planning Projects of Changsha, China (K1406018-21 to ZY).

Compliance with Ethical Standards

Conflict of interest Xilin Xu, Aiping Wu, Xinlei Zhang, Mingming Su, Taijiao Jiang, and Zheming Yuan declare that they have no conflict of interest.

Human and Animal Rights and Informed Consent This article does not contain any studies with human or animal subjects performed by any of the authors.

References

Belda-Ferre P, Alcaraz LD, Cabrera-Rubio R, Romero H, Simon-Soro A, Pignatelli M, Mira A (2012) The oral metagenome in health and disease. ISME J 6:46–56

Ben-Hur A, Noble WS (2005) Kernel methods for predicting protein-protein interactions. Bioinformatics 21(Suppl 1):i38–i46

Bolger AM, Lohse M, Usadel B (2014) Trimmomatic: a flexible trimmer for Illumina sequence data. Bioinformatics 30:2114–2120

Bottou L (2007) Large-scale kernel machines. The MIT Press, Cambridge

Caporaso JG, Bittinger K, Bushman FD, DeSantis TZ, Andersen GL, Knight R (2010a) PyNAST: a flexible tool for aligning sequences to a template alignment. Bioinformatics 26:266–267

Caporaso JG, Kuczynski J, Stombaugh J, Bittinger K, Bushman FD, Costello EK, Fierer N, Pena AG, Goodrich JK, Gordon JI et al (2010b) QIIME allows analysis of high-throughput community sequencing data. Nat Methods 7:335–336

Chang CC, Lin CJ (2011) LIBSVM: a library for support vector machines. ACM Trans Intell Syst Technol 2:27

Ciccarelli FD, Doerks T, von Mering C, Creevey CJ, Snel B, Bork P (2006) Toward automatic reconstruction of a highly resolved tree of life. Science 311:1283–1287

DeSantis TZ, Hugenholtz P, Larsen N, Rojas M, Brodie EL, Keller K, Huber T, Dalevi D, Hu P, Andersen GL (2006) Greengenes, a chimera-checked 16S rRNA gene database and workbench compatible with ARB. Appl Environ Microbiol 72:5069–5072

Edgar RC (2010) Search and clustering orders of magnitude faster than BLAST. Bioinformatics 26:2460–2461

Fan RE, Chang KW, Hsieh CJ, Wang XR, Lin CJ (2008) LIBLINEAR: a library for large linear classification. J Mach Learn Res 9:1871–1874

Gevers D, Kugathasan S, Denson LA, Vazquez-Baeza Y, Van Treuren W, Ren B, Schwager E, Knights D, Song SJ, Yassour M, Morgan XC, Kostic AD, Luo C, González A, McDonald D, Haberman Y, Walters T, Baker S, Rosh J, Stephens M, Heyman M, Markowitz J, Baldassano R, Griffiths A, Sylvester F, Mack D, Kim S, Crandall W, Hyams J, Huttenhower C, Knight R, Xavier RJ (2014) The treatment-naive microbiome in new-onset Crohn's disease. Cell Host Microbe 15:382–392

Ghoshal UC, Shukla R, Ghoshal U, Gwee KA, Ng SC, Quigley EM (2012) The gut microbiota and irritable bowel syndrome: friend or foe? Int J Inflam 2012:151085

Goecks J, Nekrutenko A, Taylor J, Galaxy T (2010) Galaxy: a comprehensive approach for supporting accessible, reproducible, and transparent computational research in the life sciences. Genome Biol 11:R86

Handelsman J, Rondon MR, Brady SF, Clardy J, Goodman RM (1998) Molecular biological access to the chemistry of unknown soil microbes: a new frontier for natural products. Chem Biol 5:R245–R249

Hsiao EY, McBride SW, Hsien S, Sharon G, Hyde ER, McCue T, Codelli JA, Chow J, Reisman SE, Petrosino JF, Patterson PH, Mazmanian SK (2013) Microbiota modulate behavioral and physiological abnormalities associated with neurodevelopmental disorders. Cell 155:1451–1463

Human Microbiome Project C (2012) A framework for human microbiome research. Nature 486:215–221

Liao L, Noble WS (2003) Combining pairwise sequence similarity and support vector machines for detecting remote protein evolutionary and structural relationships. J Comput Biol 10:857–868

Lozupone C, Lladser ME, Knights D, Stombaugh J, Knight R (2011) UniFrac: an effective distance metric for microbial community comparison. ISME J 5:169–172

Magoc T, Salzberg SL (2011) FLASH: fast length adjustment of short reads to improve genome assemblies. Bioinformatics 27:2957–2963

McDonald D, Clemente JC, Kuczynski J, Rideout JR, Stombaugh J, Wendel D, Wilke A, Huse S, Hufnagle J, Meyer F, Knight R, Caporaso JG (2012) The Biological Observation Matrix (BIOM) format or: how I learned to stop worrying and love the ome-ome. Gigascience 1:7

Navas-Molina JA, Peralta-Sanchez JM, Gonzalez A, McMurdie PJ, Vazquez-Baeza Y, Xu Z, Ursell LK, Lauber C, Zhou H, Song SJ, Huntley J, Ackermann GL, Berg-Lyons D, Holmes S, Caporaso JG, Knight R (2013) Advancing our understanding of the human microbiome using QIIME. Methods Enzymol 531:371–444

Pavlidis P, Weston J, Cai J, Noble WS (2002) Learning gene functional classifications from multiple data types. J Comput Biol 9:401–411

Price MN, Dehal PS, Arkin AP (2009) FastTree: computing large minimum evolution trees with profiles instead of a distance matrix. Mol Biol Evol 26:1641–1650

Pushalkar S, Mane SP, Ji X, Li Y, Evans C, Crasta OR, Morse D, Meagher R, Singh A, Saxena D (2011) Microbial diversity in saliva of oral squamous cell carcinoma. FEMS Immunol Med Microbiol 61:269–277

Qin J, Li Y, Cai Z, Li S, Zhu J, Zhang F, Liang S, Zhang W, Guan Y, Shen D, Peng Y, Zhang D, Jie Z, Wu W, Qin Y, Xue W, Li J, Han L, Lu D, Wu P, Dai Y, Sun X, Li Z, Tang A, Zhong S, Li X, Chen W, Xu R, Wang M, Feng Q, Gong M, Yu J, Zhang Y, Zhang M, Hansen T, Sanchez G, Raes J, Falony G, Okuda S, Almeida M, LeChatelier E, Renault P, Pons N, Batto JM, Zhang Z, Chen H, Yang R, Zheng W, Li S, Yang H, Wang J, Ehrlich SD, Nielsen R, Pedersen O, Kristiansen K, Wang J (2012) A metagenome-wide association study of gut microbiota in type 2 diabetes. Nature 490:55–60

Qin N, Yang F, Li A, Prifti E, Chen Y, Shao L, Guo J, Le Chatelier E, Yao J, Wu L, Zhou J, Ni S, Liu L, Pons N, Batto JM, Kennedy SP, Leonard P, Yuan C, Ding W, Chen Y, Hu X, Zheng B, Qian G, Xu W, Ehrlich SD, Zheng S, Li L (2014) Alterations of the human gut microbiome in liver cirrhosis. Nature 513:59–64

Rajilić-Stojanović M, Jonkers DM, Salonen A, Hanevik K, Raes J, Jalanka J, de Vos WM, Manichanh C, Golic N, Enck P, Philippou E, Iraqi FA, Clarke G, Spiller RC, Penders J (2015) Intestinal microbiota and diet in IBS: causes, consequences, or epiphenomena? Am J Gastroenterol 110:278–287

Ratsch G, Sonnenburg S, Scholkopf B (2005) RASE: recognition of alternatively spliced exons in C.elegans. Bioinformatics 21(Suppl 1):i369–i377

Riehle K, Coarfa C, Jackson A, Ma J, Tandon A, Paithankar S, Raghuraman S, Mistretta TA, Saulnier D, Raza S, Diaz MA, Shulman R, Aagaard K, Versalovic J, Milosavljevic A (2012) The genboree microbiome toolset and the analysis of 16S rRNA microbial sequences. BMC Bioinform 13(Suppl 13):S11

Saeys Y, Inza I, Larranaga P (2007) A review of feature selection techniques in bioinformatics. Bioinformatics 23:2507–2517

Sankar SA, Lagier JC, Pontarotti P, Raoult D, Fournier PE (2015) The human gut microbiome, a taxonomic conundrum. Syst Appl Microbiol 38:276–286

Saulnier DM, Riehle K, Mistretta TA, Diaz MA, Mandal D, Raza S, Weidler EM, Qin X, Coarfa C, Milosavljevic A, Petrosino JF, Highlander S, Gibbs R, Lynch SV, Shulman RJ, Versalovic J (2011) Gastrointestinal microbiome signatures of pediatric patients with irritable bowel syndrome. Gastroenterology 141:1782–1791

Scher JU, Sczesnak A, Longman RS, Segata N, Ubeda C, Bielski C, Rostron T, Cerundolo V, Pamer EG, Abramson SB, Huttenhower C, Littman DR (2013) Expansion of intestinal Prevotella copri correlates with enhanced susceptibility to arthritis. Elife 2:e01202

Schloss PD, Westcott SL, Ryabin T, Hall JR, Hartmann M, Hollister EB, Lesniewski RA, Oakley BB, Parks DH, Robinson CJ, Sahl JW, Stres B, Thallinger GG, Van Horn DJ, Weber CF (2009) Introducing mothur: open-source, platform-independent, community-supported software for describing and comparing microbial communities. Appl Environ Microbiol 75:7537–7541

Sonnenburg S, Schweikert G, Philips P, Behr J, Ratsch G (2007) Accurate splice site prediction using support vector machines. BMC Bioinform 8(Suppl 10):S7

Turnbaugh PJ, Hamady M, Yatsunenko T, Cantarel BL, Duncan A, Ley RE, Sogin ML, Jones WJ, Roe BA, Affourtit JP, Egholm M, Henrissat B, Heath AC, Knight R, Gordon JI (2009) A core gut microbiome in obese and lean twins. Nature 457:480–484

Wang Q, Garrity GM, Tiedje JM, Cole JR (2007) Naive Bayesian classifier for rapid assignment of rRNA sequences into the new bacterial taxonomy. Appl Environ Microbiol 73:5261–5267

A new dimethyl labeling-based SID-MRM-MS method and its application to three proteases involved in insulin maturation

Dongwan Cheng[1,2], Li Zheng[1], Junjie Hou[1], Jifeng Wang[1], Peng Xue[1], Fuquan Yang[1✉], Tao Xu[1✉]

[1] National Laboratory of Biomacromolecules, Institute of Biophysics, Chinese Academy of Sciences, Beijing 100101, China
[2] University of Chinese Academy of Sciences, Beijing 100049, China

Abstract The absolute quantification of target proteins in proteomics involves stable isotope dilution coupled with multiple reactions monitoring mass spectrometry (SID-MRM-MS). The successful preparation of stable isotope-labeled internal standard peptides is an important prerequisite for the SID-MRM absolute quantification methods. Dimethyl labeling has been widely used in relative quantitative proteomics and it is fast, simple, reliable, cost-effective, and applicable to any protein sample, making it an ideal candidate method for the preparation of stable isotope-labeled internal standards. MRM mass spectrometry is of high sensitivity, specificity, and throughput characteristics and can quantify multiple proteins simultaneously, including low-abundance proteins in precious samples such as pancreatic islets. In this study, a new method for the absolute quantification of three proteases involved in insulin maturation, namely PC1/3, PC2 and CPE, was developed by coupling a stable isotope dimethyl labeling strategy for internal standard peptide preparation with SID-MRM-MS quantitative technology. This method offers a new and effective approach for deep understanding of the functional status of pancreatic β cells and pathogenesis in diabetes.

Keywords Dimethyl labeling, Stable isotope dilution-multiple reaction monitoring (SID-MRM), Mass spectrometry, Insulin

INTRODUCTION

The development of quantitative proteomics has enabled the absolute quantification of target proteins (i.e., target proteomics). Absolute quantification typically involves stable isotope dilution coupled with multiple reaction monitoring (SID-MRM) mass spectrometry (Gallien et al. 2011). MRM experiments are typically conducted using triple quadrupole mass spectrometers to scan specific ion pairs of mass-to-charge (m/z) values associated with the peptide precursor and fragment ions, which are referred to as transitions. Intermediate or heavy stable isotope-labeled tryptic peptides from the target proteins, which are used as internal standards in the stable isotope dilution method, have the same amino acid sequence, HPLC retention time, ionization efficiency, and secondary fragment ions as the corresponding light stable isotope-labeled peptides. Due to the mass difference between the light and intermediate or heavy stable isotope-labeled peptides, the peak intensity (peak height or peak area) ratio can be obtained by MRM-MS. Based on the known quantities of stable isotope internal standards, the content of the corresponding peptides in a sample can be calculated, and the final quantities of the corresponding proteins can be determined. MRM-MS is of high sensitivity, high specificity, wide dynamic

---

Dongwan Cheng and Li Zheng contributed equally to this paper.

✉ Correspondence: fqyang@ibp.ac.cn (F. Yang), xutao@ibp.ac.cn (T. Xu)

range, high throughput, and low background and has been widely applied for the accurate quantifications of multiple proteins simultaneously, including low-abundance proteins in complex biological samples. The superiority of this technique compared to Western blotting, which is characterized by poor reproducibility, low sensitivity, low throughput, and antibody-dependence, makes MRM-MS a powerful and crucial quantitative technology in target proteomics. Thus, applications of MRM are of increasing interest.

The successful preparation of stable isotope-labeled internal standard peptides is an important prerequisite for the SID-MRM absolute quantification method. Several approaches have been applied to generate stable isotope-labeled peptides: (1) chemical synthesis of peptides with heavy stable isotope-labeled amino acids, which is costly; (2) biosynthesis of peptides in vivo via the use of heavy isotopically labeled amino acids in cell culture, which requires stable isotope-labeled amino acids and special culture medium, resulting in a long cycle and high consumption; and (3) labeling peptides with various chemical reagents at the proteolytic peptide level. Among these, commercial reagents such as iTRAQ/mTRAQ and TMT yield good labeling effects but are expensive; by contrast, ^{18}O labeling is inexpensive but has an unstable labeling efficiency. Here we employed a fast, high-efficiency labeling and inexpensive method based on dimethyl labeling, which has been used widely in relative quantitative proteomics. The dimethylation chemical reaction occurs at the N-terminus and lysine residues (α- and ε-amino groups) and involves the formation of a Schiff base, followed by the reduction of cyanoborohydride to generate a dimethylamine (Boersema et al. 2009). This method exhibits high reaction efficiency and selectivity and does not generate any significant by-products (with the exception of rarely occurring peptides containing an N-terminal proline). In comparison with other labeling methods, dimethyl labeling is reliable, cost-effective, simple, and applicable to any protein sample. Highly efficient dimethyl labeling can be performed using in-solution, online, and on-column strategies to meet the requirements of different samples. The dimethyl labeling strategy has been widely applied to different biological studies, such as stem cells, stimulation-induced phosphorylation dynamics, and interaction proteomic studies (Aye et al. 2012).

Diabetes mellitus is a common metabolic disease around the world that is accompanied by serious chronic complications and is difficult to cure. Due to its increasing incidence, diabetes (particularly type 2 diabetes) has become a threat to economic development and population health. Pancreatic islet β cells are unique endocrine cells that secrete insulin and hypoglycemic hormone. The relative or absolute deficiency of insulin secretion is an important factor in the development of diabetes. In β cells, proinsulin is processed into insulin and C-peptide by three key proteases along two routes. Type I and II prohormone convertases (PC1/3 and PC2) cleave proinsulin at the B/C chain and A/C chain junction; carboxypeptidase-E (CPE), then removes the C-terminal Lys and Arg residues exposed by endoproteolytic cleavage by the PCs. β cells lose their functions during the progression of type 2 diabetes, leading to the release of a large number of precursor molecules, namely proinsulin, into the blood, which in turn results in an increased proinsulin/insulin ratio. However, the proteins involved in these processing defect and the underlying mechanisms are unknown. Therefore, a quantitative analysis of key proteases during insulin maturation will help clarify the functional status of pancreatic islet β cells, enabling pathogenesis studies of diabetes. MRM-MS technology can simultaneously quantify multiple proteins, including low-abundance proteins (Picotti et al. 2009), which is particularly important for precious samples such as islets.

A new method for the absolute quantification of three key proteases during insulin maturation, namely PC1/3, PC2, and CPE, was developed by coupling a stable isotope dimethyl labeling preparation of internal standard peptides with SID-MRM-MS quantitative technology (for strategy scheme, see Fig. 1).

RESULTS AND DISCUSSION

Confirmation of synthetic peptides

The sequences of the 11 commercial synthetic peptides were verified by nanoLC-ESI-Q Exactive MS. Figure 2 shows the MS/MS spectrum of one of the peptides (LDLHVIPVWEK). The observed m/z of the peptide was 674.88300, consistent with the theoretical m/z (-2.74 ppm). In addition, the b- and y-ions in the MS/MS spectrum clearly supported the sequential amino acids, producing a high-confidence identification with a SEQUEST XCorr value of 3.86. The mass spectrometry results confirmed that all the 11 synthetic peptides were fully consistent with the expected sequences.

Dimethyl labeling efficiency

The efficient labeling of commercial synthetic peptides as internal standards is a prerequisite for obtaining accurate quantitative results. To avoid sample loss during the desalting procedure, in-solution dimethyl

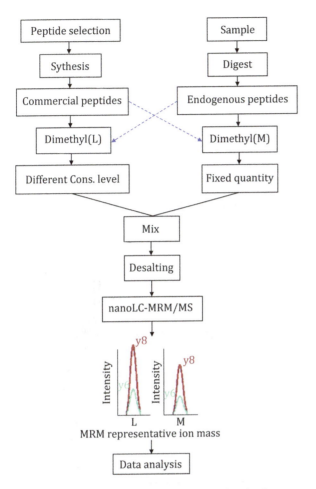

Fig. 1 Dimethyl-SID-MRM strategy. Workflow for absolute quantification by combining dimethyl labeling with SID-MRM-MS

the synthetic peptides and the islet proteotypic mixture, respectively. After a chemical reaction with dimethyl labeling reagents, all primary amines (the N-terminus and the side chain of the lysine residues) in a peptide mixture were labeled and converted to dimethylamines. The light and intermediate dimethyl label resulted in monoisotopic mass shifts of +28.0313 and +32.0564 Da per primary amine, respectively.

Figure 3 shows the representative mass spectra of synthetic peptides before and after labeling. The data in Fig. 3A, C, and E represent the label-free, light, and intermediate label spectra of peptide FGFGLLNAK, respectively. The observed m/z of the label-free peptide was 966.4 Da, consistent with the theoretical m/z of 966.548 Da. Dimethyl labeling at the N-terminus and lysine residues created mass shifts of +56.0626 Da (light) and +64.1128 Da (intermediate), which were supported by the observed mass shifts of +56.1 Da (light) and +64.2 Da (intermediate). The ion peak of the non-labeled peptide was not detected at an m/z of 966.4 Da in both the light and intermediate dimethyl-labeled sample spectra, indicating complete labeling of the peptide. Figure 3B, D and F represent the label-free, light, and intermediate label spectra of peptide YTDDWFNSHGTR, respectively. Dimethyl labeling at the N-terminus resulted in observed mass shifts of +28.2 Da (light) and +32.2 Da (intermediate), which were entirely consistent with the theoretical z values. The ion peak of the non-labeled peptide at an m/z of 1498.5 Da was also not detected in the light or intermediate dimethyl-labeled sample spectra. These MALDI-TOF-MS results demonstrate that nearly all the synthetic peptides were labeled by dimethylation, indicating the efficiency of this method for dimethyl labeling to produce an internal standard.

labeling was performed to ensure the accuracy of the quantification. MALDI-TOF-MS and nanoLC-ESI-MS/MS were performed to estimate the labeling efficiency of

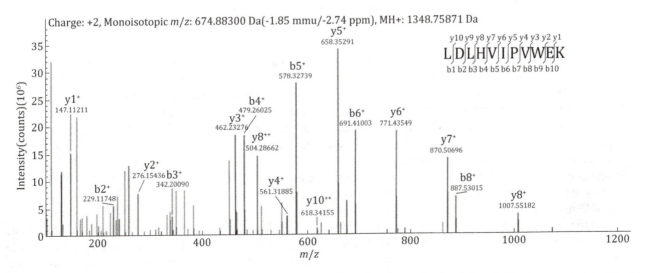

Fig. 2 The MS/MS spectrum of a commercial peptide (LDLHVIPVWEK). The observed m/z of the peptide was consistent with the theoretical m/z (−2.74 ppm), and most of the band y ions were detected in the spectrum

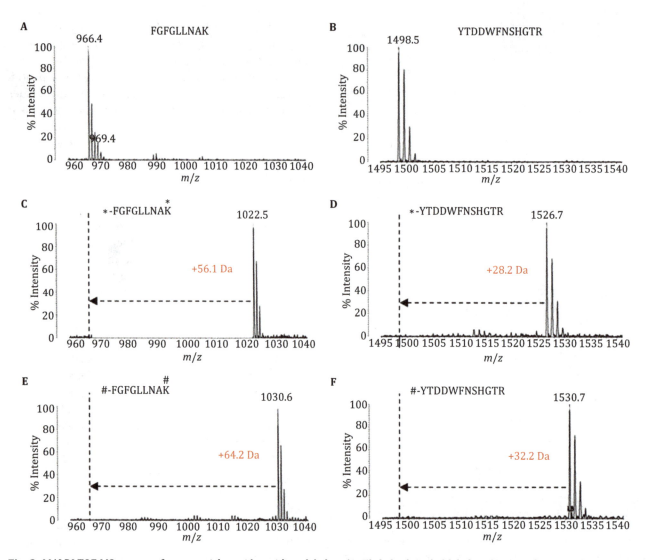

Fig. 3 MALDI-TOF-MS spectra of commercial peptides with no labeling (**A**, **B**), light dimethyl labeling (**C**, **D**) and intermediate dimethyl labeling (**E**, **F**). The peptide (FGFGLLNAK) with two labeling sites exhibited observed mass shifts of +56.1 Da (light) and +64.2 Da (intermediate), and the peptide (YTDDWFNSHGTR) with only one labeling site exhibited mass shifts of +28.2 Da (light) and +32.2 Da (intermediate). "*" represents the light dimethyl labeling sites, and "#" represents the intermediate dimethyl labeling sites

We next assessed the efficiency of protein digestion because the accuracy of the MRM assay is dependent on the completeness of the tryptic digestion reaction. The results of this assessment indicated that the percentage of confident hits with missed cleavages at the peptide level should have been less than 15%. We also evaluated the efficiency of dimethyl labeling on a complex sample by submitting 2 μg of total tryptic peptides from islets labeled by light and intermediate dimethyl reagents to nanoLC-MS analysis. The results demonstrated that the labeling efficiencies for light and intermediate dimethyl labeling were 97.7% (labeled/total: 8983/9087) and 97.5% (labeled/total: 8863/9191), respectively.

The above results demonstrate that the dimethyl labeling method ensures complete reaction and high labeling efficiency. No significant difference in labeling efficiency was observed between light and intermediate dimethyl labeling, and the labeling efficiency for both simple and complex samples reached >95%, sufficient to satisfy our next MRM-based protein absolute quantification.

Optimization of MRM-based absolute peptide quantification

Equimolar of each internal standard peptide was mixed followed by dimethyl labeling and detected by nanoLC-ESI-TSQ Vantage mass spectrometry. The LC conditions and collision energy of the fragment ions were optimized to ensure that the peptides were maximally separated by LC and that the corresponding detection parameters achieved the optimum value.

The optimized elution gradients were as follows: 0–41 min, 5%–32% phase B; 41–46 min, 32%–90% phase B; 46–49 min, 90% phase B; and after 50 min, return to 5% phase B, with a total running time of 60 min and a flow rate of 300 nL/min. Precursor ions with a charge state of +2 were selected as the parent ions, and the corresponding collision energy was optimized in a range among ± 8 of the predicted theoretical values. The collision energy that produced the most intense fragment ions was selected as the final optimized set, and the two highest intensity fragment ions were selected as the quantitative ions, with another two fragment ions as the assistant qualifier ions. The transitions and parameters of the optimized MRM-MS method are shown in Table 1, and the MRM ion chromatograms of six selected quantitative peptides are shown in Fig. 4.

Absolute quantification of target proteins in an islet sample

To maximize the quantification of the target proteins, 4 μg of an islet proteotypic sample was loaded, based on the capacity of the HPLC column. Internal standard mixtures of 2, 5, 10, 40, 100, 200, 500, and 2000 fmol with light dimethyl labeling were incorporated into 4-μg islet proteotypic samples with intermediate dimethyl labels, respectively. This series of mixed samples was analyzed using the MRM method established above. The extracted peak areas of each transition from all internal standard peptides were obtained using Pinpoint software, and linear standard measurement curves were then constructed by using the measured peak areas and theoretical sample concentrations. Of the 11 internal standard peptides, six exhibited a good MS response and an excellent linear relation between the extracted peak area and the absolute amount; these peptides were thus selected to represent the contents of the corresponding individual target proteins in the islet sample. The remaining five peptides failed to yield the desired signal intensity or an adequate curve, possibly due to interference from other components in the sample or other factors. The resulting linear fit curves corresponding to 12 "parent–daughter" ion transitions are shown in Fig. 5A. All correlation coefficients (R^2) were greater than 0.99, indicating a good linear relationship between the peak areas of the dimethyl-labeled internal standard peptides and quantity within a 2–2000 fmol range. The contents of the quantitative peptides corresponding to the target proteins of the islet samples were calculated by inputting their peak areas into the corresponding linear regression equations (see Fig. 5B).

To verify the accuracy and reliability of the above quantitative results, a reverse-labeling experiment was performed in which different concentrations of intermediate-labeled internal standard peptides were incorporated into the light labeled islet sample. The quantitative peptide contents of the islet sample were calculated based on two quantitative fragment ions using the same data analysis process, and the final content of each peptide was obtained as the mean calculated content of its two fragment ions (see Fig. 5C). The results of the forward and reverse-labeling experiments demonstrated that there was no significant difference between two experiments, confirming the accuracy of the method and its suitability for absolute quantitative analysis.

For the three target proteases in mouse islet samples, the respective protein copies per islet cell as well as the copies per granule can be estimated by calculating the number of molecules using the Avogadro constant followed by division by the number of cells or granules used in the experiment. Because 100 islets can produce approximately 30–50 μg of total protein (Petyuk et al. 2008), we used 40 μg as the calculated value. The cell number per single islet varies from a few cells to several thousand cells (Seino and Bell 2008), excluding very small or large islets and assuming an average diameter of approximately 150 μm. Based on the reported mean cell volume of islet cells, one islet of 150 μm diameter comprises approximately 1000 cells (Dean 1973); each islet cell contains 9000–13,000 granules (Dean 1973; Olofsson et al. 2002), and thus 10,000 granules were used for the calculations. On the basis of 1 molar = 6.0221367×10^{23} molecules, 415 molecules of PC1/3, 1677 molecules of PC2, and 1188 molecules of CPE were computed per single granule of each single islet cell as a rough approximation.

CONCLUSION

In this study, a new method for the absolute quantification of target proteins was developed by combining dimethyl labeling with SID-MRM mass spectrometry. High labeling efficiency was achieved for both synthetic peptides and the islet proteotypic mixtures, and high accuracy was reliably supported by forward and reverse labeling experiments. Therefore, these results demonstrate that dimethyl labeling of peptides with stable isotopes for producing internal standards and absolute quantification is a highly efficient, stable, and low-cost strategy.

With this method, an absolute quantification assay was established for three key proteases (PC1/3, PC2,

Table 1 MRM transitions and optimized collision energy of internal peptides

Protein name	Peptide sequence	Charge	m/z				CE
			Q1 (L)	Q3 (L)	Q1 (M)	Q3 (M)	
CPE	SNAQGIDLNR[a]	2	558.29	687.32[b]	560.31	687.38[b]	22
				1000.52[b]		1000.52[b]	22
				402.22		402.25	28
				886.42		886.47	22
	EGGPNNHLLK	2	567.82	288.22	571.84	292.25	24
				977.52		981.58	24
				401.32		405.34	28
				538.33		542.40	26
	SYWEDNK	2	499.23	289.12	503.26	293.21	20
				882.32		886.42	20
				719.32		723.36	20
				279.12		283.16	20
	LLAPGNYK[a]	2	466.28	606.32[b]	470.31	610.35[b]	23
				790.42[b]		794.47[b]	19
				677.32		681.39	19
				338.22		342.23	27
PC1/3	YDLTNENK[a]	2	526.77	746.42[b]	530.79	750.43[b]	21
				861.42[b]		865.46[b]	21
				307.12		311.15	21
				289.13		293.21	25
	ALVDLADPR[a]	2	499.29	272.12[b]	501.30	272.17[b]	18
				898.42[b]		898.50[b]	20
				213.12		217.18	24
				785.42		785.41	18
	LDLHVIPVWEK	2	702.92	686.33	706.94	690.41	31
				799.43		803.50	27
				898.53		902.56	25
				490.23		494.29	39
	FGFGLLNAK	2	511.81	360.22	515.83	364.25	26
				847.52		851.53	20
				246.12		250.21	28
				473.32		477.33	26
PC2	IPPTGK	2	334.72	430.21	338.75	434.29	20
				527.31		531.34	14
				232.12		236.19	26
				333.22		337.24	26
	YTDDWFNSHGTR[a]	2	763.84	1004.43[b]	765.85	1004.47[b]	29
				1335.53[b]		1335.57[b]	27
				1119.43		1119.50	27
				818.33		818.39	29
	LVLTLK[a]	2	371.77	502.32[b]	375.80	506.39[b]	17
				601.42[b]		605.45[b]	17
				288.22		292.25	23
				389.21		393.30	21

[a] The peptides selected to quantify the target protein
[b] Quantitative fragment ions

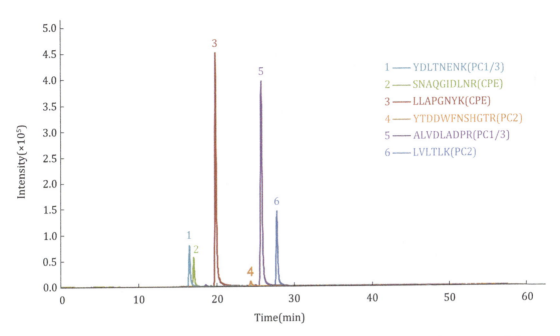

Fig. 4 The LC-MRM-MS ion chromatograms of the internal peptides were selected to represent the contents of the corresponding individual target proteins

and CPE) during insulin maturation. Based on the quantities of dimethyl-labeled peptides used as internal standards, the actual content of target proteins in the islet tissues was quantified. The linear correlation coefficient (R^2) of each standard curves was greater than 0.99 with a dynamic range of three orders of magnitude. Using this method, small samples comprising approximately 100 islets with a total protein content of less than 50 μg were analyzed for the absolute quantification of three key proteases. This method not only offers high sensitivity, high throughput, and high specificity for the absolute quantification of proteins in precious samples such as islets but also provides insights into the functions of pancreatic β cells, thus providing a robust tool for future studies of the pathogenesis of diabetes.

MATERIALS AND METHODS

Selection and synthesis of internal peptides

Internal peptides selected as precursor ions for MRM experiments must fulfill the following criteria: (1) the peptides should have 6–20 amino acids in length; (2) the peptides should not have tryptic missed cleavages; (3) the peptides should not include any amino acid residue which could be chemically modified (e.g., methionine, cysteine); and (4) the amino acid sequence of each peptide should be unique. Based on these criteria, a pilot experiment and software analysis (Maclean et al. 2010) led the selection of eleven theoretical tryptic peptides from PC1/3, PC2, and CPE (see Table 1).

The internal peptides were synthesized commercially (Scilight Biotechnology, LLC, Beijing, China) at a purity greater than 98%. Before dimethyl labeling for SID-MRM, the sequence of each synthetic peptide was verified by nanoLC-ESI-Q Exactive mass spectrometry (Thermo Scientific, Pittsburgh PA, U.S.).

Preparation of islet tryptic peptide samples

C57BL/6 male mice were purchased from the Institute of Experimental Animals, Chinese Academy of Medical Sciences. Islets were isolated from 10- to 16-week-old mice as previously described (Rorsman and Trube 1986). After rinsing and discarding the supernatant, lysis buffer (8 mol/L urea and 100 mmol/L TEAB) was added to the islet sample followed by sonication. A BCA assay was performed to quantify the total islet protein.

DTT and IAA were added successively to 200 μg total protein islet lysate to final concentrations of 10 and 40 mmol/L, respectively, to reductively alkylate cysteine. To neutralize the IAA, an additional aliquot of DTT was added to achieve a total DTT concentration of 20 mmol/L. After dilution in buffer (100 mmol/L TEAB) for obtaining a final urea concentration of 1 mol/L, the

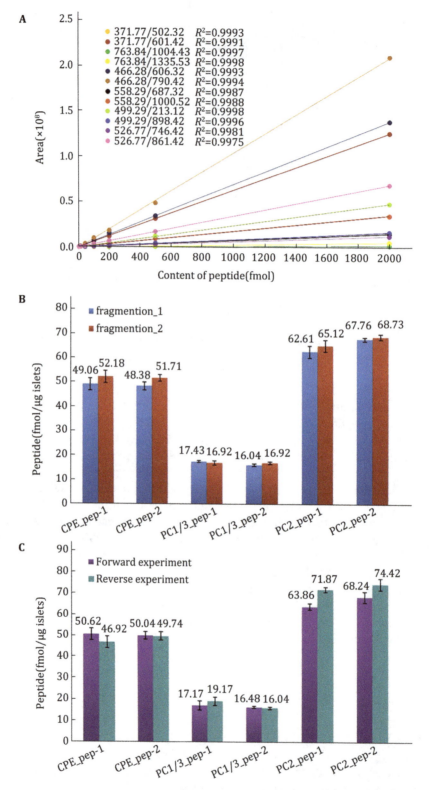

Fig. 5 **A** Linear fit curves of internal peptides corresponding to 12 "parent-daughter" ion transitions. **B** Quantitative peptide contents in samples corresponding to the top two quantitative ions in a forward experiment. **C** Quantitative peptide contents in samples determined using forward and reverse-labeling strategies

sample was digested overnight by trypsin (enzyme: substrate = 1:50 w/w). The digestion was terminated with formic acid, and 2 μg of the islet tryptic peptides mixture was submitted to LC-MS to verify the digestion efficiency.

Dimethyl labeling of synthetic peptides and islet tryptic peptide mixtures

Synthetic peptides and an islet tryptic peptides mixture were dimethyl labeled in solution with reagents containing light and intermediate isotopic formaldehyde, respectively, as described by Paul (Boersema et al. 2009). Briefly, the samples were differentially isotopically labeled in parallel by adding light or intermediate isotopic formaldehyde and $NaBH_3CN$ to the reaction. The labeling reaction was quenched via the addition of ammonia solution, and the sample was then acidified with formic acid for LC-MS analysis. Labeling efficiencies were measured on a MALDI-TOF for synthetic peptides and by nanoLC-ESI-Q Exactive MS for the islet proteotypic mixture. Only samples with a labeling efficiency greater than 95% were used for further quantitative analysis.

Mass spectrometry analysis and database search

Samples were analyzed on a Thermo Scientific Q Exactive Orbitrap MS with an ESI source. The peptides were separated on an in-house making C18 capillary column with mobile phases 0.1% formic acid in water (A) and 0.1% formic acid in acetonitrile (B) at a 78-min gradient from 0 to 8 min, 4%–8% B; 8–58 min, 8%–22% B; 58–70 min, 22%–32% B; 70–71 min, 32%–90% B; 71–78 min, and 90% B at a flow rate of 280 nL/min. MS analysis was performed in data-dependent MS/MS scan mode, spray voltage was 2.0 kV, MS full scan range was 300–1600 m/z, AGC was set as $3e^6$, resolution at 400 m/z was 70,000, the maximum ion injection time was 60 ms. Top 20 precursor ions were selected into the HCD chamber for MS^2 fragmentation analysis, MS/MS scan resolution was 17,500, AGC was set as $5e^4$, the maximum injection time was 80 ms. Dynamic exclusion time was 50 s. Normalized collision energy was 27%.

MS data were searched against Uniprot *Rattus norvegicus* proteomes datasets (ftp://ftp.uniprot.org/pub/databases/uniprot/current_release/knowledgebase/proteomes/) and 245 common protein contaminants using Proteome Discoverer 1.3 software. Parameters were set as follows: precursor mass tolerance 10 ppm, fragment ion mass tolerance 0.02 Da, two maximum missed cleavages, iodoacetamide on cysteine as fixed modification, oxidation of methionine as variable modifications, peptide false discovery rate of 1% were estimated by Percolator (Kall et al. 2007), which was implemented as a searching node in SEQUEST search engine with Proteome Discovery software.

MALDI-TOF-MS analysis of peptides was performed using an AXIMA-CFR plus MALDI-TOF mass spectrometer (Shimadzu/Kratos, Manchester, UK) equipped with a pulsed nitrogen laser operated at 337 nm. Peptide solution was mixed with 10mg/mL CHCA in 60% acetonitrile containing 0.1% TFA by 1:1 (v/v) in an Eppendorf tube, and 1 μL of the peptide/matrix solution was spotted onto the MALDI sample plate and then crystallized in the air. Positive ion MALDI mass spectra were acquired in the reflection mode under the following parameters: ion source, 20 kV; lens, 6.3 kV; pulsed extraction, −2.5 kV; reflection, 25 kV. Mass spectrometry data were processed with Launchpad 2.7.1 software (Shimadzu/Kratos, Manchester, UK).

MRM-MS analysis and transitions optimization

All the MRM-MS analyses were conducted on a nanoLC-ESI-TSQ Vantage triple quadrupole tandem mass spectrometer (Thermo Scientific, Pittsburgh PA, U.S.) with a C18 column (75 μm I.D. × 150 mm, 3 μm) and the mobile phases 0.1% formic acid in water (A) and 0.1% formic acid in acetonitrile (B) at a flow rate of 300 nL/min. Positive ion mass spectra were acquired in SRM mode under the following parameters: capillary spray voltage, 1.9 kV; capillary temperature, 250 °C; ion source discharge current, 4.0 A; S-lens, 163 V; collision gas, argon at 15 psi; Q1/Q3 peak width, 0.7 Da; and cycle time, 2 s. The mixture of dimethyl-labeled synthetic peptides was used as an internal standard to optimize the operational parameters for each targeted transition, including liquid chromatography conditions and collision energy. Only the precursor ions with a charge state of +2 were chosen as the parent ions, of which the two highest abundant fragment ions in the MS/MS scan were selected as quantitative ions and an additional two relatively high-abundance fragment ions were used as assistant qualifier ions.

Quantifying target proteins

A series of different amounts of light stable isotope dimethyl-labeled synthetic peptides mixture were used as internal standards, and incorporated into the intermediate stable isotope dimethyl-labeled islet tryptic peptides sample. MRM-MS analysis was performed using the optimized method. Each sample was analyzed three times, and the data were analyzed by Pinpoint 1.2 software (Thermo Scientific, Pittsburgh PA, U.S.). The absolute amounts of the target peptides were calculated according to the linear equations acquired from the standard curve for each internal standard peptide.

A reverse-labeling experiment was also performed in the internal standards with intermediate stable isotope dimethyl labeling and the islet tryptic peptides sample with light stable isotope dimethyl labeling. The quantitative results for the target peptides in the forward and reverse experiments were compared to confirm the reliability of the absolute quantitative method.

Acknowledgments This work was supported by the National Natural Science Foundation of China (31300700).

Conflicts of interest Dongwan Cheng, Li Zheng, Junjie Hou, Jifeng Wang, Peng Xue, Fuquan Yang and Tao Xu declare that they have no conflict of interest.

References

Aye TT, Low TY, Bjorlykke Y et al (2012) Use of stable isotope dimethyl labeling coupled to selected reaction monitoring to enhance throughput by multiplexing relative quantitation of targeted proteins. Anal Chem 84(11):4999–5006

Boersema PJ, Raijmakers R, Lemeer S et al (2009) Multiplex peptide stable isotope dimethyl labeling for quantitative proteomics. Nat Protoc 4(4):484–494

Dean PM (1973) Ultrastructural morphometry of the pancreatic-cell. Diabetologia 9(2):115–119

Gallien S, Duriez E, Domon B (2011) Selected reaction monitoring applied to proteomics. J Mass Spectrom (JMS) 46(3):298–312

Kall L, Canterbury JD, Weston J et al (2007) Semi-supervised learning for peptide identification from shotgun proteomics datasets. Nat Methods 4(11):923–925

Maclean B, Tomazela DM, Shulman N et al (2010) Skyline: an open source document editor for creating and analyzing targeted proteomics experiments. Bioinformatics 26(7):966–968

Olofsson CS, Gopel SO, Barg S et al (2002) Fast insulin secretion reflects exocytosis of docked granules in mouse pancreatic B-cells. Pflugers Arch Eur J Physiol 444(1–2):43–51

Petyuk VA, Qian WJ, Hinault C et al (2008) Characterization of the mouse pancreatic islet proteome and comparative analysis with other mouse tissues. J Proteome Res 7(8):3114–3126

Picotti P, Bodenmiller B, Mueller LN et al (2009) Full dynamic range proteome analysis of *S. cerevisiae* by targeted proteomics. Cell 138(4):795–806

Rorsman P, Trube G (1986) Calcium and delayed potassium currents in mouse pancreatic beta-cells under voltage-clamp conditions. J Physiol 374:531–550

Seino S, Bell G (2008) Pancreatic beta cell in health and disease. Springer, Tokyo

Permissions

All chapters in this book were first published in BR, by Springer; hereby published with permission under the Creative Commons Attribution License or equivalent. Every chapter published in this book has been scrutinized by our experts. Their significance has been extensively debated. The topics covered herein carry significant findings which will fuel the growth of the discipline. They may even be implemented as practical applications or may be referred to as a beginning point for another development.

The contributors of this book come from diverse backgrounds, making this book a truly international effort. This book will bring forth new frontiers with its revolutionizing research information and detailed analysis of the nascent developments around the world.

We would like to thank all the contributing authors for lending their expertise to make the book truly unique. They have played a crucial role in the development of this book. Without their invaluable contributions this book wouldn't have been possible. They have made vital efforts to compile up to date information on the varied aspects of this subject to make this book a valuable addition to the collection of many professionals and students.

This book was conceptualized with the vision of imparting up-to-date information and advanced data in this field. To ensure the same, a matchless editorial board was set up. Every individual on the board went through rigorous rounds of assessment to prove their worth. After which they invested a large part of their time researching and compiling the most relevant data for our readers.

The editorial board has been involved in producing this book since its inception. They have spent rigorous hours researching and exploring the diverse topics which have resulted in the successful publishing of this book. They have passed on their knowledge of decades through this book. To expedite this challenging task, the publisher supported the team at every step. A small team of assistant editors was also appointed to further simplify the editing procedure and attain best results for the readers.

Apart from the editorial board, the designing team has also invested a significant amount of their time in understanding the subject and creating the most relevant covers. They scrutinized every image to scout for the most suitable representation of the subject and create an appropriate cover for the book.

The publishing team has been an ardent support to the editorial, designing and production team. Their endless efforts to recruit the best for this project, has resulted in the accomplishment of this book. They are a veteran in the field of academics and their pool of knowledge is as vast as their experience in printing. Their expertise and guidance has proved useful at every step. Their uncompromising quality standards have made this book an exceptional effort. Their encouragement from time to time has been an inspiration for everyone.

The publisher and the editorial board hope that this book will prove to be a valuable piece of knowledge for researchers, students, practitioners and scholars across the globe.

List of Contributors

Zhou Gong, Xu Dong and Chun Tang
CAS Key Laboratory of Magnetic Resonance in Biological Systems, State Key Laboratory of Magnetic Resonance and Atomic Molecular Physics, Wuhan Institute of Physics and Mathematics, Chinese Academy of Sciences, Wuhan 430071, China

Yue-He Ding, Na Liu, E. Erquan Zhang and Meng-Qiu Dong
National Institute of Biological Sciences, Beijing 102206, China

Chang Huang
National Laboratory of Biomacromolecules, CAS Center for Excellence in Biomacromolecules, Institute of Biophysics, Chinese Academy of Sciences, Beijing 100101, China

Bing Zhu
National Laboratory of Biomacromolecules, CAS Center for Excellence in Biomacromolecules, Institute of Biophysics, Chinese Academy of Sciences, Beijing 100101, China
College of Life Sciences, University of Chinese Academy of Sciences, Beijing 100049, China

Ye Yuan
National Laboratory of Biomacromolecules, CAS Center for Excellence in Biomacromolecules, Institute of Biophysics, Chinese Academy of Sciences, Beijing 100101, China
College of Life Science and Technology, Huazhong Agricultural University, Wuhan 430070, China

Ji Liu
National Laboratory of Biomacromolecules, CAS Center for Excellence in Biomacromolecules, Institute of Biophysics, Chinese Academy of Sciences, Beijing 100101, China
College of Life Science, Hubei University, Wuhan 430070, China

Jian Chen
College of Life Science, Hubei University, Wuhan 430070, China

Jibin Zhang
College of Life Science and Technology, Huazhong Agricultural University, Wuhan 430070, China

Dianbing Wang and Xian-En Zhang
National Laboratory of Biomacromolecules, CAS Center for Excellence in Biomacromolecules, Institute of Biophysics, Chinese Academy of Sciences, Beijing 100101, China

Xinmei Li and Shuangbo Zhang
National Key Laboratory of Biomacromolecules, CAS Center for Excellence in Biomacromolecules, Institute of Biophysics, Chinese Academy of Sciences, Beijing 100101, China
University of Chinese Academy of Sciences, Beijing 100049, China

Jianguo Zhang
Center for Biological Imaging, Institute of Biophysics, Chinese Academy of Sciences, Beijing 100101, China

Fei Sun
National Key Laboratory of Biomacromolecules, CAS Center for Excellence in Biomacromolecules, Institute of Biophysics, Chinese Academy of Sciences, Beijing 100101, China
University of Chinese Academy of Sciences, Beijing 100049, China
Center for Biological Imaging, Institute of Biophysics, Chinese Academy of Sciences, Beijing 100101, China

Shan Lu, Yong Cao and Meng-Qiu Dong
National Institute of Biological Sciences, Beijing, Beijing 102206, China

Sheng-Bo Fan, Zhen-Lin Chen, Run-Qian Fang and Si-Min He
Key Lab of Intelligent Information Processing of Chinese Academy of Sciences (CAS), University of CAS, Institute of Computing Technology, CAS, Beijing 100190, China
University of Chinese Academy of Sciences, Beijing 100049, China

Fei Sun
National Key Laboratory of Biomacromolecules, CAS Center for Excellence in Biomacromolecules, Institute of Biophysics, Chinese Academy of Sciences, Beijing 100101, China
Center for Biological Imaging, Institute of Biophysics, Chinese Academy of Sciences, Beijing 100101, China
University of Chinese Academy of Sciences, Beijing 100049, China
Sino-Danish Center for Education and Research, Beijing 100190, China

List of Contributors

Yang Shi
National Key Laboratory of Biomacromolecules, CAS Center for Excellence in Biomacromolecules, Institute of Biophysics, Chinese Academy of Sciences, Beijing 100101, China
University of Chinese Academy of Sciences, Beijing 100049, China
Sino-Danish Center for Education and Research, Beijing 100190, China

Yujia Zhai
National Key Laboratory of Biomacromolecules, CAS Center for Excellence in Biomacromolecules, Institute of Biophysics, Chinese Academy of Sciences, Beijing 100101, China
University of Chinese Academy of Sciences, Beijing 100049, China

Li Wang and Jianguo Zhang
Center for Biological Imaging, Institute of Biophysics, Chinese Academy of Sciences, Beijing 100101, China

Chengyan Dong
Interdisciplinary Laboratory, Institute of Biophysics, Chinese Academy of Sciences, Beijing 100101, China

Zhaofei Liu
Medical Isotopes Research Center and Department of Radiation Medicine, School of Basic Medical Sciences, Peking University, Beijing 100191, China

Fan Wang
Interdisciplinary Laboratory, Institute of Biophysics, Chinese Academy of Sciences, Beijing 100101, China
Medical Isotopes Research Center and Department of Radiation Medicine, School of Basic Medical Sciences, Peking University, Beijing 100191, China

Yangyu Huang, Haotian Li and Yi Xiao
Biomolecular Physics and Modeling Group, School of Physics, Huazhong University of Science and Technology, Wuhan 430074, China

Yiran Li
Department of Biological Science and Biotechnology, School of Biological Science and Medical Engineering, Beihang University, Beijing 100191, China
National Laboratory of Biomacromolecules, Institute of Biophysics, Chinese Academy of Sciences, Beijing 100101, China

Shimeng Xu
National Laboratory of Biomacromolecules, Institute of Biophysics, Chinese Academy of Sciences, Beijing 100101, China
University of Chinese Academy of Sciences, Beijing 100049, China

Xuelin Zhang
Capital University of Physical Education and Sport, Beijing 100191, China

Simon Cichello
School of Life Sciences, La Trobe University, Melbourne, VIC 3086, Australia

Zongchun Yi
Department of Biological Science and Biotechnology, School of Biological Science and Medical Engineering, Beihang University, Beijing 100191, China

Zhou Gong, Xu Dong and Chun Tang
CAS Key Laboratory of Magnetic Resonance in Biological Systems, State Key Laboratory of Magnetic Resonance and Atomic Molecular Physics, and National Center for Magnetic Resonance at Wuhan, Wuhan Institute of Physics and Mathematics of the Chinese Academy of Sciences, Wuhan 430071, China
National Center for Magnetic Resonance at Wuhan, Wuhan Institute of Physics and Mathematics of the Chinese Academy of Sciences, Wuhan 430071, China

Zhu Liu
Department of Pharmacology, Institute of Neuroscience, Key Laboratory of Medical Neurobiology of the Ministry of Health of China, Zhejiang University School of Medicine, Hangzhou 310057, China

Yue-He Ding
RNA Therapeutics Institute, University of Massachusetts Medical School, 368 Plantation Street, Worcester, MA 01605, USA

Meng-Qiu Dong
National Institute of Biological Sciences, Beijing 102206, China

Xiaowu Li and Hongrong Liu
College of Physics and Information Science, Synergetic Innovation Center for Quantum Effects and Applications, Hunan Normal University, Changsha 410081, China

Lingpeng Cheng
School of Life Sciences, Tsinghua University, Beijing 100084, China

Xingguo Liu, Liang Yang and Qi Long
Key Laboratory of Regenerative Biology, Guangdong Provincial Key Laboratory of Stem Cell and Regenerative Medicine, South China Institute for Stem Cell Biology and Regenerative Medicine, Guangzhou Institutes of Biomedicine and Health, Chinese Academy of Sciences, Guangzhou 510530, China

David Weaver and György Hajnóczky
Department of Pathology, MitoCare Center, Anatomy and Cell Biology, Thomas Jefferson University, Philadelphia, PA 19107, USA

Ming Liu
College of Biotechnology, Tianjin University of Science and Technology, Tianjin 300457, China

Jie Heng, Yuan Gao and Xianping Wang
National Laboratory of Macromolecules, National Center of Protein Science - Beijing, Institute of Biophysics, Chinese Academy of Sciences, Beijing 100101, China

Peng Cao and Mei Li
National Laboratory of Biomacromolecules, CAS Center for Excellence in Biomacromolecules, Institute of Biophysics, Chinese Academy of Sciences, Beijing 100101, China

Xin Sheng, Xiuying Liu and Zhenfeng Liu
National Laboratory of Biomacromolecules, CAS Center for Excellence in Biomacromolecules, Institute of Biophysics, Chinese Academy of Sciences, Beijing 100101, China
University of Chinese Academy of Sciences, Beijing 100049, China

Wei Li
National Laboratory for Condensed Matter Physics and Key Laboratory of Soft Matter Physics, Institute of Physics, Chinese Academy of Sciences, Beijing 100190, China

Ping Chen
National Laboratory of Biomacromolecules, CAS Center for Excellence in Biomacromolecules, Institute of Biophysics, Chinese Academy of Sciences, Beijing 100101, China

Xue Xiao, Yi-Zhou Wang, Peng-Ye Wang and Ming Li
National Laboratory for Condensed Matter Physics and Key Laboratory of Soft Matter Physics, Institute of Physics, Chinese Academy of Sciences, Beijing 100190, China
National Laboratory of Biomacromolecules, CAS Center for Excellence in Biomacromolecules, Institute of Biophysics, Chinese Academy of Sciences, Beijing 100101, China

Liping Dong and Guohong Li
National Laboratory of Biomacromolecules, CAS Center for Excellence in Biomacromolecules, Institute of Biophysics, Chinese Academy of Sciences, Beijing 100101, China
University of Chinese Academy of Sciences, Beijing 100049, China

Thor C. Møller, David Moreno-Delgado, Jean-Philippe Pin and Julie Kniazeff
Institut de Génomique Fonctionnelle (IGF), CNRS, INSERM, Univ. Montpellier, 34094 Montpellier, France

Jiasong Li, Qi Liu, Xudong Wang, Weimin Gong, Xiaohong Liu and Haiping Liu
Institute of Biophysics, Chinese Academy of Sciences, Beijing 100101, China

Li Wang, Xuzhen Guo, Fuying Kang, Cheng Hu and Jiangyun Wang
Institute of Biophysics, Chinese Academy of Sciences, Beijing 100101, China
College of Life Sciences, University of Chinese Academy of Sciences, Beijing 100049, China

Xian Chen
Key Laboratory of Physics and Technology for Advanced Batteries (Ministry of Education), Department of Physics, Jilin University, Changchun 130012, China

Wei Zhuang
State Key Laboratory of Structural Chemistry, Fujian Institute of Research on the Structure of Matter, Chinese Academy of Sciences, Fuzhou 350002, China

Xianjin Xu and Xiaoqin Zou
Dalton Cardiovascular Research Center, University of Missouri, Columbia, MO 65211, USA
Department of Physics and Astronomy, University of Missouri, Columbia, MO 65211, USA
Informatics Institute, University of Missouri, Columbia, MO 65211, USA
Department of Biochemistry, University of Missouri, Columbia, MO 65211, USA

Marshal Huang
Dalton Cardiovascular Research Center, University of Missouri, Columbia, MO 65211, USA
Informatics Institute, University of Missouri, Columbia, MO 65211, USA

Jun Wang and Yi Xiao
Institute of Biophysics, School of Physics and Key Laboratory of Molecular Biophysics of the Ministry of Education, Huazhong University of Science and Technology, Wuhan 430074, China

Yunjie Zhao
Institute of Biophysics, School of Physics and Key Laboratory of Molecular Biophysics of the Ministry of Education, Huazhong University of Science and Technology, Wuhan 430074, China
Institute of Biophysics and Department of Physics, Central China Normal University, Wuhan 430079, China

List of Contributors

Chen Zeng
Department of Physics, The George Washington University, Washington, DC 20052, USA

Dou Yeon Youn and Jeffrey E. Pessin
Division of Endocrinology, Department of Medicine, Diabetes Research Center, Albert Einstein College of Medicine, Bronx, NY 10461, USA
Department of Molecular Pharmacology, Albert Einstein College of Medicine, Bronx, NY 10461, USA

Alus M. Xiaoli and Fajun Yang
Division of Endocrinology, Department of Medicine, Diabetes Research Center, Albert Einstein College of Medicine, Bronx, NY 10461, USA
Department of Developmental and Molecular Biology, Albert Einstein College of Medicine, Bronx, NY 10461, USA

Zhe-Ming Yuan
Hunan Provincial Key Laboratory for Biology and Control of Plant Diseases and Insect Pests, Hunan Agricultural University, Changsha 410128, China

Xilin Xu
Hunan Provincial Key Laboratory for Biology and Control of Plant Diseases and Insect Pests, Hunan Agricultural University, Changsha 410128, China
Center for Systems Medicine, Institute of Basic Medical Sciences, Chinese Academy of Medical Sciences & Peking Union Medical College, Beijing 100005; Suzhou Institute of Systems Medicine, Suzhou 215123, China

Aiping Wu and Taijiao Jiang
Center for Systems Medicine, Institute of Basic Medical Sciences, Chinese Academy of Medical Sciences & Peking Union Medical College, Beijing 100005; Suzhou Institute of Systems Medicine, Suzhou 215123, China

Xinlei Zhang
Suzhou Geneworks Technology Company Limited, Suzhou 215123, China

Mingming Su
Institute of Basic Medical Sciences, Chinese Academy of Medical Sciences & Peking Union Medical College, Beijing 100005, China

Li Zheng, Junjie Hou, Jifeng Wang, Peng Xue, Fuquan Yang and Tao Xu
National Laboratory of Biomacromolecules, Institute of Biophysics, Chinese Academy of Sciences, Beijing 100101, China

Dongwan Cheng
National Laboratory of Biomacromolecules, Institute of Biophysics, Chinese Academy of Sciences, Beijing 100101, China
University of Chinese Academy of Sciences, Beijing 100049, China

Index

A
Apex2, 52-57, 59
Auto-inhibitory Mechanism, 14

B
Base-pairing Consistency, 189-191, 193, 195

C
Chemical Cross-linking, 1-3, 5, 7, 9-11, 82, 89
Chemical Cross-linking Coupled With Mass Spectrometry, 1, 89
Chromatin Fibers, 36, 130-140
Chromophore, 21, 23-28, 159-167, 169-171
Clonable Tags Localization, 52
Cryo-electron Diffraction, 29, 31-32, 35
Crystal Cryo-lamella, 29, 35
Cyanobacteria, 115-116, 128

D
Diacylglycerol, 75, 79-81, 128-129
Dimethyl Labeling, 39, 51, 217-221, 223, 225-226
Disulfide Bonds, 38-41, 43-47, 49-51

E
Electron Microscopy, 8, 10, 29, 36, 52-53, 57-59, 89, 91, 98, 115, 117, 128, 131
Electron-transfer Dissociation, 39

F
Fluorescent Proteins, 21-23, 25, 27-28, 99-100, 102, 105-106, 160, 165, 170-171
Focus Ion Beam Scanning Electron Microscopy, 52

G
G Protein-coupled Receptors, 152-153, 155, 157
Gas-phase Vertical Absorption, 160-161
Gene Transcription, 18, 139, 198-201
Genetic Code Expansion, 159-160, 166
Glucose-fatty Acid Cycle, 74

H
High Throughput Screening, 67, 180
Higher-energy Collisional Dissociation, 39
Histone Acetylation, 13
Histone H3k36 Methylation, 13, 20
Histone Methyltransferases, 13, 15, 17, 19
Homology Modelling, 108

I
Icosahedral Viral Capsids, 91
Insulin Resistance, 73-75, 77-81, 151, 198, 202
Intramuscular Triglyceride, 73-74
Intramyocellular Lipid, 73, 75, 77, 79-81
Inverse Virtual Screening, 172-173, 175, 177, 179, 181, 183-187
Isotope-labeled Peptides, 217-218

L
Light-harvesting Complexes Ii, 115-116
Liquid Chromatography-mass Spectrometry (LC-MS), 38-39, 41

M
Magnetic Tweezers, 130-140
Mass Spectrometry, 1-2, 10-12, 38-39, 41, 51, 85, 89-90, 142, 146, 149, 151, 180, 217-218, 220-221, 223, 225
Mediator Complex, 198-206
Metadp, 207-213, 215
Metagenomics, 207-208
Micro-electron Diffraction, 29-30
Mitochondrial Dynamics-related Proteins, 52, 56
Mitochondrial Fission, 53, 55-56, 58-59, 81, 99, 107, 199, 204, 206
Mitochondrial Fission Factor, 53, 58
Monogalactosyldiacylglycerol, 128-129

N
Natural Isotope Abundance, 6, 8-9, 84

O
On-grid Crystallization, 29, 33

P
Photoswitching, 99, 102-103
Photosystem Ii, 115, 128-129
Plant Chloroplast, 115
Polychlorinated Biphenyls, 180
Positive Predictive Value, 190
Protein Complex Structures, 9, 68
Protein Localization, 52-53, 55-59, 160
Protein-protein Complex, 68, 82

R
Radioligand Binding, 60-61, 67
Raman Spectra, 21-28

Raman Spectra Bands, 21-23
Ribosomal Rna, 189-192
Rna-protein Complexes, 68, 72
Root Mean Square Deviation, 68-69
Rprank, 68-70

S
Saturation Binding Assays, 60
Scanning Electron Microscopy, 52-53, 58-59
Sea Anemone, 21
Serum, 57, 66, 89, 141-145, 147-151, 201-204
Sid-mrm-ms, 217-219, 221, 223, 225
Signal Recognition Particle, 189-190
Single-molecule Magnetic Tweezers, 130-131, 133, 135, 137, 139
Single-side Blotting, 29, 34

Small Orf-encoded Peptides, 141, 143, 145, 147, 149, 151
Symmetry-mismatch Reconstruction Method, 92

T
Triacylglycerol, 73-75, 79, 81

U
Unnatural Amino Acids, 160

V
Venus Flytrap (VFT) Domain, 152-153
Vesicular Neurotransmitter Transporters, 108, 113
Viral Genome, 91-92

W
Watson-crick Base Pairs, 190, 195

CPSIA information can be obtained
at www.ICGtesting.com
Printed in the USA
BVHW011522180820
586704BV00004B/134